Microbial Bioactive Compounds

Ravindra Soni • Deep Chandra Suyal • Lourdes Morales-Oyervides
Editors

Microbial Bioactive Compounds

Industrial and Agricultural Applications

Springer

Editors
Ravindra Soni
Department of Agricultural Microbiology
Indira Gandhi Krishi Vishwavidyalaya
Raipur, India

Deep Chandra Suyal
Vidyadayini Institute of Science, Management and Technology
Bhopal, India

Lourdes Morales-Oyervides
Facultad de Ciencias Quimicas
Universidad Autonoma de Coahuila
Saltillo, Mexico

ISBN 978-3-031-40081-0 ISBN 978-3-031-40082-7 (eBook)
https://doi.org/10.1007/978-3-031-40082-7

This Springer imprint is published by the registered company Springer Nature Switzerland AG
The registered company address is: Gewerbestrasse 11, 6330 Cham, Switzerland

Paper in this product is recyclable.

Preface

Microbial natural products, either as pure compounds or as standardized extracts, provide unlimited opportunities for new drug leads because of the unmatched availability of chemical diversity and ease of simplicity. A variety of organisms, such as bacteria and fungi, produce diverse bioactive metabolites. These microorganism-derived compounds have been utilized in medicine, agriculture, food industry, cosmetics, textiles, and scientific research. Moreover, they are also being investigated for their role in green chemistry and various other sustainable development strategies. However, much research is still needed to be done in this field as most of the soil and marine microbes are not easily accessible due to cultural biases. Besides these, the available scientific information in this field needs to be compiled and shared effectively among the beneficiaries. From this perspective, this book is a perfect documentation of primary and secondary data-based information on the latest research findings, case studies, experiences, and innovations regarding microbial bioactive compounds.

We acknowledge the suggestions and encouragement made by colleagues and well-wishers. Moreover, we are grateful to all the authors who have contributed to this book. Suggestions for the improvement of the book will be highly appreciated and incorporated in the subsequent editions.

Raipur, India — Ravindra Soni
Bhopal, India — Deep Chandra Suyal
Saltillo, Mexico — Lourdes Morales-Oyervides

Contents

Chapter 1
Microbial Production of Bioactive Compounds: Recent Advancements and Trends

Juan Pablo Ruiz-Sanchez, Miguel Angel Villegas-Mendez, Julio Montañez, Juan Roberto Benavente-Valdés, and Lourdes Morales-Oyervides

Abstract The production of bioactive compounds through microbial sources has gained considerable attention in recent years. The use of microorganisms for producing a wide range of complex molecules with different biological activities for various applications has become increasingly popular. Researchers have recognized a vast number of microorganisms as producers of bioactive compounds with industrial applications. In addition, the use of microorganisms to produce bioactive compounds is considered to be an environmentally friendly and sustainable approach. However, finding the optimal conditions for producing these compounds remains a challenge, and exploring new niches with new microorganisms expands the possibility of discovering novel bioactive compounds. The chapter provides an overview of various applications of bioactive compounds, including food, cosmeceutical/cosmetic products, and environmental and agricultural applications. Overall, this chapter provides valuable insights into the recent advancements and trends in the microbial production of bioactive compounds and identifies the challenges and opportunities for future research in this field.

Keywords Bioactive compounds · Pigments · Polysaccharides · Cosmeceuticals

1.1 Introduction

Throughout centuries, humans have harnessed the potential of bioactive compounds (BCs) to improve their lives [1]. The scientific community has thoroughly researched BCs, which are characterized by their ability to interact with living tissue

J. P. Ruiz-Sanchez · J. Montañez · J. R. Benavente-Valdés · L. Morales-Oyervides (✉)
Facultad de Ciencias Quimicas, Universidad Autonoma de Coahuila, Unidad Saltillo, Saltillo, Coahuila, Mexico
e-mail: lourdesmorales@uadec.edu.mx

M. A. Villegas-Mendez
Departamento de Ciencias Básicas, Universidad Autónoma Agraria Antonio Narro, Saltillo, Coahuila, Mexico

R. Soni et al. (eds.), *Microbial Bioactive Compounds*,
https://doi.org/10.1007/978-3-031-40082-7_1

and cause various biological effects and reactions. Although the exact definition of a BC may be unclear, its influence on biological systems is undeniable [2]. It is widely agreed among authors that these compounds have distinct advantageous qualities that differentiate them from detrimental compounds, like toxic or carcinogenic substances [3, 4].

Recently, there has been a growing interest in using bioactive compounds for various applications, including food, pharmaceuticals, cosmetics, as well as environmental and agricultural applications [3–7].

For instance, in the food industry, BCs serve as additives or can be utilized to create health-promoting products such as food supplements, nutraceuticals, and functional foods [3]. The pharmaceutical industry is researching BCs in order to discover new drugs and therapies, specifically new antibiotics that can combat resistant pathogens [4]. Additionally, BCs are being investigated for the treatment of other disorders such as genetic (cancer), neurological (Alzheimer or Parkinson disease), and metabolic (diabetes and obesity) [8].

As for the cosmetic sector, BCs can be used to formulate skin antiaging, hydrating, whitening, and brightening products. Also, the cosmetic industry is increasingly invested in finding new BCs that provide a range of skin benefits, including protection against UV radiation and treatment of various skin conditions [5, 9]. These compounds can be used in the formulation of various cosmetic products such as moisturizers, antiaging creams, and sunscreens.

When it comes to environmental and agricultural applications, microorganisms have the potential to produce BCs for various applications, including bioremediation, biofertilizers, and biopesticides. Mainly, BCs are being explored for their potential to replace the use of agrochemicals. They can improve the growth, yield, and quality of crops while also reducing the impact on the environment and human health [7].

Certainly, BCs can come from either natural or synthetic sources. In this regard, the worldwide exigencies for natural products have been boosted during the last decade. Furthermore, naturally occurring biological compounds created through microbial cell factories are gaining popularity in both industrial and academic fields [4, 10]. A range of microorganisms, including bacteria, yeast, fungi, and algae, possess the ability to produce a wide assortment of BCs that exhibit diverse biological properties. These BCs have found utility across various industries as well (Fig. 1.1).

The utilization of microbial-produced BCs can provide numerous benefits as compared to other natural sources. One of the major advantages is its higher efficiency, which enables the production of larger amounts of the desired complex molecules. Moreover, the microbial-produced BCs offer greater versatility in terms of the range of molecules that can be produced, making them a more versatile option for various applications. Additionally, the production process is easier to scale, ensuring more efficient and cost-effective production of the desired molecules. Additionally, microbial cell factories can be engineered to consume renewable resources, making them a more sustainable approach to obtaining BCs. Indeed, microbial production of BCs can face many bottlenecks in reaching a commercial

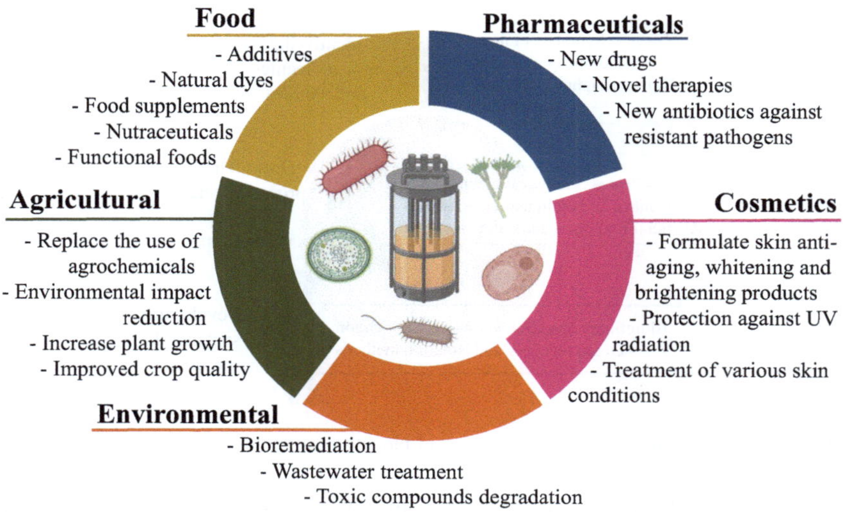

Fig. 1.1 Applications of microbial bioactive compounds

stage, such as low manufacturing yields, high processing costs, and challenging recovery methods.

This chapter explores the current trend of including bioactive compounds in various industries and explains how microbial production of these compounds is being used in different sectors.

1.2 Recent Trends Toward Bioactive Compounds Incorporation into the Market

It has been estimated that the market for bioactive ingredients on a global scale has reached an impressive size of USD $45.5 billion in the year 2022. The projections indicate that this market will continue to grow at a compound annual growth rate (CAGR) of 7.42% from 2022 to 2028, culminating in a forecasted value of USD $69.9 billion by the year 2028 [11]. The fact that there is a growing demand for bioactive ingredients in different industries and their potential to drive innovation and progress is evidenced by this growth. According to the report, the largest segment was food supplements, but personal care and animal nutrition were also included. The report also covered natural BCs extracted from plants.

Regarding the market of microbial products, according to recent projections, the microbial products industry is expected to experience a notable CAGR of 7.6% between 2022 and 2027. This growth is predicted to increase the market's overall value from its current standing of $231.5 billion to $334.2 billion [12]. These numbers indicate significant advancements in this industry and suggest a promising

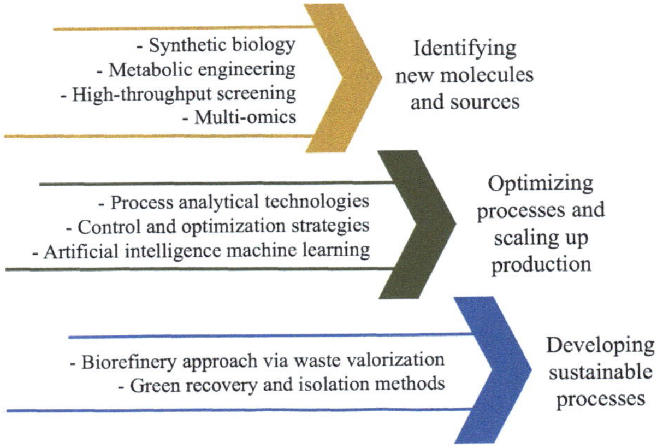

Fig. 1.2 Recent trends for microbial production of bioactive compounds

future for microbial products. It is worth noting that consumers' increasing awareness of healthy lifestyle choices and their growing demand for products made with natural ingredients are the driving forces behind the growth of both markets. It is crucial to consider these factors when analyzing the market's expansion. On the other hand, the time required to obtain regulatory approval, such as from the FDA, may limit the growth of the microbial BCs market due to high processing costs.

In this regard, the latest trends in microbial-produced compounds aim to enhance process efficiency, improve product quality, reduce costs, and find new sources and compounds. In addition, due to the environmental crisis, both industry and academia have suggested bioprocesses that follow the ideals of cleaner production and a circular economy.

Recent trends in microbial-produced compounds depend on the process stage, whether upstream or downstream, as well as the research and development stage. Therefore, efforts can be targeted toward identifying new molecules and sources, optimizing processes, scaling up production, and developing environmentally friendly recovery methods (Fig. 1.2).

For example, synthetic biology, metabolic engineering, high-throughput screening, and a multi-omics approach have contributed to advancing the development of new microbial cell factories, improving the efficiency of existing ones, and engineering microorganisms to achieve specific objectives [13]. On the other hand, within the processing and scale-up trends, the utilization of process analytical technologies has enabled the application of bioprocess control and optimization strategies to achieve maximum yields and consistency [14, 15]. Artificial intelligence and machine learning are most certainly concepts that have emerged in industrial microbiology to improve the efficiency, consistency, and quality of bioprocesses while reducing the time and cost of developing new bioprocesses [16, 17].

Regarding developing sustainable processes, two main trends are followed, a biorefinery approach for the upstream and a design of a green recovery method for the downstream. In this sense, waste valorization allows converting waste materials into BCs, thus improving economic viability while reducing the amount of waste that ends up contaminating the environment [18, 19]. Meanwhile, green recovery methods enable isolating BCs while reducing the utilization of organic solvents and improving the energy efficiency of the recovery process [20].

Further sections will describe the applications of microbial BCs within distinct sectors and purposes, as shown in Fig. 1.1.

1.3 Food Applications and Health Benefits

The extraction and use of secondary BCs in food is a practice that dates back centuries. In ancient Asia, mold-fermented rice was utilized as both a food and traditional medicine [21]. Today, microbial BCs are employed in a variety of food applications, including preservation, color, flavor, texture, and nutritional enhancement, as outlined in Table 1.1.

Lactic acid bacteria (LAB) are a group of bacteria that produce lactic acid as a metabolic product [23]. They are commonly used as starter cultures in the production of fermented foods such as yogurt, cheese, and sauerkraut. Additionally, LAB strains can produce antimicrobial compounds that can serve as natural preservatives in various food products. Juodeikiene et al. [28] employed the extracellular metabolites present in the supernatant broth fermented by LAB to wash off mycotoxins and spores from wheat grains before malting, preventing the undesirable growth of other microorganisms or off-flavors. LAB are recognized not only as an essential group of microorganisms for the dairy industry but also for their catabolic and anabolic metabolism and the benefits that they bring to other fields in the food

Table 1.1 Applications of bioactive compounds in food and health industry

Microorganism	Bioactive compound	Activity/Application	References
Lactobacillus rhamnosus B103	Lactic acid	Texture and flavor modification in dairy products, meat products preservation.	[22, 23]
Pichia pastoris	α-Amylases	Starch liquefaction, saccharification, maltose syrup production.	[24]
Talaromyces leycettanus JCM12802	Glucoamylases	Starch saccharification, production of high fructose and glucose syrups.	[25]
Saccharomyces cerevisiae	Glucans	Noncaloric food thickener, fat substitute, emulsifier, foam stabilizer, source of dietary fiber.	[26]
Rhodobacter sphaeroides HY01	CoQ10	Food supplement, energy boosting, antioxidant.	[27]

industry [22]. Regarding this, the BCs obtained from LAB can be produced either by degradation or synthesis.

LAB have the ability to degrade polysaccharides, providing unique flavors and odors to sourdough, as well as proteins and amino acids, effectively hydrolyzing proteins present in milk and other non-nutritive and harmful substances such as phytic acid or undesirable peptides. Conversely, LAB also produce essential secondary BCs with a significant impact on the food industry, including lactic acid and other organic acids, bacteriocin, vitamins, extracellular polysaccharides, gamma-aminobutyric acid, flavor substances, and antioxidant substances [29].

Filamentous fungi can serve as natural producers of food colorants, with an incredible variety of pigments such as carotenoids, melanins, flavins, phenazines, quinones, monascins, atrorosins, violacein, and indigo [30]. To date, *Monascus* sp. is one of the most extensively studied filamentous fungi, with over 50 different pigments examined. However, *Monascus*-like pigments have also been reported in species of *Talaromyces* and *Penicillium* [31]. Similarly, microbial pigments can also be produced by bacteria. Unlike other microorganism-produced pigments, bacterial pigments have the advantages of a short life cycle and ease of genetic modification. Nevertheless, it is important to note that most bacterial pigments are still in the research and development stage, unlike those produced by fungi.

Enzymes play a crucial role in the food industry, as they can catalyze a wide range of reactions and processes that are important for food production and processing. Raveendran et al. [32] explained the most relevant enzymes and their application, including α-amylase, glucoamylases, proteases, lactase, lipases, phospholipases, esterases, lipoxygenases, cellulases, xylanases, pectinases, glucose oxidase, laccase, catalase, and peroxidase.

Some of the applications of these enzymes are as follows: α-amylase is capable of hydrolyzing glycosidic bonds, resulting in the production of short-chain dextrins. It has a wide range of uses in the baking industry for flavor enhancement, starch liquefaction, brewing, etc.

The major microorganisms used for the industrial production of α-amylase are *Bacillus amyloliquefaciens*, *Bacillus stearothermophilus*, or *Bacillus licheniformis* [33]. Glucoamylases hydrolyze polysaccharide starch, releasing β-glucose. They have major applications in the production of high glucose and fructose syrups, bread quality improvement, and beer production. Glucoamylases are primarily produced by *Aspergillus niger*, *Aspergillus awamori*, and *Rhizopus oryzae*.

Proteases are hydrolytic enzymes that release peptides and amino acids from proteins. They are commonly used for meat tenderization, coagulation of milk, brewing, etc. An important producer of this enzyme due to its low pH tolerance is *Aspergillus usamii* [34]. Lactase is principally used in dairy products to reduce lactose intolerance but is also used as a prebiotic food ingredient. Lactase for industrial use is principally obtained from bacteria like *Bifidobacterium infantis* CCRC 14633, *B. longum* CCRC 15708, *B. longum* CCRC15708, and *Lactobacillus* spp. [35]. For a better understanding of enzyme usage in the food industry, please refer to the work of Raveendran et al. [32].

In addition to enzymes, polysaccharides produced by microbes are also used as food additives to improve the texture of foods. For instance, xanthan gum produced by bacteria such as *Xanthomonas campestris* is used as a thickening agent in foods like salad dressings and sauces [36]. Other examples of microbial polysaccharides, their producers, and applications are glucans from *Saccharomyces cerevisiae*, which can be used as a noncaloric food thickener, gellan from *Pseudomonas elodea*, which is used as a gelling agent, levan from *Alcaligenes viscosus*, which has prebiotic and hypocholesterolemic effects, and emulsan from *Acinetobacter caloaceticus*, which is used as an emulsifying agent [37].

Likewise, microbial BCs are increasingly being recognized for their beneficial properties, which make them attractive for nutritional enhancement and the development of functional foods. These compounds possess various properties such as antioxidants, immune system enhancers, and enzymes that aid in digestion. Carotenoids, for example, are microbial BCs that have been shown to scavenge free radicals and prevent oxidative damage to cells [38], making them potential candidates for preventing chronic diseases such as cancer, diabetes, and cardiovascular disease.

One microbial BC that has gained significant attention in recent years is CoQ10. CoQ10, also known as ubiquinone, is an essential compound that plays a crucial role in the production of ATP, the main energy source for cells. It also acts as an antioxidant, protecting cells from damage caused by free radicals [39]. Although CoQ10 can be found in small amounts in some foods, it can also be produced by microorganisms.

Food supplement products containing CoQ10 produced by microorganisms are available in various forms, such as capsules, tablets, and soft gels. These supplements are marketed as an aid in supporting heart health and improving energy levels. CoQ10 produced by microorganisms has also been added to certain foods, such as beverages, yogurt, and energy bars, to increase their nutritional value. Researchers have explored the potential of purple non-sulfur bacteria (PNSB), a photosynthetic bacterium, for the production of CoQ10. He et al. [40] explained the development and future prospects of CoQ10 production by PNSB. They discussed the possibility of using nontoxic wastewater effluent as a nutrient source for the production of BCs by PNSB, specifically CoQ10. They also explained the bioreactor configuration and important factors that influence the production of CoQ10, such as light, oxygen, and C/N source and ratio. To compensate for the cost of production in their analysis, they must produce 1.4 g/L of biomass and 49.65 mg/g of CoQ10 content. In addition, Zhang et al. [27] demonstrated that a strategy of phosphate limitation along with glucose-fed batch fermentation with the industrial strain *Rhodobacter sphaeroides* HY01 was a positive strategy for CoQ10 production.

Further research is needed to explore the full potential of microbial BCs in food applications and their impact on human health.

1.4 Potential Use in Cosmeceutical/Cosmetic Products

Cosmeceutical is a term used to describe a range of products that combine cosmetic and pharmaceutical properties. That is, it describes products that serve cosmetic purposes but also contain active ingredients that may have medicinal or drug-like effects on skin health. These products have become increasingly popular in the marketplace in recent years. The cosmetic industry has shown considerable interest in microbial BCs because of their natural, safe, and effective properties. Additionally, the use of BCs in pharmaceutical/cosmetic fields is of major interest due to their high demand, low required quantity, and high sale price [42]. However, the utilization of BCs entails a rigorous purification and refinement process [43], which may potentially lead to higher processing costs. As previously mentioned, the cost of pure isolated BCs may be too high and could outweigh any benefits. The BCs and their different applications in the cosmeceutical and cosmetic industries are summarized in Table 1.2.

BCs can be utilized as physical agents in cosmetic formulations in order to enhance the stability, thickness, and overall gel-like texture of a diverse array of cosmetic products [52]. For instance, cosmetic products can use microbially produced biosurfactants for diverse functions, including acting as detergents, creating foam, and emulsifying [53].

Moreover, BCs are versatile and can be used in various ways in the cosmetic industry, such as reducing skin aging, brightening the skin, protecting it from UV

Table 1.2 Applications of bioactive compounds in cosmeceutical/cosmetic industry

Microorganism	Bioactive compound	Activity/Application	References
Amorphotheca resinae	Melanin	Sunscreen, UV protection, antioxidant, antiproliferative effect.	[44]
Serratia marcescens	Prodiogiosin	Dye, antimicrobial, antiparasitic, anti-cancer, immunosuppressive effect, sunscreen.	[45]
Nostoc sp., *LLC-10, Nostoc* sp., *CAQ-15*	Phycobiliproteins	Dye, cosmetic colorant, antioxidant.	[46]
Chromobacterium violaceum	Violacein	Dye, cosmetic colorant, UV and visible light protection.	[47]
Aspergillus oryzae	Kojic acid	Skin lightening, UV protection, collagen production.	[48]
Desmodesmus sp.	Mycosporine-like amino acids (MAAs)	UV protection, sunscreen, antioxidant, anti-inflammatory, antiaging, wound healing.	[49]
Aureobasidium pullulans	Pullulan	Drug carrier, hydrogels and films for skin hydration, photoprotective, skin whitening, antiaging, sunscreen.	[50]
Streptococcus zooepidemicus	Hyaluronic acid	Skin moisturizing, sunscreen, dermal fillers, haircare products, nails products.	[51]

radiation, and treating different skin problems. Specifically, compounds such as microbial pigments have numerous applications in the cosmetic industry due to their ability to impart color but also their beneficial properties, including sunscreen, antioxidants, antiaging agents, and skin lighteners [54]. Some of these pigments include carotenoids, a liposoluble organic pigment that absorbs light energy, thereby helping to prevent sunburn and photoaging. Melanin has antimicrobial, photoprotection, antioxidant, and thermoregulation activities [55]. Phycobiliproteins are colored proteins with antioxidant activity and exhibit red, blue, and green colors [56]. Prodiogiosin is a red pigment with immunosuppressant and anticancer activities, as well as antimicrobial and antimalarial activities [21]. Indigoidine is a water-soluble blue pigment that provides resistance to oxidative stress, and Violacein is a dark-blue pigment that possibly provides protection against UV and visible radiation, as well as antimicrobial effects [57]. The potential of pigments produced by *Talaromyces australis* and *Penicillium murcianum* as a functional cosmetic ingredient was attributed to their antioxidant properties [58]. The authors suggested using ionic gelation for encapsulation as a means to make handling dry powder easier. The method of recovery for microbial pigments holds significant importance due to their comparatively lower stability as compared to synthetic pigments [31]. Furthermore, current trends suggest sustainable and eco-friendly recovery techniques. In this regard, innovative technologies such as ultrasound and alternative solvents such as deep eutectic solvents are being studied [59].

For skin whitening and brightening, kojic acid is a promising compound derived from *Aspergillus oryzae* fermentation [48]. It acts as a natural skin-lightening agent that can help lighten dark spots and brighten the skin by filtering ultraviolet rays, thereby preventing sunburn damage. Kojic acid also inhibits tyrosinase, an enzyme involved in melanin production, making it an effective approach to reducing hyperpigmentation. In addition to its skin-related activities, kojic acid has been found to have antibacterial and antimicrobial properties, making it a potential ingredient for use as a preservative. It also possesses antioxidant activity. Furthermore, kojic acid exhibits a slight anti-inflammatory effect, expanding its range of potential applications [60]. Apart from its cosmetic applications, kojic acid can be used for collagen production, in dental care products, and as a treatment for skin disorders such as melasma and other related diseases [61].

Marine microorganisms, particularly certain strains of cyanobacteria, are capable of producing mycosporine-like amino acids (MAAs), which can absorb UV radiation and are suitable for use in sunscreens and other skincare products [62]. These naturally occurring compounds provide a safe and effective alternative to synthetic UV filters. Microbial carotenoids have also been beneficial for applications to prevent skin damage caused by excessive exposure to ultraviolet radiation [63]. The authors developed nanoemulsions with butiri oil and microbial carotenoids to provide protection against UV rays.

Furthermore, microbial BCs show promise in treating various skin conditions. For instance, *Aureobasidium pullulans*, a fungus that produces pullulan, a natural polysaccharide [64], has demonstrated the ability to improve skin hydration and elasticity and to soothe irritated skin. Hyaluronic acid (HA) is a linear polysaccharide that has become a trend in the cosmetic and cosmeceutical industry. It is widely used due to its water-retention activity, which promotes and maintains skin hydration, making it an excellent ingredient for skin care products [65]. Furthermore, HA has applications beyond esthetics, as it has been proven to be an excellent ingredient for the development of hydrogels to treat xerosis [66].

As previously mentioned, the use of BCs in cosmeceutical products requires highly purified compounds, which can increase production costs. Utilizing waste materials to extract fermentable sugars or to create a fermentation medium is a viable method to reduce costs. This strategy can effectively improve the efficiency of production while also contributing to sustainability efforts. For example, pullulan can be produced by *Aureobasidium pullulans* using soybean meal hydrolysate [41], beta-carotene can be obtained by *Rhodotorula glutinis* using orange and grape wastes [67], and cashew apple juice-based media can be used to produce hyaluronic acid [65]. Furthermore, the incorporation of metabolic pathways into genetically modified strains, such as lignocellulose degradation, has been discussed as a potential opportunity to employ low-cost natural substrates for BCs biosynthesis [68].

The demand for natural and sustainable cosmetic products is increasing, and as a result, the cosmetic industry is expected to use more microbial bioactive compounds in the future.

1.5 Environment and Agricultural Applications

One of the current trends in industrial activities is to mitigate the environmental impact caused by the resources required, such as water, energy, and greenhouse gas emissions [69]. Table 1.3 displays various applications of BCs in the environment and agricultural industries.

The main concerns regarding environmental damage are related to heavy metals, petroleum hydrocarbons, synthetic dyes, and the disposal of effluents into land, air, and water bodies [69, 78, 79]. The environmental application of microbial BCs is within the bioremediation field. Bioremediation is a well-established process that involves the use of natural agents to eliminate hazardous pollutants from the environment [70]. Among these agents, microbial bioactive compounds have gained significant attention due to their ability to positively impact the growth and activity of microorganisms involved in biodegradation processes.

Microbial bioactive compounds exert their influence through various mechanisms, including stimulating the growth and activity of indigenous microorganisms and inhibiting the growth of harmful microorganisms. Within the strategies employed to mitigate environmental pollution, microbial platforms have shown the potential to remove contaminants from soil and water effluents [78, 79].

Regarding heavy metal removal, the most commonly cited mechanisms are bioleaching, biosorption, biomineralization, intracellular accumulation, and redox reactions [80]. In line with the proposed focus of this chapter, emphasis should be

Table 1.3 Applications of bioactive compounds in the environment and agricultural industry

Microorganism	Bioactive compound	Activity/Application	References
Phanerochaete chrysosporium CDBB 686	Lignin peroxidase, manganese peroxidase, lacase	Biodegradation of synthetic dyes	[70]
Trametes versicolor	Laccase	Biodegradation of bisphenol A	[71]
Aspergillus melleus	Lipase	Biodegradation of poly (ε-caprolactone)	[72]
Acinetobacter beijerinckii	Phytohormones	Enhance soybean plant growth and heavy metal resistance.	[73]
Fusarium oxysporum	Gibberellic acid	Improve tomato growth and physiological parameters under salt stress	[74]
Paenibacillus polymyxa KM2501–1	Volatile organic compounds	Biocontrol of *M. incognita* showing nematicidal, fumigant, and chemotactic activity	[75]
Streptomyces hydrogenans DH16	Indole acetic acid	Positive impact on the growth of pea seedlings	[76]
Penicillium oxalicum	Sanxiapeptin (Aminoacids)	Antimicrobial agent	[77]

placed on the principle of the bioleaching mechanism, which involves the excretion of organic acids or polymeric substances that cause mineral dissolution [80, 81].

Organic acids such as citric acid, lactic acid, gluconic acid, and oxalic acid are produced in microbial metabolism and interact with surface metal ions to form soluble metal complexes and chelate ions [82]. Bioleaching has been applied in the recovery of metal ions such as, Pb, Ni, Cu, Zn, Al, Ca, P, and Cd from mine tailings, electronic waste, and soil [81–85].

On the other hand, cell wall components of microorganisms have been exploited as bioactive compounds for the biosorption of synthetic dyes or chemical oxygen demand (COD) removal in textile and food industry effluents. Yeast cells of *Saccharomyces cerevisiae*, *Pichia pastoris*, and *Yaworria lipolytica* have shown the potential to remove red, green, blue, and orange colorants [69, 86]. Efficient COD removal of <50% has been reported in palm oil mill effluent [87], brewery wastewater [88], as well as tannery effluent [89].

Enzymes obtained from microbes or crude enzymatic extracts have been found to be effective for bioremediation [90]. The specific enzyme required for the process depends on the type of pollutant. For example, oxidoreductases can neutralize pollutants that contain free radicals, while hydrolases can assist in the decomposition of organic compounds [91].

Sosa-Martínez et al. [70] demonstrated the possibility of using the crude enzymatic extract produced by *Phanerochaete chrysosporium* CDBB 686 using only agro-industrial waste as a substrate to treat and degrade synthetic pigments in a simulated wastewater system. This approach effectively degraded and lowered the toxicity of the frequently used industrial pigment, methyl green. Even though enzymes are effective in breaking down contaminants, their use in bioremediation remains challenging due to the high costs associated with producing and purifying these biomolecules. Yet, enzymatic processes have significant implications for the management of environmental pollutants, and their potential use in various industries deserves further exploration.

The textile industry provides another example of the potential application of microbial BCs as a substitute for synthetic pigments, thus reducing the industry's impact on the environment. Venil et al. [92] demonstrated the feasibility of using fungal pigments due to their color stability, even withstanding temperature and pH variations when applied to textile fabric. Pigments produced by *Talaromyces amestolkiae* were also used to color latex gloves, replacing synthetic pigments that may cause allergies or generate wastewater effluents [93]. Microalgal by-products have also been studied for their potential application in various industrial fields. Kumar et al. [43] elaborated on the utilization of microalgae for the production of oils that can be efficiently converted into energy. This innovative approach offers a sustainable and eco-friendly alternative to conventional fossil fuel-based energy production methods.

On the other hand, the utilization of microbial BCs for agricultural applications is an emerging field that holds great promise in addressing various agricultural challenges. As mentioned in the introduction, BCs are compounds that can have a positive impact on living organisms, including plants. These compounds have shown to be effective in increasing plant growth and improving crop quality, representing a promising, safer, and more sustainable alternative to agrochemicals.

Fungi, bacteria, and yeast produce metabolites that have a synergic effect on plants to improve growth cycles and crop yields with the minimum environmental harm [94, 95]. In this respect, some mechanisms for promoting plant growth are the solubilization of phosphorous, the production of phytohormones, and the fixation of nitrogen [96].

One of the primary mechanisms by which BCs enhance plant growth is by producing phytohormones. Phytohormones are plant hormones that control specific cellular processes, promoting plant growth and development. While plants do not possess secretion glands, hormones can be located in various sections of the plant and transferred to another location [96, 97]. Nonetheless, fungi are capable of producing some phytohormones that help in the improvement of root and leaf growth [96, 98]. For instance, two of the main hormones produced by microorganisms (*Aspergillus* sp., *Lasiodiplodia theobromae*, *Gibberella fujikuroi*, *Bacillus* sp.) are gibberellic acid and jasmonic acid [98, 99]. Gibberellic acid has been found to increase thermotolerance, growth, biochemical attributes, and yield in various crops, such as tomatoes, lettuce, and chickpeas [100–102]. Such studies have shown that the application of gibberellic acid results in higher crop yields and better crop quality, making it a promising alternative to conventional agrochemicals. Jasmonic acid, on the other hand, has been found to reduce abiotic stress in wheat, cotton, and chickpea plants [103–105].

Furthermore, the production of microbial metabolites is feasible and sustainable due to the utilization of inexpensive substrates [106]. For example, lactic acid bacteria can synthesize B-group vitamins, which can be used to stimulate the growth of several fruits and vegetables as well as obtain a biofortified food crop for human consumption [95].

Microbial BCs have also been proven effective as biocontrol agents. Juveniles of *Meloidogyne incognita* were effectively controlled by the nematicidal activity of volatile organic compounds synthesized by *P. polymyxa* KM2501–1 [75] To find a comprehensive list of fungal biological controls for plant defense, refer to the detailed information review by Sikandar et al. [107].

As more research is conducted in this field, we can expect to see more sustainable and environmentally friendly agricultural practices that rely on microorganisms.

1.6 Concluding Remarks: Challenges and Opportunities

The microbial production of bioactive compounds has gained a lot of attention in recent years due to its potential to provide sustainable alternatives to traditional chemical synthesis methods. However, several challenges need to be addressed to enable the widespread adoption of microbial production technologies.

One of the primary challenges is optimizing the production yield and scaling up production to meet commercial demand. Microbial processes' yields can be affected by various factors, including microbial strain, cultivation conditions, and downstream processing. Moreover, the production of bioactive compounds on a large scale might require significant investment for process control, which can be a barrier to entry for small- and medium-sized enterprises.

Another challenge is ensuring the purity and quality of the bioactive compounds produced by microorganisms. Obtaining regulatory approval, particularly from the FDA, can be time consuming and costly, hindering the growth of the microbial BCs market.

If there are unwanted compounds or microorganisms present, it can damage the effectiveness and safety of the final product, which may also hinder their approval. Also, this makes it inappropriate for use in pharmaceuticals, nutraceuticals, food, and other areas. It is of utmost importance to enforce strict quality control measures and analytical methods in order to guarantee the safety and purity of bioactive compounds.

Although there are challenges, the current trends in microbial-produced compounds present numerous opportunities for creating new bioactive compounds. One such opportunity is using genetic engineering techniques to modify microbial strains, which can produce compounds that would be difficult or impossible to synthesize through chemical means.

Moreover, microbial production is an environmentally friendly approach to synthesizing bioactive compounds. This innovative method enables the reduction of harmful solvent use and carbon emissions while promoting a circular bioeconomy through using renewable feedstocks for microbial fermentation. By using waste streams as feedstock, microbial production technologies can help reduce the amount of organic waste sent to landfills and contribute to developing a more sustainable and efficient system. In this sense, the development and adoption of microbial production technologies can contribute to the transition to a more sustainable and circular economy, which is essential to address the environmental, social, and economic challenges of our time.

In conclusion, the microbial production of bioactive compounds offers significant opportunities for developing sustainable and innovative solutions to address various challenges in food, pharmaceutical, cosmetic, agriculture, and other industries. However, continuing research and development will require overcoming the challenges associated with microbial production, such as optimizing yield, maintaining quality and purity, and scaling up production.

References

1. Sadh PK, Kumar S, Chawla P, Duhan JS (2018) Fermentation: a boon for production of bioactive compounds by processing of food industries wastes (By-Products)
2. Guaadaoui A, Benaicha S, Elmajdoub N et al (2014) What is a bioactive compound? A combined definition for a preliminary consensus. Int J Nutr Food Sci 3:174. https://doi.org/10.11648/j.ijnfs.20140303.16
3. Martirosyan D, Miller E (2018) Bioactive compounds: the key to functional foods. Bioact Compd Heal Dis 1:36. https://doi.org/10.31989/bchd.v1i3.539
4. Pai S, Hebbar A, Selvaraj S (2022) A critical look at challenges and future scopes of bioactive compounds and their incorporations in the food, energy, and pharmaceutical sector. Environ Sci Pollut Res 29:35518–35541. https://doi.org/10.1007/s11356-022-19423-4
5. Gomes C, Silva AC, Marques AC et al (2020) Biotechnology applied to cosmetics and aesthetic medicines. Cosmetics 7:33. https://doi.org/10.3390/cosmetics7020033
6. Kawada M, Atsumi S, Wada S, Sakamoto S (2018) Novel approaches for identification of antitumor drugs and new bioactive compounds. J Antibiot (Tokyo) 71:39–44. https://doi.org/10.1038/ja.2017.97
7. Tabacchioni S, Passato S, Ambrosino P et al (2021) Identification of beneficial microbial consortia and bioactive compounds with potential as plant biostimulants for a sustainable agriculture. Microorganisms 9:426. https://doi.org/10.3390/microorganisms9020426
8. Roohbakhsh A, Karimi G, Iranshahi M (2017) Carotenoids in the treatment of diabetes mellitus and its complications: a mechanistic review. Biomed Pharmacother 91:31–42. https://doi.org/10.1016/j.biopha.2017.04.057
9. Meléndez-Martínez AJ, Stinco CM, Mapelli-Brahm P (2019) Skin carotenoids in public health and nutricosmetics: the emerging roles and applications of the UV radiation-absorbing colourless carotenoids phytoene and phytofluene. Nutrients 11:1093. https://doi.org/10.3390/nu11051093
10. Batra B, Sharma D, Bose D et al (2023) Implications of bioprospecting marine diversity and sustainable production of bioactive compounds. In: Marine antioxidants. Elsevier, pp 27–43
11. IMARC (2023) Bioactive ingredients market: global industry trends, share, size, growth, opportunity and forecast 2023–2028
12. BCC Publishing Staff (2023) Microbial products: technologies, applications and global markets BIO086E. BCC Res LLC
13. Sarnaik A, Liu A, Nielsen D, Varman AM (2020) High-throughput screening for efficient microbial biotechnology. Curr Opin Biotechnol 64:141–150. https://doi.org/10.1016/j.copbio.2020.02.019

14. Gerzon G, Sheng Y, Kirkitadze M (2022) Process analytical technologies – advances in bioprocess integration and future perspectives. J Pharm Biomed Anal 207:114379. https://doi.org/10.1016/j.jpba.2021.114379
15. Oliveira JC, Montañez JC, Méndez-Zavala A et al (2017) Selection of best conditions of inoculum preparation for optimum performance of the pigment production process by Talaromyces spp. using the Taguchi method. Biotechnol Prog 33:621–632. https://doi.org/10.1002/btpr.2470
16. Cheng Y, Bi X, Xu Y et al (2023) Artificial intelligence technologies in bioprocess: opportunities and challenges. Bioresour Technol 369:128451. https://doi.org/10.1016/j.biortech.2022.128451
17. Helleckes LM, Hemmerich J, Wiechert W et al (2022) Machine learning in bioprocess development: from promise to practice. Trends Biotechnol. https://doi.org/10.1016/j.tibtech.2022.10.010
18. Villegas-Méndez MÁ, Montañez J, Contreras-Esquivel JC et al (2022) Coproduction of microbial oil and carotenoids within the circular bioeconomy concept: a sequential solid-state and submerged fermentation approach. Fermentation 8:258. https://doi.org/10.3390/fermentation8060258
19. Villegas-Méndez MÁ, Montañez J, Contreras-Esquivel JC et al (2023) Scale-up and fed-batch cultivation strategy for the enhanced co-production of microbial lipids and carotenoids using renewable waste feedstock. J Environ Manag 339:117866. https://doi.org/10.1016/j.jenvman.2023.117866
20. Saravana PS, Ummat V, Bourke P, Tiwari BK (2022) Emerging green cell disruption techniques to obtain valuable compounds from macro and microalgae: a review. Crit Rev Biotechnol 1–16. https://doi.org/10.1080/07388551.2022.2089869
21. Narsing Rao MP, Xiao M, Li WJ (2017) Fungal and bacterial pigments: secondary metabolites with wide applications. Front Microbiol 8:1–13. https://doi.org/10.3389/fmicb.2017.01113
22. Wang Y, Wu J, Lv M et al (2021) Metabolism characteristics of lactic acid bacteria and the expanding applications in food industry. Front Bioeng Biotechnol 9:1–19. https://doi.org/10.3389/fbioe.2021.612285
23. Barcenilla C, Ducic M, López M et al (2022) Application of lactic acid bacteria for the biopreservation of meat products: a systematic review. Meat Sci 183:108661. https://doi.org/10.1016/j.meatsci.2021.108661
24. Wang YC, Hu HF, Ma JW et al (2020b) A novel high maltose-forming α-amylase from Rhizomucor miehei and its application in the food industry. Food Chem 305:125447. https://doi.org/10.1016/j.foodchem.2019.125447
25. Tong L, Zheng J, Wang X et al (2021) Improvement of thermostability and catalytic efficiency of glucoamylase from Talaromyces leycettanus JCM12802 via site-directed mutagenesis to enhance industrial saccharification applications. Biotechnol Biofuels 14:1–9. https://doi.org/10.1186/s13068-021-02052-3
26. Amer EM, Saber SH, Markeb AA et al (2021) Enhancement of β-glucan biological activity using a modified acid-base extraction method from saccharomyces cerevisiae. Molecules 26:1–17. https://doi.org/10.3390/molecules26082113
27. Zhang L, Liu L, Wang KF et al (2019) Phosphate limitation increases coenzyme Q10 production in industrial Rhodobacter sphaeroides HY01. Synth Syst Biotechnol 4:212–219. https://doi.org/10.1016/j.synbio.2019.11.001
28. Juodeikiene G, Bartkiene E, Cernauskas D et al (2018) Antifungal activity of lactic acid bacteria and their application for Fusarium mycotoxin reduction in malting wheat grains. LWT 89:307–314. https://doi.org/10.1016/j.lwt.2017.10.061
29. Xing Q, Dekker S, Kyriakopoulou K et al (2020) Enhanced nutritional value of chickpea protein concentrate by dry separation and solid state fermentation. Innov Food Sci Emerg Technol 59:102269. https://doi.org/10.1016/j.ifset.2019.102269

30. Isbrandt T, Tolborg G, Ødum A et al (2020) Atrorosins: a new subgroup of Monascus pigments from Talaromyces atroroseus. Appl Microbiol Biotechnol 104:615–622. https://doi.org/10.1007/s00253-019-10216-3
31. Morales-Oyervides L, Ruiz-Sánchez JP, Oliveira JC et al (2020) Biotechnological approaches for the production of natural colorants by Talaromyces/Penicillium: a review. Biotechnol Adv
32. Raveendran S, Parameswaran B, Ummalyma SB et al (2018) Applications of microbial enzymes in food industry. Food Technol Biotechnol 56:16–30. https://doi.org/10.17113/ftb.56.01.18.5491
33. van der Maarel MJE, van der Veen B, Uitdehaag JC et al (2002) Properties and applications of starch-converting enzymes of the α-amylase family. J Biotechnol 94:137–155. https://doi.org/10.1016/S0168-1656(01)00407-2
34. Ashraf M, Hussain N, Baqar Z et al (2023) Bioprospecting microbial proteases in various industries/sectors. In: Microbial biomolecules. Elsevier, pp 301–324
35. Movahedpour A, Ahmadi N, Ghalamfarsa F et al (2022) β-Galactosidase: from its source and applications to its recombinant form. Biotechnol Appl Biochem 69:612–628. https://doi.org/10.1002/bab.2137
36. Ramesh C, Vinithkumar NV, Kirubagaran R, et al (2019) Multifaceted applications of microbial pigments: current knowledge, challenges and future directions for public health implications
37. Jindal N, Singh Khattar J (2018) Microbial polysaccharides in food industry. Elsevier Inc.
38. Srilekha V, Gudikandula K (2023) Antioxidant potential of carotenoids derived from marine bacteria and their applications. In: Marine antioxidants. Elsevier, pp 311–315
39. Sifuentes-Franco S, Sánchez-Macías DC, Carrillo-Ibarra S et al (2022) Antioxidant and anti-inflammatory effects of coenzyme Q10 supplementation on infectious diseases. Healthc 10. https://doi.org/10.3390/healthcare10030487
40. He C, Zhang Z, Zhang Y et al (2021) Efficient pullulan production by Aureobasidium pullulans using cost-effective substrates. Int J Biol Macromol 186:544–553. https://doi.org/10.1016/j.ijbiomac.2021.07.068
41. He S, Lu H, Zhang G, Ren Z (2021) Production of coenzyme Q10 by purple non-sulfur bacteria: current development and future prospect. J Clean Prod 307:127326. https://doi.org/10.1016/j.jclepro.2021.127326
42. Alves A, Kijjoa A (2020) Marine-derived compounds with potential use as. Molecules 25:2536
43. Kumar BR, Mathimani T, Sudhakar MP et al (2021) A state of the art review on the cultivation of algae for energy and other valuable products: application, challenges, and opportunities. Renew Sust Energ Rev 138:110649. https://doi.org/10.1016/j.rser.2020.110649
44. Oh JJ, Kim JY, Son SH et al (2021) Fungal melanin as a biocompatible broad-spectrum sunscreen with high antioxidant activity. RSC Adv 11:19682–19689. https://doi.org/10.1039/d1ra02583j
45. Wang S, Nguyen VB, Doan CT, Tran TN (2020a) Bioconversion of chitin and protein-containing. Molecules 25:1–23
46. Galetovic A, Seura F, Gallardo V et al (2020) Use of phycobiliproteins from atacama cyanobacteria as food colorants in a dairy beverage prototype. Foods 9:1–13. https://doi.org/10.3390/foods9020244
47. Cheng KC, Hsiao HC, Hou YC et al (2022) Improvement in violacein production by utilizing formic acid to induce quorum sensing in chromobacterium violaceum. Antioxidants 11:1–12. https://doi.org/10.3390/antiox11050849
48. Chib S, Jamwal VL, Kumar V et al (2023) Fungal production of kojic acid and its industrial applications. Appl Microbiol Biotechnol 107:2111–2130. https://doi.org/10.1007/s00253-023-12451-1
49. Gharib R, Tabarzad M, Hosseinabadi T (2020) Effect of high salinity on mycosporine-like amino acid production in desmodesmus sp.: trends pept. Protein Sci 5:1–6 (e2). https://doi.org/10.22037/tpps.v5i0.28876

50. Singh RS, Kaur N, Singh D et al (2023) Pullulan in pharmaceutical and cosmeceutical formulations: a review. Int J Biol Macromol 231:123353
51. Ciriminna R, Scurria A, Pagliaro M (2021) Microbial production of hyaluronic acid: the case of an emergent technology in the bioeconomy. Biofuels Bioprod Biorefin 15:1604–1610. https://doi.org/10.1002/bbb.2285
52. Thiyagarasaiyar K, Goh B-H, Jeon Y-J, Yow Y-Y (2020) Algae metabolites in cosmeceutical: an overview of current applications and challenges. Mar Drugs 18:323. https://doi.org/10.3390/md18060323
53. Gupta PL, Rajput M, Oza T et al (2019) Eminence of microbial products in cosmetic industry. Nat Products Bioprospect 9:267–278. https://doi.org/10.1007/s13659-019-0215-0
54. de Mejia EG, Zhang Q, Penta K et al (2020) The colors of health: chemistry, bioactivity, and market demand for colorful foods and natural food sources of colorants. Annu Rev Food Sci Technol 11:145–182. https://doi.org/10.1146/annurev-food-032519-051729
55. Singh S, Nimse SB, Mathew DE et al (2021) Microbial melanin: recent advances in biosynthesis, extraction, characterization, and applications. Biotechnol Adv 53:107773. https://doi.org/10.1016/j.biotechadv.2021.107773
56. Ji L, Qiu S, Wang Z et al (2023) Phycobiliproteins from algae: current updates in sustainable production and applications in food and health. Food Res Int 167:112737
57. Dodou HV, Batista AHM, Medeiros SC et al (2020) Violacein antimicrobial activity on Staphylococcus epidermidis biofilm. Nat Prod Res 34:3414–3417. https://doi.org/10.1080/14786419.2019.1569654
58. Contreras-Machuca PI, Avello M, Pastene E et al (2022) Chemical characterization and microencapsulation of extracellular fungal pigments. Appl Microbiol Biotechnol 106:8021–8034. https://doi.org/10.1007/s00253-022-12255-9
59. Colorado Gómez VK, Ruiz-Sánchez JP, Méndez-Zavala A et al (2023) Biotechnological production of microbial pigments: recent findings. In: Handbook of natural colorants. Wiley, pp 439–457
60. Phasha V, Senabe J, Ndzotoyi P et al (2022) Review on the use of kojic acid—a skin-lightening ingredient. Cosmetics 9:64. https://doi.org/10.3390/cosmetics9030064
61. Saeedi M, Eslamifar M, Khezri K (2019) Kojic acid applications in cosmetic and pharmaceutical preparations. Biomed Pharmacother 110:582–593. https://doi.org/10.1016/j.biopha.2018.12.006
62. Dextro RB, Delbaje E, Geraldes V et al (2023) Exploring the relationship between biosynthetic gene clusters and constitutive production of mycosporine-like amino acids in brazilian cyanobacteria. Molecules 28. https://doi.org/10.3390/molecules28031420
63. Mansur MCPPR, Campos C, Vermelho AB et al (2020) Photoprotective nanoemulsions containing microbial carotenoids and buriti oil: efficacy and safety study. Arab J Chem 13: 6741–6752. https://doi.org/10.1016/j.arabjc.2020.06.028
64. Wani SM, Mir SA, Khanday FA, Masoodi FA (2021) Advances in pullulan production from agro-based wastes by Aureobasidium pullulans and its applications. Innov Food Sci Emerg Technol 74:102846. https://doi.org/10.1016/j.ifset.2021.102846
65. Pires AMB, Macedo AC, Eguchi SY, Santana MHA (2010) Microbial production of hyaluronic acid from agricultural resource derivatives. Bioresour Technol 101:6506–6509. https://doi.org/10.1016/j.biortech.2010.03.074
66. Ha NG, Lee SH, Lee EH et al (2022) Safety and efficacy of a new hydrogel based on hyaluronic acid as cosmeceutical for xerosis. J Cosmet Dermatol 21:6840–6849. https://doi.org/10.1111/jocd.15368
67. Uğurlu Ş, Günan Yücel H, Aksu Z (2023) Valorization of food wastes with a sequential two-step process for microbial β-carotene production: a zero waste approach. J Environ Manag 340. https://doi.org/10.1016/j.jenvman.2023.118003
68. Xu S, Gao S, An Y (2023) Research progress of engineering microbial cell factories for pigment production. Biotechnol Adv 65:108150. https://doi.org/10.1016/j.biotechadv.2023.108150

69. Danouche M, El Arroussi H, Bahafid W, El Ghachtouli N (2021) An overview of the biosorption mechanism for the bioremediation of synthetic dyes using yeast cells. Environ Technol Rev
70. Sosa-Martínez JD, Balagurusamy N, Montañez J et al (2020) Synthetic dyes biodegradation by fungal ligninolytic enzymes: process optimization, metabolites evaluation and toxicity assessment. J Hazard Mater 400:123254. https://doi.org/10.1016/j.jhazmat.2020.123254
71. Taghizadeh T, Talebian-Kiakalaieh A, Jahandar H et al (2020) Biodegradation of bisphenol A by the immobilized laccase on some synthesized and modified forms of zeolite Y. J Hazard Mater 386:121950. https://doi.org/10.1016/j.jhazmat.2019.121950
72. Amin M, Bhatti HN, Bilal M (2020) Kinetic and thermodynamic characterization of lipase from Aspergillus melleus and its biocatalytic performance for degradation of poly(ε--caprolactone). J Chem Technol Biotechnol 0–1. https://doi.org/10.1002/jctb.6649
73. Husna HA, Shah M et al (2023) Phytohormones producing rhizobacteria alleviate heavy metals stress in soybean through multilayered response. Microbiol Res 266:127237. https://doi.org/10.1016/j.micres.2022.127237
74. Ben Rhouma M, Kriaa M, Ben Nasr Y et al (2020) A new endophytic fusarium oxysporum gibberellic acid: optimization of production using combined strategies of experimental designs and potency on tomato growth under stress condition. Biomed Res Int 2020. https://doi.org/10.1155/2020/4587148
75. Cheng W, Yang J, Nie Q et al (2017) Volatile organic compounds from Paenibacillus polymyxa KM2501-1 control Meloidogyne incognita by multiple strategies. Sci Rep 7:1–11. https://doi.org/10.1038/s41598-017-16631-8
76. Kaur T, Manhas RK (2022) Evaluation of ACC deaminase and indole acetic acid production by Streptomyces hydrogenans DH16 and its effect on plant growth promotion. Biocatal Agric Biotechnol 42:102321. https://doi.org/10.1016/j.bcab.2022.102321
77. Yang YC, Li K, Liu CX et al (2022) Sanxiapeptin, a linear pentapeptide from Penicillium oxalicum, inhibited the growth of citrus green mold. Food Chem 366:130541. https://doi.org/10.1016/j.foodchem.2021.130541
78. Feng L, Jiang X, Huang Y et al (2021) Petroleum hydrocarbon-contaminated soil bioremediation assisted by isolated bacterial consortium and sophorolipid. Environ Pollut. https://doi.org/10.1016/j.envpol.2021.116476
79. Massoud R, Hadiani MR, Hamzehlou P, Khosravi-Darani K (2019) Bioremediation of heavy metals in food industry: application of *Saccharomyces cerevisiae*. Electron J Biotechnol
80. Sharma R, Talukdar D, Bhardwaj S et al (2020) Bioremediation potential of novel fungal species isolated from wastewater for the removal of lead from liquid medium. Environ Technol Innov 18:100757. https://doi.org/10.1016/j.eti.2020.100757
81. Ye M, Liang J, Liao X et al (2021) Bioleaching for detoxification of waste flotation tailings: relationship between EPS substances and bioleaching behavior. J Environ Manag. https://doi.org/10.1016/j.jenvman.2020.111795
82. Dusengemungu L, Kasali G, Gwanama C, Mubemba B (2021) Overview of fungal bioleaching of metals. Environ Adv 5:100083
83. Alavi N, Partovi K, Majlessi M et al (2021) Bioleaching of metals from cellphones batteries by a co-fungus medium in presence of carbon materials. Bioresour Technol Rep. https://doi.org/10.1016/j.biteb.2021.100768
84. Castro L, Blázquez ML, González F, Muñoz JA (2020) Bioleaching of phosphate minerals using aspergillus niger: recovery of copper and rare earth elements. Metals (Basel). https://doi.org/10.3390/met10070978
85. Mendes G d O, Dyer T, Csetenyi L, Gadd GM (2022) Rock phosphate solubilization by abiotic and fungal-produced oxalic acid: reaction parameters and bioleaching potential. Microb Biotechnol. https://doi.org/10.1111/1751-7915.13792
86. Saravanan P, Kumaran S, Bharathi S et al (2021) Bioremediation of synthetic textile dyes using live yeast *Pichia pastoris*. Environ Technol Innov. https://doi.org/10.1016/j.eti.2021.101442

87. Karim A, Islam MA, Bin KZ et al (2021) Microbial lipid accumulation through bioremediation of palm oil mill effluent using a yeast-bacteria co-culture. Renew Energy. https://doi.org/10.1016/j.renene.2021.05.055
88. Dias C, Gouveia L, Santos JAL et al (2020) Using flow cytometry to monitor the stress response of yeast and microalgae populations in mixed cultures developed in brewery effluents. J Appl Phycol. https://doi.org/10.1007/s10811-020-02236-8
89. Okoduwa SIR, Igiri B, Udeh CB et al (2017) Tannery effluent treatment by yeast species isolates from watermelon. Toxics. https://doi.org/10.3390/toxics5010006
90. Sosa-Martínez J, Balagurusamy N, Gadi SK et al (2022) Critical process parameters and their optimization strategies for enhanced bioremediation. In: Bioremediation of environmental pollutants. Springer International Publishing, Cham, pp 75–110
91. Bhandari S, Poudel DK, Marahatha R et al (2021) Microbial enzymes used in bioremediation. J Chem 2021:1–17. https://doi.org/10.1155/2021/8849512
92. Venil CK, Velmurugan P, Dufossé L et al (2020) Fungal pigments: potential coloring compounds for wide ranging applications in textile dyeing. J Fungi 6. https://doi.org/10.3390/jof6020068
93. Mussagy CU, Oshiro AA, Lima CA et al (2023) Journal of Industrial and Engineering Chemistry Natural fluorescent red colorants produced by Talaromyces amestolkiae as promising coloring agents for custom-made latex gloves. J Ind Eng Chem 119:357–366. https://doi.org/10.1016/j.jiec.2022.11.056
94. Strobbe S, Verstraete J, Fitzpatrick TB et al (2022) A novel panel of yeast assays for the assessment of thiamin and its biosynthetic intermediates in plant tissues. New Phytol. https://doi.org/10.1111/nph.17974
95. Viscardi S, Marileo L, Barra PJ, et al (2020) From farm to fork: it could be the case of Lactic Acid Bacteria in the stimulation of folates biofortification in food crops. Curr Opin Food Sci
96. Rehman F, Kalsoom M, Adnan M et al (2020) Plant growth promoting rhizobacteria and their mechanisms involved in agricultural crop production: a review. SunText Rev Biotechnol 01. https://doi.org/10.51737/2766-5097.2020.010
97. Mukherjee A, Gaurav AK, Singh S et al (2022) The bioactive potential of phytohormones: a review. Biotechnol Rep 35:e00748
98. Salvatore MM, Alves A, Andolfi A (2020) Secondary metabolites of *Lasiodiplodia theobromae*: distribution, chemical diversity, bioactivity, and implications of their occurrence. Toxins (Basel) 12:457
99. Keswani C, Singh SP, Cueto L et al (2020) Auxins of microbial origin and their use in agriculture. Appl Microbiol Biotechnol 104(20):8549–8565
100. Guo T, Gull S, Ali MM et al (2022) Heat stress mitigation in tomato (*Solanum lycopersicum* L.) through foliar application of gibberellic acid. Sci Rep 12:11324. https://doi.org/10.1038/s41598-022-15590-z
101. Miceli A, Moncada A, Sabatino L, Vetrano F (2019) Effect of gibberellic acid on growth, yield, and quality of leaf lettuce and rocket grown in a floating system. Agronomy. https://doi.org/10.3390/agronomy9070382
102. Rafique M, Naveed M, Mustafa A, et al (2021) The combined effects of gibberellic acid and rhizobium on growth, yield and nutritional status in chickpea (*Cicer arietinum* l.). Agronomy. https://doi.org/10.3390/agronomy11010105
103. Ahmad P, Raja V, Ashraf M et al (2021) Jasmonic acid (JA) and gibberellic acid (GA3) mitigated Cd-toxicity in chickpea plants through restricted cd uptake and oxidative stress management. Sci Rep. https://doi.org/10.1038/s41598-021-98753-8
104. Ma C, Wang ZQ, Zhang LT, et al (2014) Photosynthetic responses of wheat (*Triticum aestivum* L.) to combined effects of drought and exogenous methyl jasmonate. Photosynthetica. https://doi.org/10.1007/s11099-014-0041-x
105. Nazim M, Ali M, Shahzad K et al (2021) Kaolin and Jasmonic acid improved cotton productivity under water stress conditions. Saudi J Biol Sci. https://doi.org/10.1016/j.sjbs.2021.07.043

106. Werle LB, Abaide ER, Felin TH et al (2020) Gibberellic acid production from *Gibberella fujikuroi* using agro-industrial residues. Biocatal Agric Biotechnol 25. https://doi.org/10.1016/j.bcab.2020.101608
107. Sikandar S, Saqib AY, Afzal I (2020) Fungal secondary metabolites and bioactive compounds for plant defense. In: Agriculturally important fungi for sustainable agricultur, pp 149–179

Chapter 2
Amazing Potential and the Future of Fungi: Applications and Economic Importance

Priya Arya, Sunidhi Shreya, and Amit Gupta

Abstract In the new bioeconomy, fungi play a decisive role in addressing major global challenges by upgrading resource efficiency, fabricating renewable replacements for products made from fossil fuels, upgrading waste materials to valuable food and feed ingredients, combating lifestyle diseases and antibiotic resistance by strengthening the gut biota, enhancing crop plants' resilience to climate change conditions, and acting as hosts. Through more effective and productive use of natural resources, the usefulness of fungal techniques and goods can increase sustainability. They are employed as biofertilizers and food sources due to their high protein content, and they have antibacterial properties. One of nature's most lucrative areas for discovering novel medication candidates and antimicrobials is fungi. Future resources can be found in the variety of fungi. The fungal species have an extensive economic value ranging from positive aspects such as weed killers, usage in baking, alcohol, beverages, paper and pulp, and the textile industry to negative aspects such as spreading diseases in both plants and animals, deteriorating the quality of textiles, producing hallucinogens, and secreting harmful toxins, which can result in agricultural losses and the loss of lives. Fungi also produce many antibiotics that help eradicate widespread diseases and infections. The loss of habitat, which also results in the extinction of animals and other forms of biodiversity, puts fungal diversity at risk. Therefore, we need to preserve this art and use it for the betterment of humankind.

Keywords Fungi · Economic value · Antibiotics · Biofertilizers · Humankind

P. Arya · S. Shreya · A. Gupta (✉)
Department of Microbiology, Graphic Era Deemed to be University, Dehradun, Uttarakhand, India
e-mail: dr.amitgupta.bt@geu.ac.in

R. Soni et al. (eds.), *Microbial Bioactive Compounds*,
https://doi.org/10.1007/978-3-031-40082-7_2

2.1 Introduction

Fungi (plural form of fungus) belong to eukaryotic organisms and have a heterotrophic mode of nutrition. Due to their unique features, fungi hold great promise and provide some useful applications and benefits in the pharmaceutical industry. Moreover, fungi are reported in the form of black and white dots that appear on bread and mustard leaves. This may happen due to the presence of yeast, which belongs to the ascomycetes group of fungi. In general, to date, more than two lakh species have been reported, and these fungi may grow under warm and moist conditions [1, 2]. Since ancient times, fungi have contributed to the manufacturing of bread and beverages (alcoholic), among other foods and drinks. In contrast, both yeasts and moulds play a significant role in contemporary industrial fermentation for the fabrication of food components. They are carried out as biological transformation agents for the large-scale production of substances including fatty acids, organic acids, enzymes, vitamins, and pigments. But as time has progressed, the usage of fungus and its application has become versatile. It now facilitates a number of products and deals with a number of areas. For example, in fermentation, fungus plays an important role as it breaks down important enzymes and converts solid and liquid into suitable form for fermentation. Similarly, antibiotic fungi naturally produce antibiotics to inhibit and kill microorganisms and because of which many commercial products can be derived from it [1–4]. The use of fungal (and bacterial) enzymes in lieu of chemical processes has allowed industries including textile, leather, paper, and pulp to transition from chemical to biological processing, greatly reducing their environmental impact. The production of food and feed from biological raw materials, including animal feed, baked goods, beer, wine, and juice, has been greatly enhanced by the use of enzymes. Recently, researchers still collected some fungal species from different locations and reported some diversity having economic importance in living organisms and also being able to develop some novel products that are beneficial for humans [2–5]. In this chapter, our objective is to show fungal diversity and its importance in the pharmaceutical industry, and these are:

2.1.1 Immunosuppressant Drugs

Inhibiting or reducing the severity of the body's immune response is what an immunosuppressant does. The majority of these drugs are used to reduce the likelihood that the body will reject a transplanted organ. To activate early-stage immunosuppression, control late-stage immunosuppression, or maintain organ rejection, immunosuppressive drugs are required. Fungal species produce many kinds of these drugs, e.g. Cyclosporin A is a secondary metabolite produced by the fungus *Trichoderma polysporum* and used in organs like the kidney, liver, bone marrow, and pancreas. It also aids in autoimmune diseases like AIDS because of its high specificity towards T-cells and low levels of myelotoxicity. It is made by the submerged fermentation of the fungus. It is one of the first metabolites derived from microbes to be used for clinical regulation of mammalian cells. Besides the *T. polysporum* species, it is also produced by *Aspergillus terreus, Fusarium solani,* and *Fusarium oxysporum* [6, 7].

2.1.2 *Enzymes*

Enzymes are utilized to treat and alter fibres, especially while producing textiles and handling them subsequently. For instance, cotton fibres are treated with enzymes referred to as catalases in order to get them ready for the dyeing procedures. Numerous enzymes, including certain cellulases and xylanases, are employed to degrade surface fibres in order to finish fabrics, aid in the tanning of leathers, or impart a stonewashed appearance to denim. For example, cellulase enzymes are mainly produced by the *Trichoderma* species of fungi, which gives the jeans their washed appearance and softens the fibre. Species like *Aspergillus* sp. and *Trichoderma* sp. are industrially used for the production of xylanase, which is used for the hydrolysis of xylan, the second most abundant polysaccharide in nature [8, 9]. In addition, fungal enzymes are also used in the food and beverage industry. By removing, adding, or altering ingredients like vitamins, nutritional components, colours, and flavours, enzymes may also render food more palatable or appetizing, e.g. for the production of red wine, *Monoascus purpureus*, which gives the wine its pigment. In contrast, pulp and paper industry, enzymes from species like *Ganoderma* species, *Fomitopsis* species, and *Trametes versicolor* (Turkey Tail) are all excellent options. These tough, woody, tree-dwelling fungi have characteristics that make their fibres suitable for manufacturing paper; they are resilient and sturdy, and they take dyes and inks well and provide a substitute for chemical bleaching. By generating enzymes like peroxidase and xylanase, *Bjerkandera is* used to bleach hardwood cellulose [8–10].

Numerous enzymes, including cellulolytic, proteolytic, lipolytic, and pectinolytic enzymes, are secreted by *Rhizopus* species and are employed in the manufacturing of numerous foods, such as Indonesian Tempe. α-Amylase is a fungal enzyme that aids in breaking the long chains of carbohydrates, ultimately improving the quality of breads and baking goods and lifting the fermentation method [9–11]. Lactase is an enzyme produced by *Aspergillus orzyae* and *Rhizopus oryzae* that hydrolyzes lactose and is used by pharmaceutical companies to provide dietary supplements for those who are lactose intolerant. In the animal feed industry, *Aspergillus niger* and *Rhizopus oligosporus* produce the enzyme phosphatase that hydrolyzes phytase acid, an indigestible form of phosphorus, which enhances the availability of dietary phosphorus [10–12]. In the literature, we collect some information about fungal enzymes along with source and its applications as shown in Table 2.1.

Table 2.1 Sources and applications of fungal enzymes [8–12]

Fungal enzymes	Source	Applications
Xylanase	*Trichoderma. inhamatum*	Used in breaking down xylan. In the paper and pulp industry, which removes trapped lignin and aids in bleaching.
Amylase	*Aspergillus niger*	In the baking industry, it is used to improve the quality of breads.
Lipase	*Aspergillus tamarii*	Used in the manufacturing of detergents and in baking.
Lactase	*Aspergillus orzyae* *Rhizopus oryzae*	Used by pharmaceutical companies to provide dietary supplements for those who are lactose intolerant.
Phytase	*Rhizopus oligosporus* *Aspergillus niger*	In the animal feed industry, it enhances the availability of dietary phosphorus.
Glucose oxidase	*Aspergillus tubingensis*	In the food industry, it improves the colour and taste of food materials by removing glucose from dried eggs.
Laccase	*Thievia* sp.	It is used for the treatment of effluents from pulp mills or other industries containing chlorolignins or phenolic compounds.
Pectinase	*Aspergillus niger*	Used in the textile industry for a pretreatment of cotton that aids in increased absorbance.
Invertase	*Aspergillus terreus*	Used in industries for the hydrolysis of sucrose.

2.1.3 Antioxidants

Numerous efforts were made to give more attention to antioxidant molecules from fungi. These antioxidant molecules are responsible for the inactivation and decline in the concentration of reactive nitrogen and oxygen species and have shown their effectiveness against neurodegenerative and cardiovascular diseases. In the literature, researchers have reported some antioxidant candidates from marine fungi (mycelium and culture broth). Various strategies were applied with reference to the extraction and purification of novel antioxidants, which are totally dependent on the cultivation (solid or liquid) and, most importantly, on class molecules. Small metabolites were extracted from fungi mainly through the liquid extraction method, frequently using ethyl acetate. Similarly, when fungi grow on solid media and adapt different strategies in order to use different solvents with a polarity gradient, i.e. water, methanol, and butanol. The existence of two histone deacetylase (HDAC) inhibitors (nicotinamide and sodium butyrate) in marine-derived culture fungi, i.e. *Penicillium brevicompactum*, may help enhance the production of phenolic compounds [13, 14].

The most familiar example of fungal-derived antioxidant molecules is reported in the food industry, with their major contribution being antioxidant-active packaging. The requirement for active packaging of antioxidant candidates is increasing day by day because antioxidants with lower concentrations are directly added to food. But this antioxidant effect may have extended from the food matrix to the film and accumulates in the form of synthetic plastics used in food packaging. Recently, plastics have been of serious concern related to the environment and have shown a higher demand for biodegradable films and coatings. In addition, marine fungi are able to synthesize and produce polysaccharides in sufficient quantities with antioxidant properties. In short, these marine fungi species are considered one of the most potential sources of antioxidant molecules because they use waste materials and may extract the molecule at a very low cost for the fermentation of source organisms. In addition, fungal biomass was also required in large quantities for the production of large quantities of bioactive compounds, mainly through solid-state fermentation [15]. This may be one of the most valuable methods for converting by-products at a low cost into useful products. The most familiar examples of fungal metabolites as antioxidants used in food, cosmetics, etc. are cordyol (presence of phenolic compounds); aspergilol (anthraquinones); euroxanthone (xanthones), mycosporine-glutaminol-glucoside (amino acid derivatives); astaxanthin (carotenoids); AS2-1 (carbohydrates), etc. [15, 16].

2.1.4 Anticancer Agents

In the literature, several bioactive metabolites reported from marine fungi have shown their importance in the field of drug discovery because of their therapeutic properties. Numerous metabolites from marine fungi have been reported and possess several activities, like anticancer. In addition, fungus species are reported in the deep sea, a 1000 m below the surface. The conditions in the deep sea are very extreme, with a complete absence of light, low temperatures, etc. Several reports were published that clearly mentioned that there is a diversity of fungal species in this environment. One of the studies claimed that fungal species from deep sea fungi displayed anticancer activities in various cell lines. In addition, pigments from fungal species also possess anticancer properties [17, 18], and the most familiar examples are *Monascus purpureus* and *Monascus pilosus* against human colon and hepatocellular adenocarcinomas. Similarly, pigments of other fungi also showed anticancer activity, i.e. norsolorinic acid (*A. nidulans*), shiraiarin (*Shiraia bambusicola*), alterporriol (*Alternaria* sp.), and benzoquinone (*Fusarium* sp.).

2.1.5 Organic Acids

Filamentous fungi are mainly involved in the production of organic acids with low molecular weights. Today, researchers are paying more attention to these organic acids because of their industrial applications and their involvement in natural ecology. The production of organic acids is totally dependent on the type of fungi that produce them. Some organic acids are responsible for declining pH and providing an advantage to filamentous fungi (acid tolerant). The most familiar example is seen in ectomycorrhizal fungi, where declining pH is mainly involved in solubilizing the minerals of the soil and releasing the nutrient ions for plant uptake, including microorganisms that enhance the weathering of minerals. One of the familiar examples is seen in the case of saprophytic and wood-decaying fungi that produce oxalic acid because of this pH acidification, i.e. acid catalyzes the hydrolysis of holocellulose. Similarly, these organic acids may directly interact with their environment and be responsible for causing metal detoxification (metal complexion and oxalic acid for biomass degradation). Because of these properties, fungus types, especially Basidiomycota, have been extensively studied for oxalic acid production. In addition, some fungi are also able to produce organic acids in a large-scale bioprocess and have shown several industrial applications in the fields of cosmetics, pharmaceuticals, food additives, etc. Examples of organic acids are produced by species such as *Aspergillus* (e.g. citric, gluconic, malic, and itaconic acids) and *Rhizopus* genera (e.g. lactic and fumaric acids). In some of the studies, scientists worked on specific strains, and cultures were prepared according to the conditions with different complex types of media. Due to the origin and diversity of the selected strains, which may enable us to compare the potentiality of a number of fungal groups for the production of organic acids [19, 20].

2.1.6 Biofertilizers

Due to the population explosion, researchers may have expected to enhance worldwide agricultural food production in 2050 so as to feed the global enhancement in population rate. Numerous efforts were made to reduce the burden of pesticides and chemical fertilizers. For the past several years, the use of excess quantities of chemical fertilizers had a direct effect on crop production and also shown an imbalance in the soil ecosystem. This may happen due to global warming and climate change, and soil microflora may clearly indicate a healthy generation of plants, animals, and humans. One of the most serious concerns in today's world is soil contamination, which may be due to the use of pesticides and chemical fertilizers that directly affect the environment. Numerous efforts were taken by various researchers to control the hazard effect by using biofertilizers that are environmentally friendly, and this fertilizer was adapted in almost all countries. The major function of biofertilizers is to enhance or improve soil quality. Some of the plants may have a number of interconnections with kingdoms [21, 22] like Monera, Protista, and Fungi; the most recurrent are *Mycorrhiza*, *Rhizobium*, and *Cyanophyceae*. These biofertilizers have promising benefits in terms of their plant nutrition, including disease resistance, and reducing the burden of inauspicious soil and climatic conditions. In the literature, biofertilizers may help solve such issues due to enhancements in soil salinity and chemical discharge from agricultural fields. In short, these biofertilizers are required and are important for future generations to ensure a healthy life for several generations yet to come.

2.1.7 *Biofuels*

This is a renewable source of energy that can be derived from vegetable oils, animal fats, and alcohol through a process known as transesterification. Transesterification can be described as the conversion of an ester into fatty acids, which are constituents of oils. The carbon number ranges from C10 to C22. A tree fungus known as *Gliocladium rosea* grows on the ulmo tree in Patagonia. It produces a long chain of hydrocarbons that are similar to the existing fuels. Due to its capacity to collect more than 60% lipid of its total dry cell weight, *Rhodosporidium* sp. has attracted substantial interest in the manufacture of advanced biofuels. While *Rhodosporidium toruloides* has a high tolerance to a number of inhibitors, including 5-hydroxymethyl ester, furfural, acetic acid, and vanillin, it can assimilate lipid while growing on cellobiose, sucrose, maltose, and glycerol. *Rhodosporidium* sp.'s features, including lipid accumulation, inhibitor tolerance, lipid conversion into advanced biofuels, and increased carbon consumption rate, are now being improved by the application of genetic and metabolic engineering technologies [23, 24].

Using renewable sugars produced from lignocellulosic biomass, the current study highlights the significance of the oleaginous yeast *Rhodotorula pacifica* INDKK in the field of integrated bio-refineries. *Rhodotorula* species are among the oleaginous yeasts that could aggregate larger lipid titers and show a high innate tolerance towards inhibitory chemicals produced during lignocellulosic biomass pretreatment. The most thoroughly investigated oleaginous yeast is *Y. lipolytica*. This novel model oleaginous yeast produces lipids in large quantities and has been modified to convert this flow into a variety of lipid-derived chemicals, including fatty alcohols, alkanes, and ketones. *Y. lipolytica* has the ability to generate lipids at industrially useful rates, sometimes surpassing 1.2 g/L/h. By heterologously expressing several enzymes that may take advantage of the many fatty acid (FA) species found in yeast, *Y. lipolytica* has been exploited to create a variety of distinct alkane species. In one instance, naturally occurring linoleic acid was broken down into pentane and 13-oxo-cis-9, trans-11-tridecadienoic acid via the production of a lipoxygenase from the soybean plant [23–25].

2.1.8 *Anti-diabetics*

Diabetes (blood sugar) is a widely spreading disease that can infect any age group and may directly or indirectly affect our heart, eyes, kidneys, and nerves. Most prevalent diabetes, i.e. Type 2, is mainly affected in adults where the body is unable to produce enough insulin, or insulin-dependent diabetes, also called type 1 diabetes (chronic illness), where the pancreas is unable to produce little or no insulin in our body. Anti-hyperglycemic efficiency has been documented for medicinal mushrooms and their active components, such as polysaccharides and their protein complexes, dietary fibre, and other substances derived from cultured mycelium, fruiting bodies, or broth. Many species of mushrooms are used for the treatment and maintenance of this disease [26, 27].

- *Agaricus campestris*, a common mushroom that is identified as a traditional treatment, is an insulin releaser and also helps in the transportation of 2-deoxyglucose.
- *Agaricus bisporus* (button mushroom) is the most popularly grown edible fungus in the world. It includes bioactive substances that might benefit people with diabetes mellitus.
- *Astraeus hygrometricus* reduces blood glucose levels, triglyceride levels, and cholesterol levels.
- *Coedyceps sinensis* lowers the insulin metabolism and helps induce the secretion of insulin from the pancreas.
- *Pleurotus citrinopileatus* helps reduce fasting glucose levels.

2.1.9 Antibiotic Production

Antibiotics are natural substances secreted by a microorganism, such as a fungus, to inhibit or kill other microorganisms. Industrially, antibiotics are used to cure a range of diseases and infections. In 1929, Sir Alexander Fleming made the first discovery of the function of fungi in the synthesis of antibiotic compounds. Microbes are unable to survive penicillin, an organic chemical. A deuteromycete, or green mildew, belonging to the genus *Penicillium*, is used to make penicillin. Producing antibiotics involves the utilization of *Penicillium notatum*, *Penicillium chrysogenum*, and *Cenococcum* species. With improved *Penicillium notatum* and *Penicillium chrysogenum* strains, penicillin is now produced commercially all over the world, including in India (Table 2.2). From *Streptomyces griseus,* we can obtain streptomycin. In medicine, it is extremely valuable and eliminates a large number of

Table 2.2 Antibiotics and their species or sources, along with their mode of action

Antibiotics	Species/ Sources	Mode of action
Penicillin	*Penicillium notatum*	Inhibiting the cell wall synthesis by blocking the transpeptidation.
Viridin	*Trichoderma viride*	Disrupts mitosis, nucleic acid synthesis
Fumagillin	*Aspergillus fumigatus*	Inhibitor of parasitic RNA synthesis and binds to methionine aminopeptidase to inactivate it.
Cephalosporin	*Acremonium* sp.	Inhibition of cell wall synthesis
Citrinin	*P. citrinum*	Rapid induction of antioxidants and drug extrusion properties
Fusidic acid	*Fusidium coccineum*	Interfere with the protein synthesis of cell wall by blocking the ribosomes.
Palutin	*P. palatum*	Alters the barrier function of intestinal epithelial

organisms, primarily gram-negative organisms, that penicillin is unable to kill. Also taken from *Aspergillus* cultures are a variety of drugs [28, 29].

Numerous cases of dermatophytosis (ringworm) are managed with the antifungal drug griseofulvin, produced by the fungal species *Penicillium griseofulvum*, discovered in 1939 in soil. This includes skin fungal infections after antifungal treatments have failed, as well as nail and scalp fungal infections. It is consumed orally.

Acremonium (formerly known as *Cephalosporium*), a mould, is the source of a broad class of medicines known as cephalosporins. Cephalosporins function similarly to penicillins and are bactericidal (kill bacteria). The enzymes that produce peptidoglycan, a crucial part of the bacterial cell wall, are bound to them, and their activity is inhibited. Following the discovery of the first cephalosporin in 1945, scientists altered the composition of cephalosporins to increase their potency against a wider variety of bacteria. A new "generation" of cephalosporins was produced every time the structure altered. Cephalosporins are divided into five generations. FA/PHA, everything else, ONE/TEN/IME, PI & QUI, and ROL, prefix is used to indicate a cephalosporin [30, 31]. A broad-spectrum cephalosporin antibiotic called cefixime is frequently used to treat bacterial infections of the upper respiratory tract, urinary tract, and ears. Various bacterial infections can be treated with ceftriaxone, a third-generation cephalosporin antibiotic that is marketed under the trade name Rocephin. Among them are infections of the middle ear, endocarditis, meningitis, pneumonia, and infections of the bones and joints.

2.2 Biocontrol of Insects Using Fungi

Utilizing natural enemies to reduce or mitigate insect pests and their effects is known as myco-biocontrol. This method is both efficient and environmentally friendly. One of the earliest species to be utilized for the biological management of pests was an entomopathogenic fungus. A number of entomopathogens can offer efficient long- and short-term management when introduced inundatively into a range of environments [27–33].

- *Verticillium lecanii*: In monocultures of vulnerable crops, the main parasite *Verticillium lecanii* was thought to be responsible for a drastic reduction in cereal-cyst nematode populations. For usage on greenhouse chrysanthemums, many decades' prior, *Verticillium lecanii* was created to suppress whiteflies and various aphid species, notably the green peach aphid (*Myzus persicae*).
- *Nomuraea* species: There are many species of *Nomuraea*, such as a dimorphic hyphomycete called *Nomuraea rileyi*, that has been linked to epizootic fatalities in a variety of insects. Numerous insect species, including *Spodoptera litura* and some Coleoptera, have been demonstrated to be vulnerable to *N. rileyi*.
- *Beauveria* species: These are filamentous fungi that belong to the group *Deuteromycetes*. The fungus *Beauveria bassiana* naturally develops in soils all

over the world and causes white muscardine sickness in a number of insect species. It is highly host specific and causes pest suppression.
- *Paecilomyces* species: Nematophagous fungi of the genus *Paecilomyces* destroy dangerous nematodes through pathogenesis, which results in illness in the worms. So, by applying the fungus to the soil, it can be utilized as a biocontrol agent for managing nematodes. Specifically, *Paecilomyces lilacinus* infects and ingests the eggs of root knot and cyst nematodes.

2.2.1 Biofilm Inhibitors

A surface-associated microbial colony is known as a biofilm. Biofilms may develop on a variety of fungi. This growth type is important for understanding infection biology because biofilm development on devices that are implanted is a key contributor to recurring infections. Device-associated infections are very challenging to treat since biofilms are also poorly responsive to drugs [27–30].

2.3 Economical Importance of Fungi

Fungi has a vast range of activities which can be both useful and disastrous.

2.3.1 Fungi as a Harmful [1–5]

2.3.1.1 Production of Harmful Toxins

Mycotoxins are harmful secondary metabolites produced by some fungi and play a part in the spread of some illnesses in both humans and other animals. Mycotoxins, such as patulin, aflatoxin, ergot alkaloids, and ochratoxin, can have detrimental health effects ranging from immediate poisoning to long-term consequences including immunological deficiencies, liver and kidney fibrosis, and cancer.

2.3.1.2 Fungi Causing Animal and Plant Diseases

Several minor and serious plant diseases are brought on by fungi. Some of them also contribute to famine in various regions of the world. Such as downy mildew caused by white rust caused by the families *Albuginaceae* and *Peronosporaceae*, late blight of potato diseases caused by *Phytophthora infestans*, and damping of seedling diseases caused by *Pythium debaryanum*. Some fungi are parasitic on both humans and animals, causing infections of the skin, hair, and nails. Malassezia species and dermatophytes that can use keratin as a food source have a special enzymatic capacity called keratinase.

2.3.1.3 Production of Hallucinogenic Substances

The ergot disease of rye's causative agent, *Claviceps purpurea*, produces LSD (d-lysergic acid diethylamide), a well-known hallucinogenic substance, from its sclerotia. Psilocin and psilocybin, which have psychedelic characteristics, are produced by other fungi like *Psilocybe mexicana*. The hallucinogenic drugs may harm brain cells and alter a person's ability to perceive reality.

2.3.1.4 Bio-deterioration of Textiles

Any unwanted modification to a substance caused by an organism's essential functions is referred to as "bio-deterioration." It can be achieved by many methods, such as the penetration of microorganisms into the cavity of fibres, the occurrence of spots, bubbles on the surface of textiles, and the deterioration of mechanical properties by *Aspergillus* species.

2.3.2 *Benefits of Fungi [1–5, 8–12]*

2.3.2.1 Alcohol Production

The brewing business is built on the fungi that produce alcohol. The unicellular fungus *Saccharomyces cerevisiae* is often known as yeast. Fermentation is the process through which food, such as starch and sugar, is turned into carbon dioxide and alcohol. It is employed in the creation of alcoholic drinks as well as bakery products, including cake, bread, and other baked goods. *Saccharomyces ellipsoideus* produces wines with an alcohol content of around 14% from grapes or other fruits. *Saccharomyces cerevisiae* uses barley malt to produce beer that has 3–8% alcohol.

2.3.2.2 Bread Making

In the baking industry, some strains of fungi, such as *Saccharomyces cerevisiae*, are used to make bakery goods such as bread. These strains are grown on molasses and are known as "baker's yeast. Yeast is added to the flour, kneaded, and kept at a warm temperature until it rises. This process is called leavening. Enzymes like amylase, maltase, and zymase are used in this process.

2.3.2.3 Food

Soy sauce is a thick, salty liquid with a meat-like taste that is abundant in amino acids. The unpleasant soybeans were fermented using microbial cultures in Japan, where they were initially manufactured. There are two phases to fermentation. Soybeans are first soaked, then boiled to eliminate impurities, and then combined with toasted wheat. *Aspergillus oryzae* is added to the mixture, which is then maintained aerobically at 25 °C for 20–40 h. The soy paste is broken down by the fungus' invertases, amylases, and cellulases. After mixing, the mixture is introduced to the second fermentation stage.

2.3.2.4 Weed Killer

Bioherbicides, or "microbial weed killers," In contrast to synthetic pesticides, fungi are recognized for their extremely precise and effective activity and minimal residual effects. Fungi are used as bioherbicides, and a few examples with their intended targets, which are *Wallrothiella arecuthobii*, *Septagloeum gillis*, and *Colletotrichum gloeosporiordes*, are all mistletoe species. *Phyllosticta* (*Glycosmis*), Leptosphaerulina trifolia (Passiflora's), *Puccinia chondrillina* (Rush weed), and the pamakani weed, *Cercospora ageratinae* are used as bioherbicides and show higher benefits as compared with synthetic pesticides.

2.4 Future Prospects of Fungi

Fungal species had made their mark in the past and continued to do so in the present via their endless properties, advantages, and applications. And its future remains bright and prosperous. Due to coronaviruses, many species of fungi have emerged as agents of infection. These can be identified and thoroughly studied, understanding fungal illnesses like mucormycosis that are linked to COVID-19 might benefit from more investigation. A Global Virome Project has also been started in order to identify zoonotic viral dangers and prevent future pandemics. This project will sample bats and other animals in order to find viruses. In addition to screening for viruses and fungi in animals, this project poses a concern since human mobility while sampling might transfer illness between wildlife groups. Metagenomics, a new application of next-generation sequencing (NGS), involves the direct sequencing of whole communities from clinical samples using a whole-genome shotgun technique in conjunction with computer-based reconstruction of the entire or part genome sequences of the species present in the sample. Many antifungal medications have developed resistance over time and because there are only a few systemic antifungal medications still in use to treat IFIs, the prevalence of resistance to drugs in *Candida* and *Aspergillus* presents a grave risk to human health. The rising incidence of resistance to azole among invasive NAC species, particularly *C. parapsilosis* and *C. tropicalis*, emphasizes the urgent need for a deeper understanding of the underlying resistance mechanisms. This needs the isolation of more novel species that can tackle the new and existing strains.

2.5 Conclusion

While the fungal kingdom offers tremendous potential for applications in biotechnology, medicine, and environmental sustainability, it also poses grave risks to the health of people, plants, and animals. In order to prevent them from posing an even greater threat to humanity, we need to give them special attention. But many of its benefits can still be investigated and used in different contexts. Such development is necessary to fully utilize fungi in the bioeconomy. More ideas can be drawn from the kingdom of fungi. To prevent microbiological contamination of products and equipment, the appropriate antimicrobial compounds are widely utilized in a variety of sectors. Building support for mycology worldwide may be accomplished through raising understanding of and appreciation for the function of fungi.

References

1. Cerimi K, Akkaya KC, Pohl C, Schmidt B, Neubauer P (2019) Fungi as source for new bio-based materials: a patent review. Fungal Biology and Biotechnology 6:17
2. Sankaran S, Khanal SK, Jasti N, Jin B, Pometto AL, van Leeuwen JH (2010) Use of filamentous fungi for wastewater treatment and production of high value fungal byproducts: a review. Crit Rev Environ Sci Technol 40:400–449
3. Pessoa MG, Paulino BN, Mano MCR, Neri-Numa IA, Molina G, Pastore GM (2017) Fusarium species-a promising tool box for industrial biotechnology. Appl Microbiol Biotechnol 101: 3493–3511
4. Wosten HAB (2019) Filamentous fungi for the production of enzymes, chemicals and materials. Curr Opin Biotechnol 59:65–70
5. Meyer V, Basenko EY, Benz JP, Braus GH, Caddick MX, Csukai M et al (2020) Growing a circular economy with fungal biotechnology: a white paper. Fungal Biol Biotechnol 7:5
6. Ianiri G, Applen Clancey S, Lee SC, Heitman J (2017) FKBP12-dependent inhibition of calcineurin mediates immunosuppressive antifungal drug action in alassezia. MBio 8: e01752–e01717
7. Noori S, Naderi GA, Hassan ZM, Habibi Z, Bathaie SZ, Hashemi SM (2004) Immunosuppressive activity of a molecule isolated from *Artemisia annua* on DTH responses compared with Cyclosporin A. Int Immunopharmacol 4:1301–1306
8. Da Luz JMR, Nunes MD, Paes SA, Torres DP, da Silva MDCS, Kasuya MCM (2012) Lignocellulolytic enzyme production of *Pleurotus ostreatus* growth in agroindustrial wastes. Braz J Microbiol 43:1508–1515
9. Green BJ, Beezhold DH (2011) Industrial fungal enzymes: an occupational allergen perspective. J Allergy 2011:682574
10. Sutay Kocabaş D, Grumet R (2019) Evolving regulatory policies regarding food enzymes produced by recombinant microorganisms. GM Crop Food 10:191–207

11. Fuglsang CC, Johansen C, Christgau S, Adler-Nissen J (1995) Antimicrobial enzymes: applications and future potential in the food industry. Trends Food Sci Technol 6:390–396
12. Santacruz-Juarez E, Buendia-Corona RE, Ramirez RE, Sanchez C (2021) Fungal enzymes for the degradation of polyethylene: molecular docking simulation and biodegradation pathway proposal. J Hazard Mater 411:125118
13. Smith H, Doyle S, Murphy R (2015) Filamentous fungi as a source of natural antioxidants. Food Chem 185:389–397
14. Schmidt CG, Gonçalves LM, Prietto L, Hackbart HS, Furlong EB (2014) Antioxidant activity and enzyme inhibition of phenolic acids from fermented rice bran with fungus *Rizhopus oryzae*. Food Chem 146:371–377
15. Hur SJ, Lee SY, Kim YC, Choi I, Kim GB (2014) Effect of fermentation on the antioxidant activity in plant-based foods. Food Chem 160:346–356
16. Arora DS, Chandra P (2010) Assay of antioxidant potential of two *Aspergillus* isolates by different methods under various physio-chemical conditions. Braz J Microbiol 41:765–777
17. Bladt TT, Frisvad JC, Knudsen PB, Larsen TO (2013) Anticancer and antifungal compounds from *Aspergillus, Penicillium* and other Filamentous Fungi. Molecules 18:11338–11376
18. Giridharan P, Verekar SA, Khanna A, Mishra PD, Deshmukh SK (2012) Anticancer activity of sclerotiorin, isolated from an endophytic fungus *Cephalotheca faveolata* Yaguchi, Nishim. and Udagawa. Indian. J Exp Biol 50(7):464–468
19. Jarosz-Wilkolazka A, Graz M (2006) Organic acids production by white rot Basidiomycetes in the presence of metallic oxides. Can J Microbiol 52(8):779–785
20. Liu AM, Chen JF, Zhang JJ, Fang Q, Wang M, Zha Q (2010) Effect of organic acid and amino avcid on the mycelial growth of Agaricus bisporus 2796. Chinese Agric Sci Bull 26(3):46–51
21. Rodriguez A, Sanders IR (2015) The role of community and population ecology in applying mycorrhizal fungi for improved food security. ISME J 9:1053–1061
22. Pellegrino E, Bedini S (2014) Enhancing ecosystem services in sustainable agriculture: biofertilization and biofortification of chickpea (*Cicer arietinum* L.) by arbuscular mycorrhizal fungi. Soil Biol Biochem 68:429–439
23. Ranganathan A, Smith OP, Youssef NH, Struchtemeyer CG, Atiyeh HK, Elshahed MS (2017) Utilizing anaerobic fungi for two-stage sugar extraction and biofuel production from lignocellulosic biomass. Front Microbiol 8:635
24. Sharma HK, Xu C, Qin W (2019) Biological pretreatment of lignocellulosic biomass for biofuels and bioproducts: an overview. Waste Biomass Valori 10:235–251
25. Zabed HM, Akter S, Yun J, Zhang G, Awad FN, Qi X, Sahu JN (2019) Recent advances in biological pretreatment of microalgae and lignocellulosic biomass for biofuel production. Renew Sust Energ Rev 105:105–128
26. Agrawal S, Samanta S, Deshmukh SK (2022) The antidiabetic potential of endophytic fungi: future prospects as therapeutic agents. Biotechnol Appl Biochem 69:1159–1165
27. Khan R, Tahira S, Naqvi Q, Fatima N (2019) Study of antidiabetic activities of endophytic fungi isolated from plants. Pure Appl Biol 8:1287–1295
28. Sanchez JF, Somoza AD, Keller NP, Wang CC (2012) Advances in Aspergillus secondary metabolite research in the post-genomic era. Nat Prod Pep 29:351–371
29. Payne GA et al (2006) Whole genome comparison of *Aspergillus flavus* and *A. oryzae*. Med Mycol 44:S9–S11
30. Pichichero ME, Casey JR (2007) Safe use of selected cephalosporins in penicillin-allergic patients: a meta-analysis. Otolaryngol Head Neck Surg 136:340–347

31. Yang MS, Kang DY, Seo B, Park HJ, Park SY, Kim MY, Park KH, Koo SM, Nam YH, Kim S et al (2018) Incidence of cephalosporin-induced anaphylaxis and clinical efficacy of screening intradermal tests with cephalosporins: a large multicenter retrospective cohort study. Allergy 73:1833–1841
32. Rana IS, Kanojiya A, Sandhu SS (2008) Mosquito larvicidal potential of fungi isolated from larval mosquito habitats against *Aedes aegypti*. J Biol Control 22:179–183
33. Nunez E, Iannacone J, Gómez H (2008) Effect of two entomopathogenic fungi in controlling *aleurodicus cocois* (Curtis, 1846) (Hemiptera: *Aleyrodidae*) Chilean. J Agric Res 68(1):21–30

Chapter 3
Commercial Compounds from Algae

J. R. Benavente-Valdés, D. Rodríguez-Zúñiga, V. Cepeda-Tovar, and O. Solís-Quiroz

Abstract Microalgae are microscopic unicellular organisms, with the ability to convert solar energy into chemical energy and fix inorganic molecules such as CO_2 to obtain organic molecules through a set of chemical reactions known as photosynthesis. Because microalgae are more efficient in the photosynthetic process than higher plants, there has been great interest in microalgae as organisms with great potential to produce valuable compounds of industrial value. Currently, microalgae have a wide range of commercial applications, ranging from the animal feed industry, cosmetics, water bioremediation, agriculture, nutraceutical food, and drug applications. Metabolites obtained from microalgae can improve the nutritional value of food and feed due to their chemical composition. These microorganisms can be used to produce a variety of metabolites such as proteins, polyunsaturated fatty acids, carbohydrates, carotenoids, vitamins, minerals, and phytohormones for application in medicine, food, and feed additives, cosmetics, and energy production. The current consumer preference for organic and naturally sourced products is driving the expansion of the market for bioactive compounds from microalgae. Industry growth with microalgae products is focused on dietary supplements and the developing nutraceutical industry. The main microalgae strains used with the highest current and future projected use are *Spirulina, Chlorella, Dunaliella salina, Haematococcus pluvialis,* and *Nannochloropsis*. Due to the great importance of biomolecules obtained from natural sources such as microalgae, this chapter will analyze the main bioactive compounds derived from microalgae, the main generating strains, and the commercial value of each of the metabolites of interest.

Keywords Biocompounds · Algae · Commercial value · Industrial interest · Applications

J. R. Benavente-Valdés (✉) · D. Rodríguez-Zúñiga · V. Cepeda-Tovar · O. Solís-Quiroz
Faculty of Chemical Sciences, Department of Chemical Engineering, Autonomous University of Coahuila, Saltillo, Mexico
e-mail: roberto.benavente@uadec.edu.mx

R. Soni et al. (eds.), *Microbial Bioactive Compounds*,
https://doi.org/10.1007/978-3-031-40082-7_3

3.1 Introduction

Microalgae are photosynthetic organisms with the capacity to accumulate a diversity of biocompounds of industrial interest, for example, the natural pigments present in microalgae such as chlorophyll or carotenoids, responsible for capturing solar energy for photosynthesis and act as photoprotectors in the cell; they are used at the industrial level for their antioxidant, anticarcinogenic, anti-inflammatory, and neuroprotective activity; with applications for the food, cosmetic, and pharmaceutical industries. Microalgae have the capacity to accumulate 60–80% of proteins, which can be incorporated into the industry as functional ingredients [1].

The production of polysaccharides in microalgal cells occurs during the process of photosynthesis and accumulates in the cytosol and chloroplasts, and these organisms are considered a sustainable source of these compounds; the carbohydrate content of microalgae can reach up to 50% of their dry weight; they are used in the pharmaceutical, cosmetic, and food industries [2] and can accumulate between 20% and 50% of their dry weight in lipids, this content can reach 80% depending on nutritional and environmental conditions [3]. Lipids are used for biodiesel production, as food supplements or in the pharmaceutical industry. The lipids of most interest are eicosapentaenoic acid (EPA), saturated and polyunsaturated fatty acids (PUFA), and docosahexaenoic acid (DHA) [3].

The growth of the market for bioactive compounds from microalgae is driven by the current consumer inclination toward organic and naturally sourced products, industry growth with products from microalgae is inclined toward dietary supplements, and growing nutraceuticals industry. The global microalgae market by 2027 is expected to reach $4.6 billion, with a compound annual growth rate (CAGR) of 4.3% [4]. By the year 2023, the microalgae species most commonly used at the industrial level are *Spirulina, Chlorella, Dunaliella salina, Haematococcus pluvialis y Nannochloropsis* [5]. Spirulina microalgae species will account for the largest growth in the global microalgae market by 2023 [4].

Microalgae can be grown in a variety of environments, require less soil, water, and have a shorter growth cycle than traditional crops, allowing for faster and simpler cultivation. Microalgae are a sustainable source of bioactive compounds; however, the amount and type of metabolites obtained varies depending on cultivation conditions. When these processes are better understood and allow obtaining biomolecules at lower costs, The metabolites produced by microalgae give these organisms the potential to become sustainable and economical sources of new chemical compounds with applications in different industries.

In this chapter, we will discuss the main bioactive compounds obtained from microalgae, the main producing strains, and the commercial value of each of the metabolites of interest.

3.2 Microalgae as a Source of High-Value Compounds

Microalgae are unicellular photosynthetic microorganisms ranging in size from 2 to 200 μm, commonly found in freshwater, saline, and hypersaline environments, they are photosynthetic organisms with the ability to tolerate different temperatures, pH values, and light intensities [6]. Due to their high metabolic flexibility, they are adaptable to various growing conditions, as well as the possibility of rapid growth [7].

Microalgae are classified into five phylum, which are Chlorophyta, known as green algae, Chlorophylls a and b are found in a single chloroplast surrounded by two enveloping membranes. *Haematococcus pluvialis, Botryococcus brauniis, Chlorella vulgaris, Dunaliella salina,* and *Parietochloris incisa* are the main strains of the phylum Chlorophyta and are widely used in commercial production [8, 9]. Rhodophyta, known as red algae, include mainly multicellular marine species. They are spherical cells and contain a single chloroplast that is surrounded by two enveloping membranes *Rhodella reticulata* and *Porphyridium cruentum* are currently under investigation with promising characteristics for phycobilin production [6].

The species belonging to the phylum Haptophyta are characterized by being mainly marine, unicellular colonial species, although several freshwater species have been reported. *Isochrysis aff. galbana (T-ISO)* and *Pavlova salina* are the most commonly used in aquaculture and are recognized for their high carbohydrate content [8, 9]. Stramenopiles are characterized by containing only chlorophyll a, and *Nannochloropsis oculata* is frequently used as live food in aquaculture. The class Bacillariophyceae is characterized as unicellular organisms, although some are colonial. They are used as feed in aquaculture. The diatoms *Skeletonema costatum, Chaetoceros muelleri*, and *Thalassiosira pseudonana* are cultivated on a commercial scale. Finally, the microalgae of the class Labyrinthulomycetes are identified as heterotrophic, filamentous protists. Most are saprotrophic decomposers and some act as parasites. *Species* such as *Ulkenia, Schizotrichium, Auranthiochytrium,* and *Thraustochytrium* are of great commercial interest due to the amount of fatty acids and pigments they accumulate [6, 9].

Dinophyta are unicellular microalgae with some species from marine and freshwater areas. It is known that 50% of this species are photosynthetic and the other half lacks chloroplasts and their nutrition is by heterotrophy. The species *Crypthecodinium cohnii* is of great commercial interest for its heterotrophic production of DHA [9].

Microalgae capture sunlight through photosystems I and II present in the thylakoid membrane of chloroplasts, to obtain chemical energy in the form of ATP and NADPH, then these molecules enter the Calvin cycle so that through redox reactions in which CO_2 is converted into carbohydrates, as a by-product of the photosynthesis process O_2 is obtained [10]. The microalgae possess many molecules of special interest, such as carbohydrates, proteins, ω-3 and ω-6 fatty acids, pigments, and various types of vitamins, for which they have been widely used as food supplements due to their diverse bioactive health benefits, also have wide application in the agricultural, energy, food and pharmaceutical industries. The diversity of these molecules will depend mainly on the species of microalgae and the environmental and nutritional conditions to which the strain is subjected [11].

The rate of microalgal biomass production and the accumulation of the metabolites of interest in a culture system depend on the process conditions, such as agitation, salinity, temperature, light intensity, photoperiod, carbon nitrogen (C/N) ratio, gas exchange, and the design and operation of an algae culture system plays a decisive role in both photosynthetic productivity and economic efficiency. The design of the cultivation system must allow the entry of light, the optimum volume of liquid for adequate agitation, and light dispersion [12].

Several large-scale cultivation systems have been developed to produce microalgae, the cultivation systems are as follows open raceway ponds and closed photobioreactors (PBR). However, closed systems allow continuous operation and are equipped with agitation, aeration, pH control, heat exchange, addition of medium and CO_2, which effectively control cultivation conditions and reduce contamination, allowing high purity biomass to be obtained. The use of hybrid systems, which are designed in two stages, is also reported, the first stage of algae growth occurs in a closed system (photobioreactor), and the second stage takes place in an open system, exposing the cells to nutritional stress, which stimulates the synthesis of desired metabolites [7, 13].

When microalgae culture is limited with nutrients such as phosphorus and nitrogen, an increase in the accumulation of lipids, especially triacylglycerides, has been evidenced. Glycolipids and phospholipids in conditions of low light intensity are associated with cell membranes [13], also when there is a decrease of nutrients, there is a stress on the cells causing the formation of free radicals and a change in the content of antioxidants such as primary carotenoids (violaxanthin, β-carotenoids chlorophylls, and voheriaxanthin) [14].

The most widely used microalgae at the industrial level are *Spirulina*, it is used in the production of food supplements, due to its high protein and nutrient content. *Haematococcus pluvialis*, used in the production of anthoxanthin, a natural antioxidant used in food supplements, skin care products, and the aquaculture industry, and *Dunaliella salina*, used in the production of beta-carotene in the food industry.

3.3 Pigments

In algae as photosynthetic organism has three main types of photosynthetic pigments: chlorophylls, carotenoids (carotenes and xanthophylls), and phycobilins (Table 3.1). Chlorophylls and carotenoids typically fat soluble, whereas phycobilins water soluble [22]. They also possess advantageous biological properties as neuroprotective, anti-angiogenic, anti-inflammatory, anticancer, and antioxidant [23, 24]. This section will describe the importance of these pigments in the market, their benefits, and their commercial value.

Table 3.1 Main commercially important microalgae pigments, source strain and market price

Pigment	Principal commercial strain	Color	Yield (% w/w)	Price range (USD per kg)	References
Astaxanthin	*Haematococcus pluvialis and Chlorella zofingiensis*	Red	0.15–4.0	2500–10,000	[15, 16]
β-carotene	*Dunaliella salina*	Red orange	12	300–3000	[15, 16]
Lutein	*Dunaliella salina*	Yellow orange	0.5	1866–2800	[17]
Phycocyanin	*Spirulina Platensis*	Blue	25	50–900	[18]
[19]	*Chlorella vulgaris*	Green	4.5	12–30	[20]
Phycoerythrin	*Porphyridium* sp.	Red	8.5	230–5600	[21]

3.3.1 Carotenoids

Carotenoids are fat-soluble pigments derived from tetraterpenes with a 40-carbon polyene structure, and they have the capacity to absorb light with a wavelength of 400–550 nm, which is advantageous for photosynthesis [25, 26]. These carotenoids are divided into xanthophylls that include oxygen, such astaxanthin and zeaxanthin, and those that do not, such as carotene and lycopene. However, the actions of the two groups of carotenoids in photosynthesis are what separates them. β-carotene and lutein, the two main carotenoids, are essential for the energy transfer to chlorophyll, which supports photosynthesis and maintains cellular viability [27]. On the other hand, two secondary carotenoids, astaxanthin and canthaxanthin, function as protective elements under stressful conditions by constructing protective layers and insulating cells from oxidative damage [15, 16].

As for the main commercial uses of carotenoids, there is the food industry, animal feed, cosmetics, and supplements for therapeutic purposes. Of these, the use of β-carotene stands out for coloring food and beverages, providing color and preservation properties, as well as in the formulation of vitamin compounds [28]. Astaxanthin is used as an additive in animal feed with the purpose of improving the color of salmon meat or egg yolk, as well as being an important source of antioxidant with neuroprotective and anticancer properties, and its use has recently been demonstrated in the treatment of gastrointestinal diseases against the bacterium *Helicobacter* [29]. Lutein, in addition to being a widely used pigment in food, stands out for its application in the treatment of macular degeneration and cataract formation in people over 40 years of age [29].

3.3.2 Chlorophylls

Chlorophylls are green pigments found in algae, bacteria, and higher plants. Chlorophylls are tetrapyrroles with centrally bound magnesium that are gaining economic attention due to their color and biological features. Chlorophyll a (blue-green color), b (bright-green), c (yellow-green), d (brilliant-forest green), and f (emerald-green) are examples of chlorophyll structures. Photosynthetic organisms primarily include chlorophyll a and b, but chlorophyll c, d, and f are found only in a few microalgae species, algae, and photosynthetic bacteria [30, 31]. This pigment is a fat-soluble molecule having bioactive capabilities like antioxidant and antimutagenicity. Furthermore, this pigment can be isolated in varying amounts from microalgae biomass 0.5% to 1.5% (on a dry weight basis). This natural green pigment is used in the cosmetic, pharmaceutical, and food industries. Although chlorophylls are acknowledged as natural pigments, however, is chemically unstable in pH settings, and is heat and light sensitive. *Chlorella* is the main commercial source of chlorophyll and has two forms of this pigment chlorophyll a and chlorophyll b, which can collect in significant levels (4.5% of dry weight) and are the most abundant pigments. *Chlorella* is regarded as the "emerald food" because of its chlorophyll content. It does, however, contain additional pigments, such as astaxanthin, c-astaxanthin, beta-carotene, and lutein, even in little concentrations [32].

Chlorophyll is a pigment of great abundance in plant sources. The main commercial uses of this pigment are the coloring of food, beverages, and other products such as chewing gum, candies, and pills [33]. It is widely used in oral hygiene products to combat halitosis, as well as therapeutic uses in gut microbiota rebalance and anemia treatment [34, 35].

3.3.3 Phycobiliproteins

Phycobiliproteins are light-harvesting protein pigments. The two most well-known cyanobacteria species for commercial production of phycobiliproteins are *Arthrospira* sp. (phycocyanin), *Spirulina*, and *Porphyridium* sp. (phycoerythrin) [36]. The pigments are found in supramolecular phycobilisomes on the thylakoid membrane's external surface and account for 40–50% of total soluble proteins [37]. They are water-soluble molecules divided into three groups: phycoerythrin (red pigment), allophycocyanin (light-blue pigment) and phycocyanin (blue pigment). The chemical structure, color, and absorption spectra differ [38].

Due to their fluorescent properties, phycobiliproteins are also used as molecular markers and immunoassays, being phycocyanin and phycoerythrin extracted from cyanobacteria tested as protein markers in electrophoresis [39, 40]. In addition, the phycocyanin extracted from Spirulina, as well as chlorophylls, has been used as a colorant in foods, beverages, cakes, and candies, as well as an additive in dairy products such as yogurt, demonstrating that its use is safe in the food industry [41].

3.3.4 Polyunsaturated Fatty Acids

Fatty acids (FA) are the most important components of active compounds obtained from microalgae, as they have structural diversity and great importance at an industrial level. These organisms have an average FA content that varies between 1 and 70% of dry weight; under certain conditions, some species can reach up to 90% of dry weight [42, 43] and are classified according to the number of double bonds present in the carbon chain: saturated (SFA, no double bonds), monounsaturated (MUFA, single double bonds), and polyunsaturated (PUFA, more than 2 double bonds). PUFAs are represented by two families: $n-3$ u $\omega-3$, eicosapentaenoic acid EPA ($C_{20:5}$, $n-3$), (α-linolenic acid (α-ALA, $C_{18:2}$, $n-3$), and docosahexaenoic acid (DHA) ($C_{22:6}$, $n-3$) and $n-6$ u $\omega-6$ ($n-6$ PUFAs include γ-linoleic acid (LA, $C_{18:3}$, $n-6$) and arachidonic acid ARA ($C_{20:4}$, $n-6$)). $n-3$ and $n-6$ represent the first double bond at the third or sixth carbon of the carbon chain end and are synthesized from linoleic acid (LA) and α-linolenic acid (ALA), as shown in Fig. 3.1 [44].

In microalgal cells, FA have three basic functions: to be structural components of cell membranes, to store energy, and to act as signaling molecules. Polar lipids such as PUFAs are important structural components of cell membranes, provide a selective, permeable barrier that protects the cell from the outside, and aid in the separation of different intracellular organelles. These lipids have specific functions in membranes and have the ability to mediate cell signaling pathways and play a role in response to environmental changes, as well as being directly involved in membrane fusion [45]. Fatty acid synthesis in algae takes place in chloroplasts and generally produces 16C and 18C fatty acid chains, which are used by the cell for

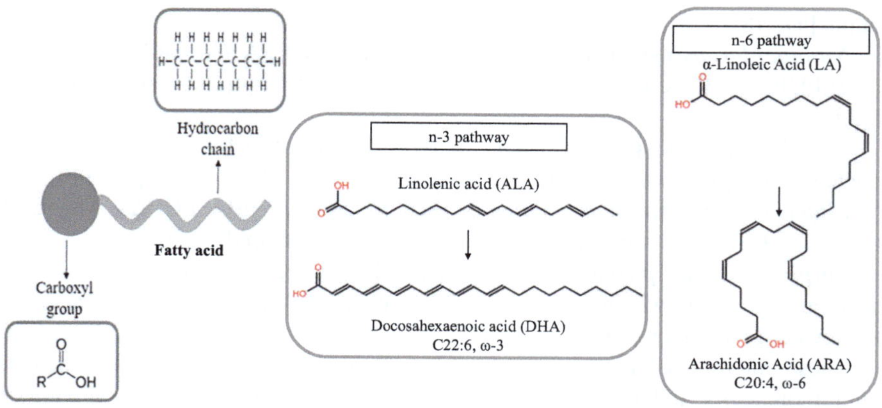

Fig. 3.1 Biosynthesis of PUFAs in microalgae, $n-3$ and $n-6$ pathway involved with different types of desaturases and elongases genes. The $n-3$ pathway has α-linolenic acid as synthesis precursor, which through four desaturation processes and three elongation processes allows obtaining DHA, and the $n-6$ pathway has linoleic acid as precursor, which also undergoes two elongation processes and three desaturations through the desaturase enzyme to obtain ARA

membrane synthesis and neutral lipid storage and desaturase enzymes are responsible for the sequential addition of double bonds to saturated fatty acids for the de novo synthesis of $n - 3$ PUFA [44, 46].

The most commercially important PUFAs are eicosapentaenoic acid (EPA) and docosahexaenoic acid (DHA). These PUFAs are omega-3 fatty acids and are mainly obtained from fish oil, in recent years, the use of microalgae as a source of EPA and DHA has gained popularity as it offers a more sustainable and environmentally friendly alternative to obtain these compounds of interest compared to animal sources such as fish and krill, which are often overexploited and can contain high levels of contaminants and heavy metals. Microalgae can be grown in controlled environments, which allows for consistent and high-quality production of PUFAs. In addition, microalgae can be grown using wastewater or other waste streams, which further reduces their environmental impact. Overall, the production of PUFAs with microalgae offers a promising solution to meet the growing demand for these essential nutrients [44].

3.4 Species of Microalgae Commonly Used for Production of PUFAs

There are several species of microalgae that are used worldwide by companies to obtain PUFAs. Some of the commonly used species include:

***Schizochytrium* sp.** This species is widely used to produce DHA and EPA PUFAs, is a marine oleaginous microalga, which has become the major commercial source of DHA. With a high concentration of lipids representing between 36% and 84% of the biomass, in which the concentration of DHA exceeds 62% of the total lipids [47, 48].

***Nannochloropsis* sp.** This species is used to produce EPA and ARA (arachidonic acid) PUFAs. EPA content in *Nanochloropsis* can vary from 20% to 30% of total lipids.

***Crypthecodinium* sp.** This species is used to produce DHA PUFAs. The DHA content in *Crypthecodinium* can vary from 20% to 50% of the total lipids [49].

***Phaeodactylum* sp.** This species is used to produce EPA PUFAs. The EPA content in *Phaeodactylum* can vary from 20% of the total lipids, C20:5 ($n - 3$) [50].

Porphyridium purpureum, it is characterized for being one of the microalgae that accumulates ARA, which can exceed 30% of its total fatty acids [51], also the freshwater green microalgae.

These microalgae are cultivated in large-scale production systems, such as photobioreactors and open ponds, to produce PUFAs that are used in various applications, including dietary supplements, functional foods, and animal feed [52].

According to recent market research reports, the microalgae species expected to account for the largest share of the global market for polyunsaturated fatty acid production by 2023 is *Nannochloropsis* sp. This is due to its ability to grow rapidly and produce high amounts of EPA and DHA, which are popular omega-3 fatty acids commonly used in dietary supplements and functional foods [53].

3.5 Fatty Acids Market

Grand View Research, in its report evidenced that the global fatty acids market size was valued at USD 37.1 billion in 2019 and is expected to grow at a compound annual growth rate (CAGR) of 5.2% from 2020 to 2027. Increasing demand for fatty acids in various applications such as food and beverages, personal care, pharmaceuticals, and animal feed is driving the market growth, the fatty acids industry is projected to grow from USD 24.5 million in 2023 to USD 34.2 million by 2030 [54].

Some of the leading companies in the market of acid grades using microalgae worldwide are DSM Nutritional Products, BASF SE, Royal Dutch Shell, Dow Chemical Company, Corbion NV, Archer Daniels Midland Company, Cargill, Inc., Roquette Freres, Novozymes A/S, and Evonik Industries. These companies oversee generating microalgae-based products, such as PUFA-rich oils, biofuels, and other value-added products.

DSM Nutritional Products, the holland multinational company uses the microalgae of the genus *Schizochytrium* sp. to obtain oil rich in EPA and DHA, with an initial cell culture of microalgae in fermenters of between 80,000 and 260,000 l. The products are available in spray dried powder, free-flowing powder with 10% al 1% of DHA [55].

Corbion NV, a company located in Amsterdam the Netherlands, offers the product AlgaPrime™ DHA an efficient, clean, and sustainable source of long-chain omega-3. Develop flexible solutions in liquid and powder form to meet the needs of different food manufacturing systems. AlgaPrime™ DHA is produced sustainably using renewable energy and cane sugar as raw material. This results in a low carbon, water, and land use impact [56].

Evonik Industries AG, located in Germany,. Evonik and DSM's highly concentrated algae oil meets the growing demand for essential omega-3s without endangering fish species, thus contributing to ecological balance and the maintenance of ocean biodiversity. Researchers have developed a process to obtain omega-3 fatty acids from natural seaweeds [57]. The companies have state-of-the-art technologies and patented methodologies that allow them to obtain PUFAs from microalgae at competitive prices in the world market.

The market for acidic grades derived from microalgae is expected to increase significantly in demand over the next 5 years, driven by the growing demand for sustainable and environmentally friendly products in the beverage, pharmaceutical, personal care, and food industries. The use of microalgae as a source of PUFAs offers several advantages over traditional sources, such as higher yields, lower environmental impact, and greater versatility in terms of product customization.

Key players in the market include DSM Nutritional Products, BASF SE, Royal Dutch Shell, and Corbion NV. The Asia-Pacific region is expected to be the fastest growing market for microalgae-derived PUFAs, owing to the increasing demand for natural and organic products in countries such as China and India. However, high production cost and limited scalability of microalgae cultivation remain major challenges to market growth. To support the industrial production of polyunsaturated fatty acids from algae, studies in recent years have devoted numerous efforts to screening algal strains, exploring biosynthetic mechanisms, optimizing induction conditions, and improving the mode of algal cultivation. Although microalgae-based products/coproducts are recognized as nutritional supplements, only a limited number of species have been successful in the market.

3.6 Polysaccharides

Microalgae contain large amounts of carbohydrates, between 15 and 75% of the dry biomass, and are mainly found as structural polymers forming part of the cell wall or storage polymers for energy in various metabolic processes. Microalgae polysaccharides have advantages over other polysaccharide sources (land plants, crustaceans, squids, or fungal cell walls), such as safety, stability, biocompatibility, and biodegradability. These characteristics contribute to its application for different industries such as pharmaceutical, nutraceutical, and biomedical due to its promising bioactive properties. In addition, it has become a potential source for energy applications, such as biofuels [58].

Carbohydrates are macromolecules structured by chains of carbon, hydrogen, and oxygen atoms (C, H, and O), covalently bonded. By glycosidic bonds of up to 40 monosaccharides to form the different types of carbohydrates, according to IUPAC-IUB carbohydrates are classified into oligosaccharides and polysaccharides taking into account their degree of polymerization (see Fig. 3.2) [59]. Polysaccharides consist of long chains of monosaccharide units and in some cases glucuronic acid and sulfate groups. Microalgal polysaccharides can be distinguished into three main groups: intracellular, structural, and extracellular carbohydrates (EPS) [60]. Extracellular polysaccharides (EPSs) are the molecules with the greatest industrial interest, since they have different chemical properties that confer biological activity, these macromolecules could find application in the industrial, pharmaceutical, and medical fields.

Carbohydrate synthesis occurs in the chloroplast during the CO_2 fixation allows the production of carbohydrates in microalgae, and this occurs through the Calvin cycle in the dark phase of photosynthesis, the Calvin cycle consists of three phases: fixation, reduction, and regeneration.

In fixation, CO_2 taken up by the stomata is added to ribulose-1,5-bisphosphate, a five-carbon sugar by the enzyme ribulose-1,5-bisphosphate carboxylase oxygenase (RuBisCo), forming two three-carbon molecules or phosphoglycerate. In the reduction phase, the two phosphoglycerate molecules are converted into two glyceraldehyde-3-phosphate molecules. At the end of the cycle, the regeneration phase takes

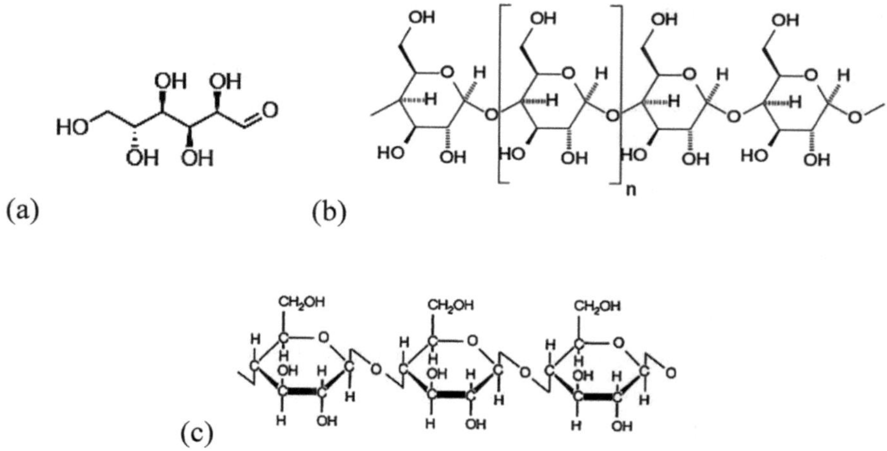

Fig. 3.2 Chemical structure of the main polysaccharides extracted from algae: (**a**) D-glucose; (**b**) starch; (**c**) cellulose (Own Authorship)

place, where one of the glyceraldehyde-3-phosphate molecules continues in the Calvin cycle, while the other is used as a substrate to form carbohydrates [61].

Glyceraldehyde-3-phosphate that is not immediately utilized as an energy source is transformed into starch in the chloroplast. The Golgi complex is responsible for sulfation and biosynthesis of polysaccharides, especially exopolysaccharides (EPS).

Currently, polysaccharides are used as dietary prebiotics (dietary fibers) and applications in the food industry as thickeners and food additives due to the high viscosity of polysaccharides over a wide range of pH, temperature, and salinity [62]. In animal feed, the use of β-glucans as antibiotic substitutes in chicken feed has been studied due to the antibacterial properties of this type of polysaccharides. In the research carried out by Koçer et al. [63], broiler chickens fed dry biomass of the *Euglena gracilis* with 55% β-glucan content showed better protection against coccidiosis. In addition, biostimulant activity has been demonstrated for the protection against biotic and abiotic stress of plants [64]. The cultivation of microalgae for the production of exopolysaccharides is receiving more attention as a result of their high hydrocarbon biosynthesis capacity [58, 63].

The main microalgae strains used for the production and extraction of polysaccharides are known to be *Chlorella* sp., *Spirulina* sp., *Porphyridium* sp., and *Nostoc* sp. [62]. *Spirulina platensis* produces various polysaccharides with anticancer, antioxidant, immunomodulatory, hypolipidemic and hypoglycemic, antithrombotic, antiviral, and intestinal microbiota-regulating properties [65].

Environmental conditions have been shown to affect the production of polysaccharides from microalgae, to obtain polysaccharides, cultivation is usually optimized by adjusting growth conditions such as light intensity, temperature, nutrient availability, and culture density. In addition, to obtain high-purity polysaccharides with the desired properties, various downstream processing techniques such as

Table 3.2 Main commercially important microalgae pigments, source strain and market price

Pigment	Strain	Yield (% w/w)	Price (USD per Kg)	References
Astaxanthin	*Haematococcus pluvialis and Chlorella zofingiensis*	0.15–4.0	2500–10,000	[15]
β-carotene	*Dunaliella salina*	12	300–3000	[15]
Lutein	*Dunaliella salina*	0.5	1866–2800	[20]
Phycocyanin	*Spirulina Platensis*	25	50–900	[18]
Chrophylls	*Chlorella vulgaris*	4.5	12–30	[19]
Phycoerythrin	*Porphyridium* sp.	8.5	230–5600	[21]

Table 3.3 Main costs of polysaccharides extracted from microalgae (Own Authorship)

General application of polysaccharide of microalgae	Precio (€)
Biostimulants from plants	10,000
Immunostimulants	200,000
Biofuels: Bioethanol, Biohydrogen	2500
Moisturizing Agents	400

harvesting, extraction, and purification can be applied [66]; therefore, conventional techniques for the extraction of polysaccharides are described below.

Table 3.2 summarizes the main selling prices (for wholesalers of polysaccharides derived from microalgae biomass).

3.7 Conventional Techniques for the Extraction of Polysaccharides from Algal Biomass

In conventional experimental techniques to extract polysaccharides from microalgal cell rupture, the extraction process is considered the first step to isolate polysaccharides from algal sources. To extract these biocomposites from algal biomass species, efforts have been focused on the management of explicit techniques to achieve the best extraction and recovery conditions. The traditional or conventional method of extraction of polysaccharides from algae mainly uses solvents that allow the rupture of the cell wall of the microalgae, listing hot water, strong acids, bases, or alkalis or a combination of the above. The process is simple as it involves subjecting pretreated algae biomass to different solvents at different temperatures for defined periods of time to obtain polysaccharides with minimal impurities [67]. For example, Abraham et al. [68] extracted 43.57% (by weight/weight ratio) of the total polysaccharide, composed of fucoidan, alginate and laminarin, from the algae *Durvillaea potatorum*, using 0.05 mol/L of acid medium (HCl) at a temperature of 60 °C for a period of 3 h. The yield of carrageenan polysaccharide reported in the literature was 32.95 ± 1.43% from *Kappaphycus alvarezii*, and the results were provided by Meinita et al. [69] use 60% KOH alkaline medium at 80 °C for 3 h. Table 3.3 lists

Table 3.4 Microalgae and main bioproducts considering culture medium and system, extraction procedure, detection, and content of bioproducts (Own Authorship)

Strain	Biomolecules	Extraction method	Characterization method	References
Chlorella fusca	Fatty acids, carbohydrates, pigments and proteins.	Bradford, Dubois, Bligh y Dyer y Jeffrey–Humphrey, respectively	Spectrophotometric	[70]
Dunaliella salina	Fatty acids, carbohydrates, pigments and proteins.	Bradford, Dubois, Bligh y Dyer y Jeffrey–Humphrey, respectively	Spectrophotometric	[70]
Kirchneriella sp.	Fatty acids, carbohydrates, pigments and proteins.	Solvent extraction for all by-products	GC–MS, HPLC–UV Spectrophotometric, respectively	[71]
Spirulina sp.	Fatty acids, carbohydrates, pigments and proteins.	Bradford, Dubois, Bligh y Dyer y Jeffrey–Humphrey, respectively	Spectrophotometric	[70]
Chlorella pyrenoidosa	Polysaccharides: Arabinogalactan	Aqueous extraction at 80°C, 1 h, precipitation with ethanol.	GLC–MS; SEC; NMR spectroscopy	[72]

the main extraction techniques for hydrocarbons (polysaccharides), different working strains, and characterization methods from macroalgae or microalgae.

Although polysaccharides have a variety of applications and health benefits, production at the industrial level faces obstacles such as achieving high microalgae growth rate and high polysaccharide production simultaneously during cultivation (Table 3.4). Some problems related to polysaccharide production could be solved by optimizing cultivation variables such as temperature, photoperiod time, light intensity, aeration, and nutrients.

3.8 Vitamins and Antioxidants

Microalgae are recognized as natural sources of vitamins and pigments with antioxidant activity and have become the most promising and innovative source of functional foods (Table 3.5). Among the various natural sources of antioxidants, microalgae have become of great interest due to their ease of cultivation, they are easily adapted to environmental conditions and do not compete with food production, adding to this the demand for antioxidants of natural origin, in fact it has been reported that many synthetic antioxidants (butylated hydroxyanisole (BHA), butylated hydroxytoluene (BHT)) have a carcinogenic and/or toxic effect in animal models, although most natural antioxidants available on the market globally are from plants, microalgae are considered as organisms with great potential for the

Table 3.5 Main antioxidant compounds and vitamins obtained from microalgae

Antioxidant/ Vitamins	Strains	Field/application	Commercial value	References
Astaxanthin	*Haematococcus pluvialis*	Food, nutraceuticals, and cosmetics industries	555.4 million US$	[73]
Beta-Carotene	*Dunaliella Salina*	Nutraceutical, pharmaceutical, cosmetical additive	224 million US$	[73]
Amino acid surfactant	Various microalgae	Cosmetics industry	30 billion US $ to 2026	[73]
Phytosterols variety	Various microalgae	Cosmetics industry	935 million US$	[74]
A/provitamin A	*Spirulina platensis*	Nutraceuticals	520 million US$	[74]
B1	*Spirulina platensis*	Nutraceuticals	170 million US$	[74]
B2	*Spirulina platensis*	Nutraceuticals	1.3 billion US $	[74]
B3	*Spirulina platensis*	Nutraceuticals	370 million US$	[74]
B5	*Spirulina platensis*	Nutraceuticals	460 million US$	[74]
B7	*Spirulina platensis*	Nutraceuticals	1.9 billion US $	[74]
C	*Spirulina platensis*	Nutraceuticals	1.3 billion US $	[74]
D	*Spirulina platensis*	Nutraceuticals	1.3 billion US $	[74]
E	*Spirulina platensis*	Nutraceuticals	0.67 million US$	[74]
K	*Spirulina platensis*	Nutraceuticals	0.79 million US$	[74]

extraction of natural antioxidants for the cosmetic, food, and nutraceutical industries [26, 75].

Microalgae are integrated into the concept of biofactories for obtaining bioactive compounds such as vitamins, and vitamins are considered one of the cutting-edge compounds obtained from microalgae, these organisms have the ability to synthesize a variety of vitamins necessary for humans, even synthesize vitamins such as D and K that are not present in many plants or fruits [76].

3.8.1 Antioxidants

Antioxidants are the compounds of greatest interest around human health and the pharmaceutical industry. This wide range of molecules has the ability to protect the human body against the adverse effects caused by free radicals [77].

As microalgae are photoautotrophic organisms, they produce free radicals and other potentially oxidizing reagents when exposed to high concentrations of light or high oxygen saturation; however, it has been proven that these microorganisms lack structural damage, so it is logical to think that they can synthesize compounds that protect them against oxidation [78].

The antioxidant molecules present in microalgae are ascorbic acid or vitamin C, glutathione, tocopherols, phenolic compounds, and carotenoids. Some microalgae have also been reported to produce antioxidant molecules such as mycosporin (porphyra-334, mycosporin-glycine, shinorin, and asterin-330 are molecules that allow UV protection and have been shown to have antioxidant properties) [75].

The carotenoids are among the most abundant pigments in nature. Most of these compounds are hydrophobic tetraterpenoids, these pigments serve to absorb pigments in visible light, as well as to protect photosynthetic microorganisms from photooxidation. There are two main types of carotenoids, carotenes and xanthophylls. The use of pigments such as carotenoids has been reported to decrease the risk of cardiovascular disease, the prevalence of metabolic syndrome, adiposity, and serum triglyceride concentrations in middle-aged and elderly men, and to prevent atherosclerosis and age-related muscle degeneration and to enhance immune resistance to viral, bacterial, fungal, and parasitic infections [16].

The main carotenoids of microalgal origin are β-carotene from *Dunaliella salina* and astaxanthin from *Haematococcus pluviallis*. The β-carotene is a natural colorant with wide application in the food industry adding nutraceutical value as an antioxidant and is also widely used as an additive for cosmetics. Astaxanthin has numerous health benefits, it helps protect the eyes, improves muscle strength, and endurance also protects the skin against premature aging and UV radiation, it has been suggested in several studies that astaxanthin has an antioxidant power against free radical's equivalent to 500 times vitamin E [79].

Phenolic compounds and polyphenols are produced thanks to secondary metabolites and are recognized as antioxidants due to their chelating power toward metal ions, preventing the formation of radicals and improving the endogenous system, in this classification are grouped hundreds of molecules with the characteristic of having a benzene with a hydroxyl as a substituent, within these molecules we can find flavonoids, specifically for algae florotaninos a potent free radical scavenger [43].

3.8.2 Vitamins

Vitamins have received special attention in recent decades due to their great antioxidant power, and interest has also grown in obtaining them from natural sources, since these vitamins have proven to be more effective than synthetic vitamins [80]. Microalgae represent a valuable source of all vitamins, fat-soluble vitamins A, D, E, and K and water-soluble vitamins C and the complete B complex, on the other hand, they also have a good proportion of the minerals that humans require as nutrients, among them we can mention sodium, potassium, calcium, magnesium, iron, zinc, and some other traces. Some species such as spirulina have high levels of vitamin B12 and iron, making it ideal for use as a food supplement in vegan and vegetarian diets. The content of these materials in a microalgae depends on the genotype among other things such as the growth stage and nutritional status of the algae and light intensity, so the manipulation of culture conditions for the accumulation of these biocompounds is an efficient strategy, which can be accompanied by the choice of strain and genetic engineering modification [76]. The value of antioxidants from microalgae will depend on different variables such as composition, purity level, the application for which it will be required and, finally, whether it is integrated in the formulation of a product or whether it receives some refinement treatment or processing to improve its virtues or organoleptic properties.

Table 3.5 shows the main antioxidants and vitamins obtained from different strains of microalgae, as well as the commercial value of the biocompounds.

3.9 Other Compounds

Because microalgae produce large quantities of biomolecules, they are of interest as excellent sources of macro and micro minerals such as K, Na, Ca, Fe, Zn, Cu, and Mn [2]. It has been reported that microalgae have the ability to produce phenolic compounds, which are recognized for their biological properties to fight aging, allergies, diabetes and different types of cancer [81]. In addition, many studies have reported the presence of enzymes such as lipases, laccases, proteases, cellulases, phytases, galactosidases, and amylases, which have a high interest in various industries. Microalgae are also characterized by the production of phytohormones such as ethylene, auxin, gibberellins, and abscisic acid, and these phytohormones regulate the main phytohormone processes. These phytohormones regulate the main physiological processes of plants [2].

3.10 Conclusions

As presented in this chapter, microalgae have a diversity of bioactive compounds with applications in different industries, it is highlighted that many of these compounds have a passive role at the industrial level due to the high costs of production, extraction, and purification; therefore, the search for new and novel methodologies adapted for the efficient extraction and purification of compounds from microalgal biomass could change the perspective toward a wider industrial use. The development of cultivation systems that allow the optimization of cultivation processes and the obtaining of greater quantities of biomass could, probably in the future, significantly increase the yields of microalgal biomass. The development of bioprocesses that take full advantage of the potential of microalgae and give a bio-fabrication approach for application in large-scale production is needed. The greatest challenges in large-scale cultivation of microalgae are focused on finding strains that allow obtaining large quantities of the metabolite of interest in less time and with lower process costs, as well as limitations in the extraction and purification methodologies of biomolecules. Microalgae are a source of active ingredients with anti-inflammatory, antioxidant, antitumor and antimicrobial action, among others, with potential applications in nutraceutical, cosmetic, pharmaceutical, and agricultural foods. Thus, the production of different bioactive compounds from microalgae promotes prospects with potential for sustainable development.

References

1. Koyande AK, Chew KW, Rambabu K, Tao Y, Chu DT, Show PL (2019) Microalgae: A potential alternative to health supplementation for humans. Food Sci Human Wellness 8 (1):16–24. https://doi.org/10.1016/j.fshw.2019.03.001
2. Hachicha R, Elleuch F, Ben Hlima H, Dubessay P, de Baynast H, Delattre C, Pierre G, Hachicha R, Abdelkafi S, Michaud P, Fendri I (2022) Biomolecules from microalgae and cyanobacteria: applications and market survey. Appl Sci 12(4):1924. https://doi.org/10.3390/app12041924
3. Fawcett CA, Senhorinho GNA, Laamanen CA, Scott JA (2022) Microalgae as an alternative to oil crops for edible oils and animal feed. Algal Res 64:102663. https://doi.org/10.1016/j.algal.2022.102663
4. Loke Show P (2022) Global market and economic analysis of microalgae technology: status and perspectives. Bioresour Technol 357:127329. https://doi.org/10.1016/j.biortech.2022.127329
5. Wong JF, Hong HJ, Foo SC, Yap MKK, Tan JW (2022) A review on current and future advancements for commercialized microalgae species. Food Sci Human Wellness 11(5):1156–1170. https://doi.org/10.1016/j.fshw.2022.04.007
6. Baudelet PH, Ricochon G, Linder M, Muniglia L (2017) A new insight into cell walls of Chlorophyta. Algal Res 25(July):333–371. https://doi.org/10.1016/j.algal.2017.04.008
7. Xiaogang, H., Jalalah, M., Jingyuan, W., Zheng, Y., Li, X., & Salama, E. (2020). Microalgal growth coupled with wastewater treatment in open and closed systems for advanced biofuel generation
8. Drews- P, Colares RG, Machado P, de Faria M, Detoni A, Tavano V (2013) Microalgae classification using semi-supervised and active learning based on Gaussian mixture models. J Braz Comput Soc 19(4):411–422. https://doi.org/10.1007/s13173-013-0121-y

9. Shaikh R, Rizvi A, Pandit S, Desai N, Patil R (2022) Microalgae: Classification, bioactives, medicinal properties, industrial applications, and future prospectives. In: An integration of phycoremediation processes in wastewater treatment. Elsevier, pp 451–486. https://doi.org/10.1016/B978-0-12-823499-0.00004-3
10. Masojídek J, Torzillo G, Koblížek M (2013) Photosynthesis in microalgae. In: Handbook of microalgal culture. Wiley, pp 21–36. https://doi.org/10.1002/9781118567166.ch2
11. Algade Amadu, A., Amoako Atta deGraft-Johnson, K., & Komla Ameka, G. (2022). Industrial applications of cyanobacteria. In: Cyanobacteria - recent advances in taxonomy and applications. IntechOpen. https://doi.org/10.5772/intechopen.98859
12. Alazaiza MYD, Albahnasawi A, Ahmad Z, Bashir MJK, Al-Wahaibi T, Abujazar MSS, Abu Amr SS, Nassani DE (2022) Potential use of algae for the bioremediation of different types of wastewater and contaminants: production of bioproducts and biofuel for green circular economy. J Environ Manag 324:116415. https://doi.org/10.1016/J.JENVMAN.2022.116415
13. Rehman M, Kesharvani S, Dwivedi G, Gidwani Suneja K (2022) Impact of cultivation conditions on microalgae biomass productivity and lipid content. Mater Today Proc. https://doi.org/10.1016/j.matpr.2022.01.152
14. Yusof NS, Yeong YS, Zakeri HA, Wahid MEA, Ghafar SNA, Yusuf N (2021) Photoperiod influenced the growth and antioxidative responses of Chlorella vulgaris, Isochrysis galbana, and Tetraselmis chuii. J Appl Pharma Sci 11(4):125–134. https://doi.org/10.7324/JAPS.2021.110415
15. Lafarga T, Clemente I, Garcia-Vaquero M (2020) Carotenoids from microalgae. In: *Carotenoids: properties, processing and applications*. Elsevier, pp 149–187. https://doi.org/10.1016/B978-0-12-817067-0.00005-1
16. Sirohi P, Verma H, Singh SK, Singh VK, Pandey J, Khusharia S, Kumar D, Kaushalendra T, P., & Kumar, A. (2022) Microalgal carotenoids: therapeutic application and latest approaches to enhance the production. Curr Issues Mol Biol 44(12):6257–6279. https://doi.org/10.3390/cimb44120427
17. Saha P, Aloui H, Yun JH, Kim HS, Kim BS (2021) Development of a novel composite film based on polyurethane and defatted Chlorella biomass: physical and functional characterization. J Appl Polym Sci 138(14):1–11. https://doi.org/10.1002/app.50152
18. Pan-utai W, Kahapana W, Iamtham S (2018) Extraction of C-phycocyanin from Arthrospira (Spirulina) and its thermal stability with citric acid. J Appl Phycol 30(1):231–242. https://doi.org/10.1007/s10811-017-1155-x
19. Khanra S, Mondal M, Halder G, Tiwari ON, Gayen K, Bhowmick TK (2018) Downstream processing of microalgae for pigments, protein and carbohydrate in industrial application: a review. Food Bioprod Process 110:60–84. https://doi.org/10.1016/j.fbp.2018.02.002
20. Saha SK, Ermis H, Murray P (2020) Marine microalgae for potential lutein production. Appl Sci 10(18):6457. https://doi.org/10.3390/app10186457
21. Huschek G, Rawel HM, Schweikert T, Henkel-Oberländer J, Sagu ST (2022) Characterization and optimization of microwave-assisted extraction of B-phycoerythrin from Porphyridium purpureum using response surface methodology and Doehlert design. Bioresource Technology Reports 19:101212. https://doi.org/10.1016/j.biteb.2022.101212
22. Garrido-Cardenas JA, Manzano-Agugliaro F, Acien-Fernandez FG, Molina-Grima E (2018) Microalgae research worldwide. Algal Res 35:50–60. https://doi.org/10.1016/j.algal.2018.08.005
23. Begum H, Yusoff FM, Banerjee S, Khatoon H, Shariff M (2016) Availability and utilization of pigments from microalgae. Crit Rev Food Sci Nutr 56(13):2209–2222. https://doi.org/10.1080/10408398.2013.764841
24. D'Alessandro EB, Antoniosi Filho NR (2016) Concepts and studies on lipid and pigments of microalgae: a review. Renew Sust Energ Rev 58:832–841. https://doi.org/10.1016/j.rser.2015.12.162
25. Ambati RR, Gogisetty D, Aswathanarayana RG, Ravi S, Bikkina PN, Bo L, Yuepeng S (2019) Industrial potential of carotenoid pigments from microalgae: current trends and future

prospects. Crit Rev Food Sci Nutr 59(12):1880–1902. https://doi.org/10.1080/10408398.2018.1432561
26. Novoveská L, Ross ME, Stanley MS, Pradelles R, Wasiolek V, Sassi J-F (2019) Microalgal carotenoids: a review of production, current markets, regulations, and future direction. Mar Drugs 17(11):640. https://doi.org/10.3390/md17110640
27. Liu J, Mao X, Zhou W, Guarnieri MT (2016) Simultaneous production of triacylglycerol and high-value carotenoids by the astaxanthin-producing oleaginous green microalga Chlorella zofingiensis. Bioresour Technol 214:319–327. https://doi.org/10.1016/j.biortech.2016.04.112
28. Fernández-Sevilla JM, Acién Fernández FG, Molina Grima E (2010) Biotechnological production of lutein and its applications. Appl Microbiol Biotechnol 86(1):27–40. https://doi.org/10.1007/s00253-009-2420-y
29. Gateau H, Solymosi K, Marchand J, Schoefs B (2017) Carotenoids of microalgae used in food industry and medicine. Mini-Rev Med Chem 17(13). https://doi.org/10.2174/1389557516666160808123841
30. Queiroz Zepka L, Jacob-Lopes E, Roca M (2019) Catabolism and bioactive properties of chlorophylls. Curr Opin Food Sci 26:94–100. https://doi.org/10.1016/j.cofs.2019.04.004
31. Silva SC, Ferreira ICFR, Dias MM, Barreiro MF (2020) Microalgae-derived pigments: a 10-year bibliometric review and industry and market trend analysis. Molecules 25(15):3406. https://doi.org/10.3390/molecules25153406
32. Koutra E, Mastropetros SG, Ali SS, Tsigkou K, Kornaros M (2021) Assessing the potential of Chlorella vulgaris for valorization of liquid digestates from agro-industrial and municipal organic wastes in a biorefinery approach. J Clean Prod 280:124352. https://doi.org/10.1016/j.jclepro.2020.124352
33. Nabi BG, Mukhtar K, Ahmed W, Manzoor MF, Ranjha MMAN, Kieliszek M, Bhat ZF, Aadil RM (2023) Natural pigments: anthocyanins, carotenoids, chlorophylls, and betalains as colorants in food products. Food Biosci 52:102403. https://doi.org/10.1016/j.fbio.2023.102403
34. Demarco M, Oliveira de Moraes J, Matos ÂP, Derner RB, de Farias Neves F, Tribuzi G (2022) Digestibility, bioaccessibility and bioactivity of compounds from algae. Trends Food Sci Technol 121:114–128. https://doi.org/10.1016/j.tifs.2022.02.004
35. Li Y, Cui Y, Hu X, Liao X, Zhang Y (2019) Chlorophyll supplementation in early life prevents diet-induced obesity and modulates gut microbiota in mice. Mol Nutr Food Res 63 (21):1801219. https://doi.org/10.1002/mnfr.201801219
36. Osório C, Machado S, Peixoto J, Bessada S, Pimentel FB, Alves C, R., & Oliveira, M. B. P. P. (2020) Pigments content (chlorophylls, fucoxanthin and phycobiliproteins) of different commercial dried algae. Separations 7(2):33. https://doi.org/10.3390/separations7020033
37. Sun H, Wang Y, He Y, Liu B, Mou H, Chen F, Yang S (2023) Microalgae-derived pigments for the food industry. Mar Drugs 21(2):82. https://doi.org/10.3390/md21020082
38. Pagels F, Salvaterra D, Amaro HM, Guedes AC (2020) Pigments from microalgae. In: Handbook of microalgae-based processes and products. Elsevier, pp 465–492. https://doi.org/10.1016/B978-0-12-818536-0.00018-X
39. Aráoz R, Lebert M, Häder D-P (1998) Electrophoretic applications of phycobiliproteins. Electrophoresis 19(2):215–219. https://doi.org/10.1002/elps.1150190213
40. Viana Carlos TA, dos Santos Pires Cavalcante KM, de Cássia Evangelista de Oliveira F, do Ó Pessoa C, Sant'Ana HB, Feitosa FX, Rocha MVP (2021) Pressurized extraction of phycobiliproteins from Arthrospira platensis and evaluation of its effect on antioxidant and anticancer activities of these biomolecules. J Appl Phycol 33(2):929–938. https://doi.org/10.1007/s10811-020-02358-z
41. Gim GH, Kim JK, Kim HS, Kathiravan MN, Yang H, Jeong SH, Kim SW (2014) Comparison of biomass production and total lipid content of freshwater green microalgae cultivated under various culture conditions. Bioprocess Biosyst Eng 37(2):99–106. https://doi.org/10.1007/s00449-013-0920-8

42. Mata TM, Martins AA, Caetano NS (2010) Microalgae for biodiesel production and other applications: a review. Renew Sust Energ Rev 14(1):217–232. https://doi.org/10.1016/j.rser.2009.07.020
43. Sudhakar MP, Kumar BR, Mathimani T, Arunkumar K (2019) A review on bioenergy and bioactive compounds from microalgae and macroalgae-sustainable energy perspective. J Clean Prod 228:1320–1333. https://doi.org/10.1016/j.jclepro.2019.04.287
44. Ramesh Kumar B, Deviram G, Mathimani T, Duc PA, Pugazhendhi A (2019) Microalgae as rich source of polyunsaturated fatty acids. Biocatal Agric Biotechnol 17(November 2018):583–588. https://doi.org/10.1016/j.bcab.2019.01.017
45. Udayan A, Pandey AK, Sirohi R, Sreekumar N, Sang BI, Sim SJ, Kim SH, Pandey A (2022) Production of microalgae with high lipid content and their potential as sources of nutraceuticals. Phytochem Rev 0123456789. https://doi.org/10.1007/s11101-021-09784-y
46. Manning SR (2022) Microalgal lipids: biochemistry and biotechnology. Curr Opin Biotechnol 74:1–7. https://doi.org/10.1016/j.copbio.2021.10.018
47. Ding J, Fu Z, Zhu Y, He J, Ma L, Bu D (2022) Enhancing docosahexaenoic acid production of Schizochytrium sp. by optimizing fermentation using central composite design. BMC Biotechnol 22(1):1–12. https://doi.org/10.1186/s12896-022-00769-z
48. Ma W, Zhang Z, Yang W, Huang P, Gu Y, Sun X, Huang H (2023) Enhanced docosahexaenoic acid production from cane molasses by engineered and adaptively evolved Schizochytrium sp. Bioresour Technol 376:128833. https://doi.org/10.1016/j.biortech.2023.128833
49. Qin Z-H, Hu X, Mou J-H, He G-H, Ye G-B, Li H-Y, Chopra SS, Dong L, Lin CSK, Wang X (2023) Environmental profiling microalgae-based eicosapentaenoic acid production along the technical advancement via life cycle assessment. J Clean Prod 397:136477. https://doi.org/10.1016/j.jclepro.2023.136477
50. Russo GL, Langellotti AL, Verardo V, Martín-García B, Oliviero M, Baselice M, Di Pierro P, Sorrentino A, Viscardi S, Marileo L, Sacchi R, Masi P (2023) Bioconversion of cheese whey and food by-products by Phaeodactylum tricornutum into fucoxanthin and n-3 Lc-PUFA through a biorefinery approach. Mar Drugs 21(3). https://doi.org/10.3390/md21030190
51. Soto-Sánchez O, Hidalgo P, González A, Oliveira PE, Hernández Arias AJ, Dantagnan P (2023) Microalgae as raw materials for aquafeeds: growth kinetics and improvement strategies of polyunsaturated fatty acids production. Aquac Nutr 2023:1–19. https://doi.org/10.1155/2023/5110281
52. Ibrahim TNBT, Feisal NAS, Kamaludin NH, Cheah WY, How V, Bhatnagar A, Ma Z, Show PL (2023) Biological active metabolites from microalgae for healthcare and pharmaceutical industries: a comprehensive review. Bioresour Technol 372:128661. https://doi.org/10.1016/j.biortech.2023.128661
53. Chen W, Li T, Du S, Chen H, Wang Q (2023) Microalgal polyunsaturated fatty acids: hotspots and production techniques. March 1–22. https://doi.org/10.3389/fbioe.2023.1146881
54. Grand View Research (2020) Fatty acids market size, share & trends analysis report by type (saturated, unsaturated), by form (oil, powder, capsule), by end-use (Industrial, F&B), by region, and segment forecasts, 2022–2030
55. DSM (2023, abril) Nutritional lipids. https://www.dsm.com/human-nutrition/en/products/nutritional-lipids.html
56. Corbion (s.f.). AlgaPrime™ DHA – sustainable, affordable, and at scale. 2023. https://www.corbion.com/Products/Algae-ingredients-products/AlgaPrimeDHA
57. *Acuicultura sostenible - Evonik Industries* (s.f.) Recuperado 11 de abril de 2023, de https://corporate.evonik.com/en/sustainable-aquaculture-648.html
58. de Carvalho Silvello MA, Severo Gonçalves I, Patrícia Held Azambuja S, Silva Costa S, Garcia Pereira Silva P, Oliveira Santos L, Goldbeck R (2022) Microalgae-based carbohydrates: a green innovative source of bioenergy. Bioresour Technol 344(September 2021). https://doi.org/10.1016/j.biortech.2021.126304

59. Depra MC, Dias RR, Maroneze MM, Nascimento TC, Severo IA, Zepka LQ, Jacob-Lopes E (2022) Bioactive polysaccharides from microalgae: a close look at the biomedical applications. Recent Pat Biotechnol 16. https://doi.org/10.2174/1872208316666220820092643
60. Prybylski N, Toucheteau C, El Alaoui H, Bridiau N, Maugard T, Abdelkafi S, Fendri I, Delattre C, Dubessay P, Pierre G, Michaud P (2020) Bioactive polysaccharides from microalgae. In: Handbook of microalgae-based processes and products. Elsevier, pp 533–571. https://doi.org/10.1016/B978-0-12-818536-0.00020-8
61. Wang K, Khoo KS, Chew KW, Selvarajoo A, Chen WH, Chang JS, Show PL (2021) Microalgae: the future supply house of biohydrogen and biogas. Front Energy Res 9:1–9. https://doi.org/10.3389/fenrg.2021.660399
62. Moreira JB, da Silva Vaz B, Cardias BB, Cruz CG, de Almeida ACA, Costa JAV, de Morais MG (2022) Microalgae polysaccharides: an alternative source for food production and sustainable agriculture. Polysaccharides 3(2):441–457. https://doi.org/10.3390/polysaccharides3020027
63. Koçer AT, İnan B, Kaptan Usul S, Özçimen D, Yılmaz MT, Işıldak İ (2021) Exopolysaccharides from microalgae: production, characterization, optimization and techno-economic assessment. Braz J Microbiol 52(4):1779–1790. https://doi.org/10.1007/s42770-021-00575-3
64. Patel AK, Vadrale AP, Singhania RR, Michaud P, Pandey A, Chen S-J, Chen C-W, Dong C-D (2022) Algal polysaccharides: current status and future prospects. Phytochem Rev. https://doi.org/10.1007/s11101-021-09799-5
65. Ai X, Yu P, Li X, Lai X, Yang M, Liu F, Luan F, Meng X (2023) Polysaccharides from Spirulina platensis: extraction methods, structural features and bioactivities diversity. Int J Biol Macromol 231:123211. https://doi.org/10.1016/j.ijbiomac.2023.123211
66. Tounsi L, Hentati F, Ben Hlima H, Barkallah M, Smaoui S, Fendri I, Michaud P, Abdelkafi S (2022) Microalgae as feedstock for bioactive polysaccharides. Int J Biol Macromol 221:1238–1250. https://doi.org/10.1016/j.ijbiomac.2022.08.206
67. He X, Fang J, Ruan Y, Wang X, Sun Y, Wu N, Zhao Z, Chang Y, Ning N, Guo H, Huang L (2018) Structures, bioactivities and future prospective of polysaccharides from Morus alba (white mulberry): a review. Food Chem 245:899–910. https://doi.org/10.1016/j.foodchem.2017.11.084
68. Abraham RE, Su P, Puri M, Raston CL, Zhang W (2019) Optimisation of biorefinery production of alginate, fucoidan and laminarin from brown seaweed Durvillaea potatorum. Algal Res 38:101389. https://doi.org/10.1016/j.algal.2018.101389
69. Meinita MDN, Marhaeni B, Jeong G-T, Hong Y-K (2019) Sequential acid and enzymatic hydrolysis of carrageenan solid waste for bioethanol production: a biorefinery approach. J Appl Phycol 31(4):2507–2515. https://doi.org/10.1007/s10811-019-1755-8
70. Díaz-Palma P, Stegen S, Queirolo F, Arias D, Araya S (2012) Biochemical profile of halophilous microalgae strains from high-andean extreme ecosystems (NE-Chile) using methodological validation approaches. J Biosci Bioeng 113(6):730–736. https://doi.org/10.1016/j.jbiosc.2012.01.015
71. Frampton DMF, Gurney RH, Dunstan GA, Clementson LA, Toifl MC, Pollard CB, Burn S, Jameson ID, Blackburn SI (2013) Evaluation of growth, nutrient utilization and production of bioproducts by a wastewater-isolated microalga. Bioresour Technol 130:261–268. https://doi.org/10.1016/j.biortech.2012.12.001
72. Suárez ER, Kralovec JA, Noseda MD, Ewart HS, Barrow CJ, Lumsden MD, Grindley TB (2005) Isolation, characterization and structural determination of a unique type of arabinogalactan from an immunostimulatory extract of Chlorella pyrenoidosa. Carbohydr Res 340(8):1489–1498. https://doi.org/10.1016/j.carres.2005.04.003
73. Aslam A, Bahadar A, Liaquat R, Saleem M, Waqas A, Zwawi M (2021) Algae as an attractive source for cosmetics to counter environmental stress. Science of The Total Environment 772:144905. https://doi.org/10.1016/j.scitotenv.2020.144905

74. Randhir A, Laird DW, Maker G, Trengove R, Moheimani NR (2020) Microalgae: a potential sustainable commercial source of sterols. Algal Res 46:101772. https://doi.org/10.1016/j.algal.2019.101772
75. Coulombier N, Jauffrais T, Lebouvier N (2021) Antioxidant compounds from microalgae: a review. Mar Drugs 19(10):549. https://doi.org/10.3390/md19100549
76. Del Mondo A, Smerilli A, Sané E, Sansone C, Brunet C (2020) Challenging microalgal vitamins for human health. Microb Cell Factories 19(1):201. https://doi.org/10.1186/s12934-020-01459-1
77. Maoka T (2020) Carotenoids as natural functional pigments. J Nat Med 74(1):1–16. https://doi.org/10.1007/s11418-019-01364-x
78. Kumar A, Ramamoorthy D, Verma DK, Kumar A, Kumar N, Kanak KR, Marwein BM, Mohan K (2022) Antioxidant and phytonutrient activities of Spirulina platensis. Energy Nexus 6:100070. https://doi.org/10.1016/j.nexus.2022.100070
79. Saini RK, Keum YS (2018) Carotenoid extraction methods: a review of recent developments. Food Chem 240
80. Verma R, Kumari KVLK, Srivastava A, Kumar A (2020) Photoautotrophic, mixotrophic, and heterotrophic culture media optimization for enhanced microalgae production. J Environ Chem Eng 8(5):104149. https://doi.org/10.1016/j.jece.2020.104149
81. Yaakob Z, Ali E, Zainal A, Mohamad M, Takriff MS (2014) An overview: biomolecules from microalgae for animal feed and aquaculture. J Biol Res-Thessaloniki 21(1):6. https://doi.org/10.1186/2241-5793-21-6

Chapter 4
Metabolic Engineering for the Biosynthesis of Terpenoids from Microbial Cell Factories

Vibha Shukla, Parul Gupta, and Suresh Chandra Phulara

Abstract Terpenes, terpenoids, or isoprenoids are the largest class of secondary metabolites produced by the natural system. Since ages the plants are being utilized as a source of flavor, fragrance, and therapeutics. Due to their higher energy density and other superior fuel properties, terpenoids are also foreseen as potential alternative to petroleum-derived fuels. The current market demand of terpenoids is fulfilled either by their plant-based extraction or by chemical synthesis. However, both the mythologies have their own set of limitations. Therefore, from the past two decades, the global research interest on terpenoid production has been shifted toward their biosynthesis from microbial routes. Microbes have provided an excellent platform for the biosynthesis of commodity chemicals due to their several advantages over plants, such as fastidious growth, low space and seasonal requirements, and simple growth medium. Moreover, the genetic modification in microbes can be done with a high success rate as compared to plants. However, microbes are not the natural producers of the majority of industrially important terpenoids. Therefore, they need to be engineered for the non-natural production of terpenoids. Advancement of synthetic and metabolic engineering approaches has enabled researchers to incorporate entire novel pathways and silence native pathways of the host microbes at the same time. Thus, provided an excellent platform not only for the non-native production of terpenoids but also improving the titers of the desired metabolites from engineered microbes. In this present chapter, we have summarized the metabolic

V. Shukla
Food, Drug and Chemical Toxicology Group, CSIR-Indian Institute of Toxicology Research, Lucknow, India

Academy of Scientific and Innovative Research (AcSIR), Ghaziabad, India

P. Gupta
Department of Respiratory Medicine, King George Medical University, Lucknow, India

S. C. Phulara (✉)
Department of Biotechnology, Govind Ballabh Pant Institute of Engineering and Technology, Ghurdauri, Pauri Garhwal, Uttarakhand, India

R. Soni et al. (eds.), *Microbial Bioactive Compounds*,
https://doi.org/10.1007/978-3-031-40082-7_4

engineering approaches for the sustainable production of terpenoid-based metabolites for microbial sources, with an emphasis on prokaryotic host organisms.

Keywords Terpenoids · Microorganisms · Biotechnology · Metabolic engineering · Homologous · Heterologous

4.1 Introduction

The molecular and functional diversity of terpenoids has always marveled researchers. The wide range of branched-chain and cyclic metabolites of terpenoid filmily are engaged in several biological activities in living system, for example, as pigments, hormones, sterols, and pheromones and several other characteristic plant compounds like limonene, menthol, and artemisisnin [1, 2]. The highly diverse nature of terpenoids provides a wealth of opportunities to encounter many social concerns such as identification of alternative nutraceutical, pharmaceutical, and biofuel resources to compete the worldwide increasing demand. There are numerous reports on the wide range of biological responses of terpenes that can be applied to the human health care sector [1, 3]. These applications are based on the antimicrobial, antiparasitic, anti-inflammatory, antiallergenic, anticancer, antihyperglycemic, immune-enhancer, neuro-therapeutic, and chemotherapeutic properties of the terpenoids [1, 3–6]. Over the past decade, terpenes are prophesied as a sustainable source of energy and an alternative to petroleum-derived fuels, due to their lower hygroscopy, higher energy density, and good fluidity at low temperatures [7]. Figure 4.1 depicts the structural and functional diversity of terpenoids.

Until date, plants are utilized as a natural source for terpenoid production; however, several environmental concerns are associated with the extraction of terpenoids from these natural sources. For example, to extract 1 kg of peclitexal, which is a highly potent chemotherapeutic agent, it requires the sacrifice of nearly 300 Yew (*Taxus* spp.) trees [9]. Due to the high demands of this drug, the Yew species are now endangered [10]. On the other hand, the extraction of resins from *Pinus roxburghii* or chir pine often causes forest fire due to the highly inflammable nature of the resin. Besides these concerns, the low-level and tissue-specific production, seasonal variations in yield, agricultural land requirement, and costly and time-consuming extraction process are other limiting factors, which limits the use of plant resources for terpenoid production [11]. To overcome these limitations, novel routes are required for the biosynthesis of terpenoid to meet the increasing demand as a raw material for the natural product-based flavors, fragrances, beauty products, and therapeutics and to maintain the ecological balance at the same time. Plant cell culture approaches have contributed to an extent in this regard; however, the lengthy culture durations, difficulties in genetic modification, and high production and extraction cost restrict its application [12]. Therefore, the global interest is now shifting to explore more sustainable approaches for the production of terpenoids.

Microbes, both the prokaryotes and eukaryotes, have shown a great potential for the biosynthesis of terpenoids [13–16]. Although majority of the terpenoids are

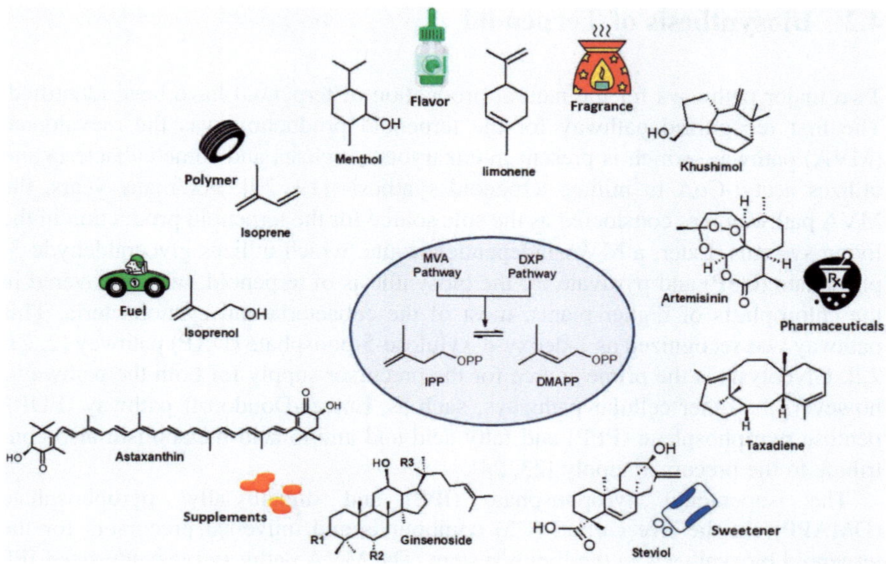

Fig. 4.1 Structural and functional diversity of terpenoids. Structural diversity of terpenoid ranges from simple five carbon atom carrying compounds (isoprenol and isoprene) to 30 and 40 carbon atom carrying compounds (ginsenoside, astaxanthin, etc.). Functionally they are rich in application that ranges from flavor, pharmaceutical to biofuel [8]

non-native to microbes; however, the advancement of synthetic biology and metabolic engineering techniques and their high success rates in microbes have enabled researcher to achieve high-yield production of terpenoids from microbial hosts [15, 17, 18]. In addition, the ease in culturing and handling, superior growth rates, and lesser nutrients and culture space requirements make microbial sources a suitable chassis for terpenoid production. Moreover, the capabilities of microbes to produce heterologous metabolites in large quantities from economical feedstock and the easy downstream processing make them ideal chassis for industrial application. Together these advantages have urged researchers to seek the production of value-added novel terpenoids from microbial sources after its discovery from animal or plant origin.

In the present chapter, we have summarized the up-to-date information on the metabolic engineering of microbial sources for the production of terpenoids with a major emphasis on prokaryotic hosts.

4.2 Biosynthesis of Terpenoid

Two major pathways for the natural production of terpenoid have been identified. The first recognized pathway for the terpenoid production was the mevalonate (MVA) pathway, which is present in eukaryotes, archaea and some eubacteria and utilizes acetyl-CoA to initiate terpenoid synthesis [19, 20]. For many years, the MVA pathway was considered as the sole source for the terpenoid production in the living systems. Later, a MVA independent route, which utilizes glyceraldehyde-3-phosphate (G3P) and pyruvate for the biosynthesis of terpenoid, was discovered in the chloroplasts of higher plants, most of the eubacteria and cyanobacteria. This pathway was recognized as 1-deoxy-d-xylulose-5-phosphate (DXP) pathway [2, 21, 22]. Glycolysis is the prime source for the precursor supply for both the pathways; however, the other cellular pathways, such as, Entner-Doudoroff pathway (EDP), pentose pyrophosphate (PPP) and fatty acid and amino acid metabolism also contribute to the precursor supply [23, 24].

The isopentenyl pyrophosphate (IPP) and dimethylallyl pyrophosphate (DMAPP) are the five carbon (C5) compounds and universal precursors for the terpenoid biosynthesis in the living system. The MVA pathway biosynthesized IPP in six enzymatic steps, which was then converted to DMAPP by an enzyme known as IPP isomerase (IDI). On the other hand, the DXP pathway utilizes seven enzymatic steps to biosynthesize both the IPP and DMAPP in a ratio 5:1 [7] and endogenous IDI enzyme maintain the balance between IPP and DMAPP. Several downstream terpenoid-precursor-specific and terpenoid-specific enzymes catalyze the conversion of IPP and DMAPP into a wide variety of terpenoids ranging from five carbon hemiterpenoids to forty or fifty carbon atoms carrying carotenoids. Thus, the terpenoid biosynthesis pathway can be divided into two modules. The upper module consists of either MVA or DXP pathway or both for the supply of IPP and DMAPP (Fig. 4.2), and the lower module carries terpenoid-precursor-specific and terpenoid-specific enzymes (Fig. 4.3).

Majority of the terpenoid-precursor-specific enzymes are heterologous to microbial hosts and need to be incorporated from plant or fungal sources. For example, plants possess geranyl pyrophosphate synthase (GPPS) enzyme convert IPP and DMAPP into GPP, which is a universal precursor for monoterpenoids (ten carbon molecules carrying terpenoids). The GPP is utilized by farnesyl pyrophosphate synthase (FPPS) to biosynthesize FPP, a common precursor to sesquiterpenoids (ten carbon molecules carrying terpenoids). Microbes on the other hand do not contain GPPS, instead they possess IspA that is responsible for direct conversion of IPP and DMAPP into FPP. FPP serves as a precursor to various vital microbial metabolites, such as ubiqinone and other membrane components. Similarly, to biosynthesize precursors for diterpenoids, triterpenoids, and carotenoids, the enzymes related to each category need to be expressed heterologously in prokaryotic hosts [11]. Like terpenoid-precursor-specific enzyme, most of the terpenoid-specific enzymes, which catalyzes the conversion of prenyl precursors into different terpenoids of a category, are also heterologous to microbial hosts [17]. However, there are

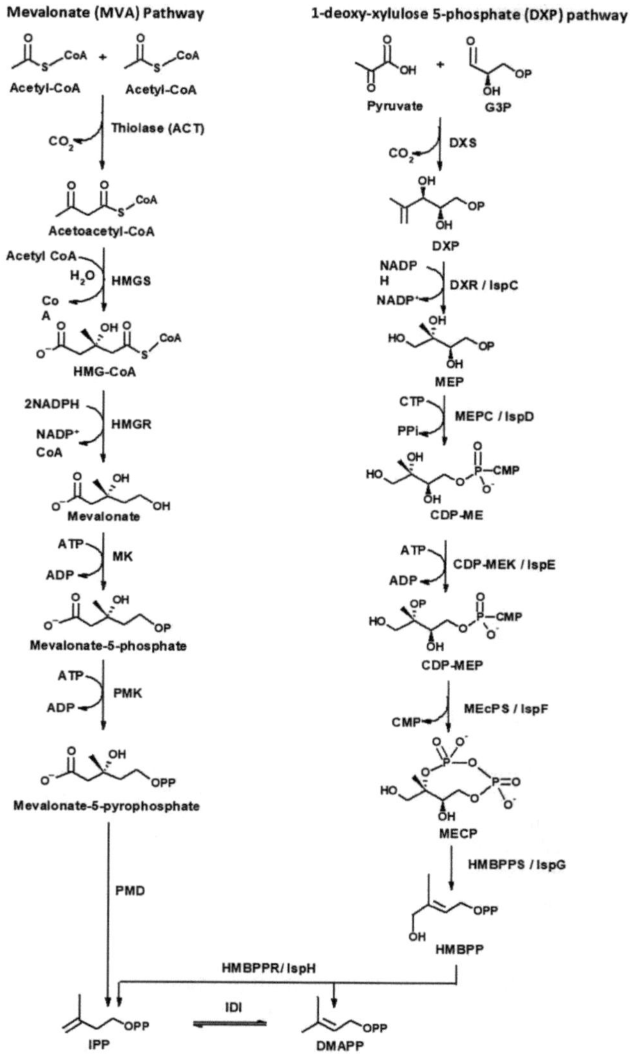

Fig. 4.2 The upper module of the terpenoid pathway for the supply of IPP and DMAPP. MVA pathway utilizes acetyl-CoA, whereas DXP pathway uses G3P and pyruvate to biosynthesize IPP and DMAPP. OP represents phosphate group, and OPP represents pyrophosphate group. *HMG-CoA* 3-hydroxy-3-methylglutaryl-CoA, *HMGS* HMG-CoA synthase, *HMGR* HMG-CoA reductase, *MK* Mevalonate kinase, *PMK* phosphomevalonate kinase, *PMD* mevalonate diphosphate decarboxylase, *G3P* glyceraldehyde-3-phosphate, *DXP* deoxyxylulose-5-phosphate, *DXS* DXP synthase, *DXR/IspC* DXP reductase, *MEP* 2-C-methylerythritol-4-phosphate, *MEPC* MEP cytidylyltransferase, *CDP-ME* 4-(cytidine-5′-diphospho)-2-C-methylerythritol, *CDP-MEK* CDP-ME kinase, *CDP-MEP* 2-phospho-4-(cytidine-5′-diphospho)-2-C-methylerythritol, *MEcP* 2-C-methylerythritol-2,4-cyclodiphosphate, *MEcPS* MEcP synthase, *HMBPP* (E)-4-hydroxy-3-methyl-but-2-enyl pyrophosphate, *HMBPPS* HMBPP synthase, *HMBPPR* HMBPP reductase, *IPP* isopentenyl pyrophosphate, *DMAPP* dimethylallyl pyrophosphate [8]

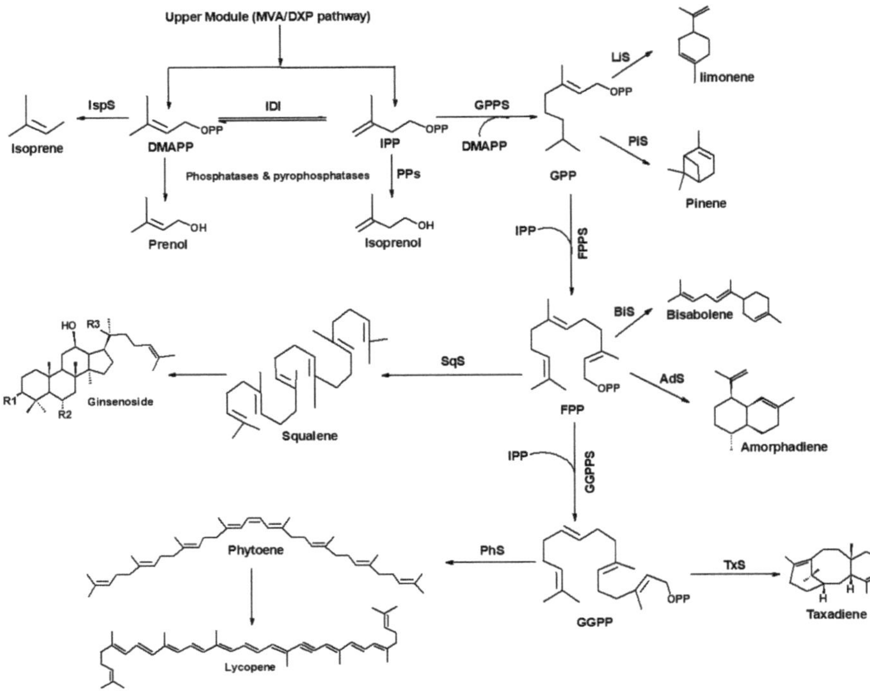

Fig. 4.3 The lower module of terpenoid biosynthetic pathway. The lower module consists of several terpenoid-precursor-specific and terpenoid-specific enzymes, which catalyze the conversion of IPP and DMAPP into a wide variety of terpenoids ranging from 5 carbon hemiterpenoids to 40 or 50 carbon atom carrying carotenoids. *IPP* isopentenyl pyrophosphate, *DMAPP* dimethylallyl pyrophosphate, *IDI* isopentenylpyrophosphate isomerase, *IspS* isoprene synthase, *GPP* geranyl pyrophosphate, *GPP* GPP synthase, *FPP* farnesyl pyrophosphate, *FPPS* FPP synthase, *PiS* pinene synthase, *LiS* limonene synthase, *BiS* bisabolene synthase, *AdS* amorphadiene synthase, *TxS* taxadiene synthase, *SqS* squalene synthase, *GGPP* geranyl pyrophosphate, *GGPPS* GGPP synthase, *PhS* phytoene synthase [8]

few enzymes, such as Nudix hydrolases and other phosphatases, which can convert prenyl precursors into their respective alcohols [25].

There is another group of enzymes that provides structural and function diversity of the terpenoids. This is cytochrome P450 monooxygenases or hydroxylases (CYP450s, CYPs). Together with the terpenoid synthases, the CYP450s govern the regio- and stereochemistry of the terpenoid molecules [26]. This category of enzymes is also heterologous to most of the microbial hosts, such as *E. coli*, and need to be expressed heterologously. Thus, the metabolic engineering of prokaryotic hosts plays an important role for the non-natural production of terpenoid-based metabolites from them.

4.3 Metabolic Engineering for Microbial Terpenoid Production

It is well understood that the production of recombinant enzymes has revolutionized modern biotechnology, which has almost ended the unethical era of sacrificing tons of animal and plant resources for the extraction of commercially important metabolites. Engineering a terpenoid biosynthesis pathway in microbial hosts can be subdivided into two modules. The first or upper module consists of either MVA pathway or DXP pathway or both and are responsible for the enhanced accumulation of IPP and DMAPP. The second or lower module comprises enzymes for the biosynthesis of the terpenoid-specific precursors and the desired terpenoid molecules. Thus, the optimization of both the modules is necessary for the non-natural production of terpenoids from engineered microbes.

As discussed earlier, only the DXP pathway enzymes are endogenous to microbial hosts and rest other enzymes are required to incorporate from foreign sources. Therefore, both the homologous and heterologous gene expression strategies are required simultaneously for the high-yield production of terpenoids from microbial sources. In the later sections, we have summarized the recent advances in both the homologous and heterologous expression strategies to achieve high-yield production of terpenoids from microbial cell factories.

4.3.1 Metabolic Engineering Strategies for Optimization of Upper Terpenoid Module

Enhancing the accumulation of IPP and DMAPP is the prime step for high-yield production of terpenoids from microbial sources, which depends upon improving the flux toward the upper module. It is well known that the endogenous DXP pathway in microbes is tightly regulated by several known and unknown regulations. The known limitations include insoluble nature of majority of the DXP pathway enzymes [27], feedback regulation of DXP synthase (DXS) by IPP and DMAPP [28], cytotoxicity to host cell by accumulation of IPP and DMAPP [29], and efflux of pathway intermediates, such as 2-C-methyl-D-erythritol-2,4-cyclodiphosphate MEcPP [30]. There might be several other regulations, which are yet to be unveiled.

Engineering *E. coli*, which is the most widely utilized microbial chassis for the production of terpenoid, has shown that the introduction of heterologous MVA pathway can improve terpenoid titers several fold in comparison to overexpression of DXP pathway [11]. The enhanced yield in MVA expressing *E. coli* might be due to the lesser or no regulatory checks, which might allow its efficient expression. Recently, expression of MVA pathway has also been utilized in other prokaryotic hosts also for the production of terpenoids. It has been observed that like *E. coli*, expression of MVA pathway in alternate hosts, such as *Corynebacterium glutamicum* and cyanobacteria, can also improve terpenoid production in respective hosts over the DXP pathway overexpression-based approaches [31, 32].

4.3.1.1 Expressing Heterologous MVA Pathway

Expression of heterologous MVA pathway in prokaryotes was imitated by Martin et al. [33] with a belief that "The DXP pathway might have unknown endogenous regulations," which were discovered later as discussed above. They utilized *E. coli* as microbial chassis for the production of a heterologous metabolite, amorphadiene (precursor to antimalarial drug, Artemisinin) [33]. The heterologous MVA pathway was further subdivided into two modules to investigate the impact of the genes present in each module on the amorphadiene yield. The upper module (MevT) was responsible for supplying mevalonate, consisted of three enzymes namely acetoacetyl-CoA synthase (AtoB) from *E. coli*, hydroxymethylglutaryl-CoA synthase (HMGS) and HMG-CoA reductase (HMGR) from *S. cerevisiae* (Fig. 4.4). The MevT module was reposible for mevalonate supply to the lower module, which they named as MevB. The MevB was constructed by mevalonate kinase (MK), phosphomevalonate kinase (PMK), and mevalonate diphosphate decarboxylase (PMD) enzymes to convert mevalonate into IPP [33] (Fig. 4.4). Further, the IDI and IspA were also added to MevB and named as MBIS. The MBIS was capable of converting mevalonate into FPP. To make the both the modules/operons functioning, the consensus Shine-Dalgarno (SD) sequences (or ribosomal binding site, RBS) of *E. coli* was added to each gene, and a spacer sequence was also added between the 5′ end of the preceding gene and SD sequence of the successor gene. Further, to regulate the expression of the modules, the operons were kept under lac promoters on compatible plasmids. For the selection of positive clones for each plasmid, the different origins of replication and antibiotic resistance genes were used [34].

Encouraged from the above study, several terpenoid-based biomolecules have been produced from microbial hosts by utilizing heterologous MVA pathway, to date. However, majority of these studies are focused on the most genetically

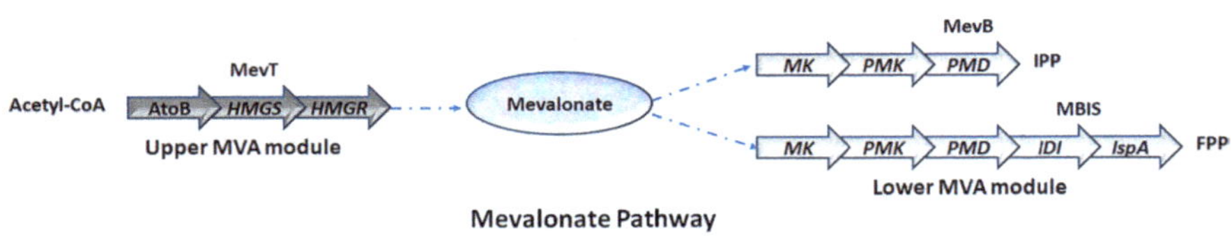

Fig. 4.4 Mevalonate pathway modules. The upper module MevT converts acetyl-CoA in mevalonate. The lower module MevB and MBIS consumes mevalonate and convert it into IPP (isopentenyl pyrophosphate) and FPP (Farnesyl pyrophosphate), respectively. *AtoB* acetoacetyl-CoA thiolase, *HMGS* 3-hydroxy-3-methylglutaryl-CoA synthase, *HMGR* 3-hydroxy-3-methylglutaryl-CoA reductase, *MK* mevalonate kinase, *PMK* phosphomevalonate kinase, *PMD* mevalonate diphosphate decarboxylase, *IDI* isopentenylpyrophosphate isomerase [8]

tractable and traceable host *E. coli* [16, 35, 36]. With the progression of the terpenoid research field, several modifications have also been made to improve the efficiency of each MVA pathway modules, which we have discussed in later sections.

Enzyme Modification

Since the MVA pathway was discovered first in the eukaryotic system, therefore the majority of the studies on terpenoid production via MVA pathway utilization used MVA enzymes from the yeast, *S. cerevisiae*. Initially, native enzymes from *S. cerevisiae* were used, which showed lower activity in prokaryotic hosts. To improve the expression of certain MVA pathway enzymes, truncated versions of respective enzymes were also used. For example, amino-terminal truncated version of HMGR was used to increase its ability for converting HMG-CoA into mevalonate [37]. It is well known that organisms have different codon usage, which is a limiting factor for the high-yield expression of heterologous proteins. Therefore, to match *E. coli* codon usage, the codons optimized variants of *S. cerevisiae* MVA pathway genes were also utilized [37]. Utilization of truncated and codon-optimized forms of the gene improved terpenoid yield in recombinant microbes over the native enzyme. However, the improved variants of *S. cerevisiae* HMGR were unable to balance the mevalonate flux inside the host. Therefore, additional copies of improved HMGR variants were expressed with heterologous MVA pathway for the efficient conversion of HMG-CoA into mevalonate [34].

Utilization of MVA Pathway Enzymes from Other Sources

Among the enzymes of the MevT module, the AtoB and HMGR have been suggested as rate-limiting steps. This is because AtoB condenses two molecules of acetyl-CoA to generate acetoacetyl-CoA and thus directs carbon flux toward the MVA pathway. The HMGR on the other hand converts HMG-CoA, whose accumulation affects fatty acid synthesis inside the host, into mevalonate [34]. On exploring enzymes from prokaryotic and eukaryotic sources, a group of enzymes have been identified that possesses AcAc–CoA synthase/acetoacetate-CoA ligase activity instead of acetoacetyl–CoA thiolase activity [38, 39]. Expression of enzymes from this subgroup, such as Acal in *E. coli*, increases the substrate range for the production of terpenoids because such enzyme can produce acetoacetyl-CoA by utilizing alternate substrates, such as lithium acetoacetate [38]. Further, expressing such enzymes together with native AtoB of the host might be beneficial to improve the mevalonate pool inside the hosts by simultaneously metabolizing wide substrate ranges.

The HMGR enzyme, for which the majority of the bottlenecks were overcome by expressing its truncated and codon-optimized version in *E. coli*. However, it was found that the improved version of HMGR was also not effective to maintain the mevalonate flux and an additional copy of it was expressed for the efficient utilization of accumulated HMG-CoA as discussed above. In an alternative way, more balance HMGR enzymes from microbial sources, especially prokaryotic origin, were also explored to improve mavolante flux inside the remobinant hosts [40–42]. The HMGR enzymes from prokaryotic origin have certain advantages over eukaryotic counterparts. For example, expression of a prokaryotic HMGR or HMGS does not require its truncated version. In some cases, codon optimization is also not required. The HMGS and HMGR enzymes from several prokaryotic hosts have been explored till now, such as from *Enterococcus faecalis* and *Staphylococcus aureus* [16, 35] and have shown better efficiency than *S. cerevisiae* HMGR [42]. This might be due to their reductase activity in addition to the acetyl-CoA transferase activity [42]. Further, exploration of such bifunctional enzymes from other microbial sources might be helpful to improve terpenoid yield beyond the currently achieved titers.

Utilization of Synthetic Scaffold

For the high-level production of terpenoids from recombinant microbes, not only the optimal expression of recombinant enzymes is required but also the stoichiometry of the participating enzymes should be balanced. Dueber et al. [43] demonstrated a promising approach, in which they tethered the MevT enzymes in synthetic scaffold to balance the metabolic flux. To prepare the scaffold, GTPase-binding domain (GBD), Src homology-3 domain (SH3), and PSD95/DlgA/Zo-1 (PDZ) binding domain from metazoans were joined together by nonstructure forming amino acids as shown in Fig. 4.5 [43]. Tethering of MevT enzymes to the scaffold was facilitated by incorporating cognate ligands of GBD, SH3, and PZD on the C-terminus of

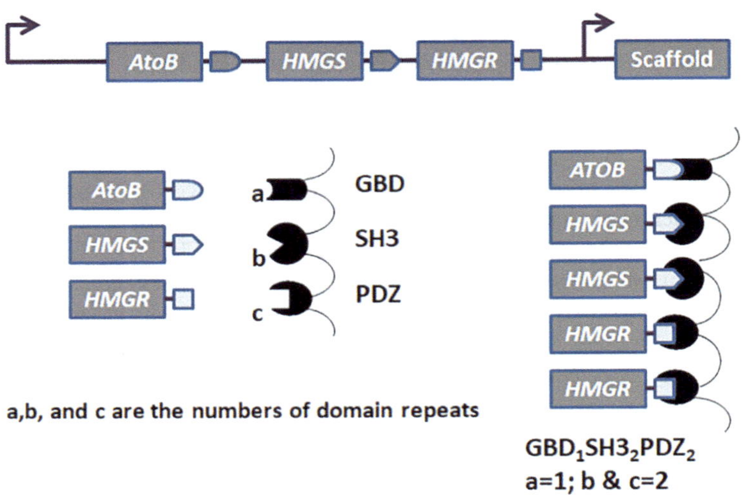

Fig. 4.5 Synthetic scaffolds to control the flux in upper MVA module flux. *GBD* GTPase-binding domain, *SH3* Src homology-3 domain, and *PDZ* PSD95/DlgA/Zo-1-binding domain [8]

AtoB, HMGS, and HMGR enzymes of MevT, respectively. The novel scaffold pathway thus formed showed better efficiency than the conventional pathway. This might be due to the co-localization of the MevT enzymes in a superlative combination, which might result in optimized stoichiometry of the enzymes at the scaffold [43]. Thus, the synthetic scaffolds might be helpful to optimize the number of the interacting domains, which would be sufficient to improve the accumulation of pathway intermediates beyond their toxic levels.

Optimization of MevT is important for the high-level accumulation of mevalonate, which is further utilized by the MBIS module to biosynthesize desired terpenoids. The first enzyme of the lower module (MBIS) of the MVA pathway is a kinase (MK), which phosphorylates mevalonate into phosphomevalonate or mevalonate-5-phosphate (MP). Like the upper module of the MVA pathway, initial studies on terpenoid production utilized MK from *S. cerevisiae*. Later it was found that the MKs are regulated by feedback inhibition from downstream prenyl precursors, such as MVPP, IPP, DMAPP, FPP, and GPP [44]. George and coworkers [44] demonstrated that the feedback regulation of MK can be overcome by expressing the feedback-resistant variants of MKs from archaeal sources. They have successfully shown the high-yield expression of isopentenol, a hemiterpene alcohol, by utilizing MK from *Methanosarcina mazei* [44]. Later, Kazieva et al. [45] discovered few more variants of feedback-resistant MKs from archaeal sources [45], which might be used in place yeast MK to improve downstream prenyl precursor flux for enhancing terpenoid production in recombinant microbial strains.

The PMK and PMD are the other two enzymes of the MBIS module, which convert MK into IPP in two successive steps. Like other MVA pathway enzymes, initially the PMK and PMD of *S. cerevisiae* were used. Later, the PMK and PMD from other microbial sources, such as *Streptococcus pneumonia*, have also been utilized [46, 47]. PMK has also been considered as a rate-limiting step in the MBIS module, as its optimal expression is challenging [48]. To improve the expression of PMK in engineered hosts, Redding-Johanson et al. [48] applied a strong promoter, which resulted in the increase in the accumulation of downstream terpenoid-based metabolites [48]. Optimization of the MBIS module has not been studied in detail as MevT, which might be the thrust area toward heterologous terpenoid production from microbial sources. Recently, the PMD enzyme, which catalyzes the biosynthesis of IPP from mevalonate pyrophosphate, has been used to construct an MVA-dependent IPP-bypass route for isopentenol production (see the later section).

Though the majority of the bottlenecks regarding the high-level expression of MVA pathway modules have been resolved; however, its expression in microbial hosts other than *E. coli* is still a challenge. Recently, expression of heterologous MVA pathway has been demonstrated in a GRAS (generally regarded as safe by FDA) status bacterium, *C. glutamicum*, for the production of isopentenol [32]. It indicates that the alternate microbes might also have potential for the expression of heterologous pathways, and they can be utilized for the production of terpenoid-based flavor and pharmaceuticals.

4.3.1.2 Tuning Endogenous DXP Pathway

Optimization of a host's endogenous pathway is always adventitious because the host's transcription and translation machinery already have tools for their expression. However, sometimes these machineries also have a tight regulation on the endogenous pathways and hinder its overexpression. The endogenous DXP pathway of the microbes is also regulated by several known and unknown factors, which restrict its optimal overexpression in prokaryotic hosts. This pathway carries seven enzymatic steps for the conversion of G3P and pyruvate into prenyl precursors. Apart from the precursors and the enzymes, the simultaneous biosynthesis of IPP and DMAPP in a ratio 5:1 [49] also makes it different from the MVA pathway of the eukaryotes. There are many endogenous regulations associated with almost each step of the DXP pathway, some of them have been identified and others are still unknown. In this section, we have summarized the metabolic engineering strategies that have been utilized for the optimal expression of DXP pathway genes. We have also discussed some of the bottlenecks in the DXP pathway and strategies to overcome those challenges for the improved production of terpenoids from prokaryotic hosts.

Until date, several bacterial chassis have been explored for production of terpenoids by engineering endogenous DXP pathway. Unlike MVA pathway-based studies, which are focused on *E. coli*, the terpenoid production via DXP pathway modulation is concentrated on alternate microbial hosts, such as *B. subtilis* [13], *C. glutamicum* [50], and cyanobacteria [14]. The other microbial communities such as *Pseudomonas putida* [51, 52] and *Streptomyces venezuelae* [53] have also been explored. Though these microbial hosts have their own set of advantages; however, the studies on these microbes are very few. This might be due to the lack of genetic tools for the multiple gene expression in *P. putida* and *Streptomyces* sp.

As discussed, the primary aim of pathway optimization is to improve the metabolic flux through the desired pathway. In the case of DXP pathway, its first enzyme, the DXP synthase (DXS), plays an important role in converging metabolic flux by condensing G3P and pyruvate to form DXP. Overexpression of DXS is the universal approach to direct metabolic flux toward the DXP pathway in engineered organisms [27, 54, 55]. Unexpectedly, its overexpression did not improve pathway flux beyond a certain extent [27, 56, 57]. It was found that the DXS enzyme has some bottlenecks, which include its low turnover number [58], feedback inhibition by IPP and DMAPP [28, 59], and insoluble nature [27, 59, 60]. Together these limitations make DXS a rate-limiting step of the DXP pathway. Kudoh et al. (2017) screened DXS enzymes from different microbial sources in order to develop a robust isoprenoid biosynthesis system. The basis of screening was resistance to feedback inhibition, solubility, and enzymatic activity. They observed that none of the DXS enzymes was fit into all the search categories. Interestingly, each parameter was satisfied by a different DXS separately. The DXS from *B. subtilis* was identified with highest resistance to feedback regulation, whereas DXS from *Paracoccus aminophilus* and *Rhodobacter capsulatus* showed highest solubility and enzymatic activity, respectively [59]. It indicates that DXS enzyme that exhibits all above-mentioned promising properties might not exist naturally. Therefore, high-throughput synthetic biology tools and techniques can be utilized to improve its functionality. A comparative scrutiny of the functionally closest set of natural variants of DXSs from each category/parameter might be helpful to decipher the evolutionary conserved loci and to select the potential motif sequences. Further, the outcomes of the scrutiny can be utilized to redesign a functionally improved variant of DXS, which may possess all the desired quality, by applying advanced synthetic biology and metabolic engineering approaches.

It has also been established that soluble variants of DXS are more catalytically active than the insoluble or less soluble enzymes [60], which might be due to their superior folding capabilities. To improve the DXS solubility, both the synthetic biology [61] and media optimization [27, 60] approaches have been utilized. Addition of the small polyionic tags, which are less likely to interfere with the core structure of a target protein, has been shown to improve protein solubility [62]. In a recent study, Han et al. [61] have demonstrated that the insertion of short peptide tags can also improve the solubility of DXS protein. For this they utilized machine learning tools, which were based on the protein sequence information, to optimize protein properties [61]. The solubility of the tagged protein was evaluated first by a support vector regression model [63] and then validated experimentally. It was found that small polypeptide tags improved the solubility of the tagged-DXS over the untagged-DXS. Although this study by Han et al. [61] was not a dedicated study on terpenoid pathway and DXS enzymes, it suggested an alternative methodology for optimizing the soluble expression of overexpressed DXP pathway proteins.

The DXP reductase (DXR or IspC) enzyme acts just downstream to DXS and catalyzes the transformation of DXP into 2*C*-methyl-D-erythritol-4-phosphate (MEP). The role of DXR in the overexpressed DXP pathway is controversial because majority of the incidences its overexpression did not enhance the downstream terpenoid titers [57, 64] and considered as a nonessential enzyme for overexpression studies. Opposed to the above-mentioned observations, there are certain studies that showed the overexpression of DXR/IspC either singly [65, 66] or as a coexpression partner with DXS [67, 68] enhanced terpenoid titers in recombinant strain.

The MEP-cytidylyltransferase (MEPC or IspD) and 4-diphosphocytidyl-2*C*-methyl-D-erythritol kinase (CDP-MEK or IspE) are the next two enzymes, which require CTP and ATP as a cofactor respectively, to catalyze the conversion of MEP into CDP-MEP and overexpression of either enzyme has not shown enhanced terpenoid titers in recombinant hosts [58, 69, 70]. The inadequate supply of the cofactors for each enzyme might be a limiting factor for the little or no improvement in terpenoid yield after their overexpression. It has been shown that optimization of culture medium by supplying high concentration of phosphate salts improved the cofactor supply and thus enhanced the accumulation of DXP pathway intermediates and final terpenoid-based metabolites in recombinant hosts [71]. Recently, coexpression of CTP synthase endogenous pyrophosphatases have been shown to restore the production of isoprene, a terpenoid-based volatile compound, in DXS and

IspD overexpression strain [69]. This suggested that balanced precursor supply with the overexpressed IspD and IspE enzyme might increase terpenoid yields in recombinant microbes.

Conversion of CDP-MEP into 2-C-methyl-D-erythritol-2,4-cyclopyrophosphate (MEcP) is catalyzed by MEcP synthase (MEcPS or IspF). In a few microbes, the IspF has been identified to be linked with IspD either in a single operon or as a fusion product (IspDF) [72]. Overexpression of the both the form of IspF has been explored in recombinant microbes to improve terpenoid yields [70, 73]. Like DXR, the overexpression of IspF did not increase terpenoid production in recombinant strains. Hence, it is also considered as a nonessential enzyme for overexpression studies. Identification of bottlenecks in the overexpression of IspF might be helpful to utilize this enzyme fullest for enhancing terpenoid yields in microbial hosts.

It has been found that after overexpression of certain DXP pathway enzyme microbial cells efflux MEcP [30], which might also be a limiting factor for the lesser or no improvement of terpenoid production in recombinant hosts after IspF overexpression. The efflux to MEcP might result in the reduction of precursor supply for the downstream enzyme, 1-hydroxy-2-methyl-2-(*E*)-butenyl-4-diphosphate synthase (HMBPPS or IspG). It was observed that overexpression of IspG can reduce the MEcP efflux in recombinant strains [30]. This indicated that the reaction catalyzed by IspG is also a rate-limiting step in the DXP pathway. Thus, its optimal and functional expression is crucial for the enhanced production of terpenoid-based metabolites in recombinant strains. Protein fusion products of two consecutive enzymes of a pathway have shown improved availability of the intermediates produced by the former enzymes to the later enzymes of the fusion product [44, 74]. Expression of IspF with the IspG could be an attractive alternative for the reduction of MEcP efflux and thus improving accumulation of downstream intermediates for terpenoid production.

The HMBPP reductase (HMBPPR or IspH) is the last enzyme of the DXP pathway that converts HMPP into IPP and DMAPP in a ratio 5:1. Like DXR, overexpression of IspH has also shown contradictory results in recombinant strains. In *B. subtilis*, overexpression of IspH improved carotenoid titer [66], whereas in *E. coli*, its overexpression resulted in reduction of squalene production [70]. Overexpression of IspH in microbial hosts has been least studied, and its endogenous regulations are still unknown. Therefore, more dedicated studies are required to understand in detail about the contribution of *ispH* in modulated DXP pathway in engineered microbes.

4.3.1.3 MVA-Dependent Novel IPP-Bypass Route

It has been established that in the absence of efficient downstream enzymes, accumulation of IPP and DMAPP (after the expression of MVA pathway components) induces toxicity in microbial hosts [29]. This prenyl precursor-induced toxicity can be removed either by the expression of more active variants of downstream enzymes or by expressing multiple copies of the same. For the production of terpenoids

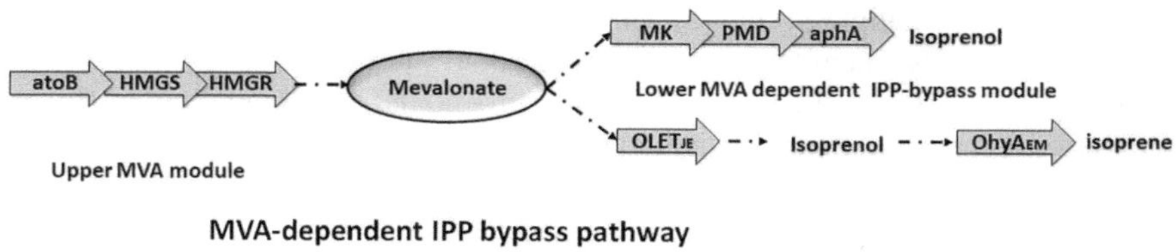

Fig. 4.6 MVA-mediated IPP bypass pathway for hemiterpene production [8]

having carbon atoms equal to ten or more, IPP and DMAPP are the essential precursors. However, for the hemiterpenoids, which possess five-carbon atoms, this requirement can be overcome by the expression of an MVA-dependent IPP-bypass route [75, 76].

Kang et al. [75] demonstrated that in the absence of PMK enzyme, PMD enzyme of the MVA pathway could catalyze the direct conversion of mevalonate and MP into isoprenol and isopentenyl monophosphate (IP), respectively. The IP can be further converted to isoprenol by endogenous phosphatases, such as AphA [75]. In the same year, Yang and coworkers [76] showed that the IPP-bypass route can also be constructed without using the MBIS module of MVA pathway. For this, they incorporated a fatty acid decarboxylase ($OleT_{JE}$) from *Jeotgalicoccus* spp. into an *E. coli* strain, which was already containing the MevT module of MVA pathway [76]. The $OleT_{JE}$ enzyme was able to catalyze the direct conversion of mevalonate into isoprenol. Further, they also demonstrated that the oleate hydratase ($OhyA_{EM}$) from *Elizabethkingia meningoseptica* could convert isoprenol into isoprene [76]. Thus, the IPP-bypass pathways overcome the requirement of truncated and codon-optimized versions of eukaryotic genes for the production of hemiterpenes and thus reduce the cost and efforts involved in train development for the same. The other advantage of the MVA-dependent routes is that they reduce cellular energetic for the production of hemiterpenoids, such as isoprenol and isoprene [75, 76] (Fig. 4.6). This is due to the reduction in the number of steps required for the hemiterpenoid production via IPP-bypass routes as compared to conventional MVA route. The reduction in enzymatic steps not only reduces the cellular energetic but also minimizes the metabolic burden on the host cell for the production of hemiterpenoids.

4.3.2 Metabolic Engineering of Downstream Terpenoid Pathway in Microbes

As discussed, the upper modules (MVA and DXP pathways) of the terpenoid biosynthesis produce the universal precursors, IPP and DMAPP, for the downstream terpenoid production. Terpenoid-based metabolites generally follow "isoprene rule," which implies that they are formed by joining multiple isoprene units and contain carbon atoms in multiple of five [5]. Based on number of carbon atoms, terpenoids are classified as hemiterpenes (five carbon atom carrying terpenoids), monoterpenoids (ten carbon atom carrying terpenoids), sesquiterpenoids (fifteen carbon atom carrying terpenoids), diterpenoids (twenty carbon atom carrying terpenoids), triterpenoids (thirty carbon atom carrying terpenoids), and carotenoids (forty or more carbon atom carrying terpenoids).

Hemiterpene-based metabolites are directly synthesized either from IPP or DMAPP. For example, isoprenol is biosynthesized from IPP, whereas, prenol and isoprene are formed from DMAPP. For the biosynthesis of the precursors for higher terpenoids, such as monoterpenoids, sesquiterpenoids and diterpenoids, the downstream enzyme utilized IPP and DMAPP. For example, precursor to all monoterpenoids, geranyl pyrophosphate (GPP), is biosynthesized via condensation of IPP and DMAPP by GPP synthase (GPPS). Similarly, farnesyl pyrophosphate (FPP), a precursor to sesquiterpenoids, is formed from the condensation of GPP and IPP by FPP synthase (FPPS). The geranylgeranyl pyrophosphate (GGPP), which is a precursor to all diterpenoids, is biosynthesized by GGPP synthase (GGPPS) via condensing FPP and IPP. Further the enzymes squalene synthase and phytoene synthase utilize FPP and GGPP to produce precursors for triterpenes and carotenoids, respectively.

Majority of the commercially important terpenoids are heterologous to microbial hosts; therefore, their biosynthesis enzymes need to be incorporated along with their precursor-specific enzymes in microbial hosts. In the later sections, we have summarized the heterologous and homologous/endogenous engineering approaches for the expression of terpenoid-specific enzymes in microbial hosts to convert accumulated IPP and DMAPP into respective terpenoids.

4.3.2.1 Pathway Engineering for Hemiterpenoid Production

Hemiterpenes are the five carbon atoms carrying simplest form of terpenoids and industrially utilized as a precursor for the synthesis of plenteous terpenoid-based flavor, pharmaceuticals, synthetic rubber, and other commodity chemicals. The IPP and its isomer DMAPP are the direct precursors for the biosynthesis of hemtiterpenoids. Among the plenteous compounds of this category, microbial production of isoprene and isopentenol (isoprenol and prenol) has been most widely studied due to the industrial importance of both the compounds. Isoprene is a volatile gas that is biosynthesized from DMAPP by isoprene synthase (IspS) enzyme. However, the IspS gene has not been identified in microbial hosts and needs to be expressed into them heterologously. The IspSs from several plant species, such as *Populus* spp., *Pueraria montana*, *Eucalyptus* spp., *Mangifera indica, Elaeocarpus photiniifolius,* and *Ipomoea batatas,* have been explored in microbial hosts for the production of isoprene [77, 78]. Majority of the studies have utilized IspS from *Populus* spp. and *P. Montana* [14, 78]; however, a more active variant of IspS was identified from *I. batatas*, which outperformed the IspSs of *Populus alba* and *P. Montana* in terms of isoprene production [77]. Further, directed evolution studies on already discovered IspS might be helpful for the development of more active variants of IspS, else, extensive studies on the discovery of microbial IspSs could provide a breakthrough in this field.

The hemiterpene alcohols, isopentenols, are synthesized from the hydrolysis of IPP and DMAPP by endogenous phosphatases and pyrophosphatases. Isoprenol is synthesized from IPP, and prenol is synthesized from DMAPP. Withers et al. [29] demonstrated that the *B. subtilis* enzymes, NudF and YhfR, could rescue the engineered host from IPP-induced toxicity by converting it into isopentenol due to their multicalatytic activity [29]. Encouraged from their findings, several phosphatases and pyrophosphates have been investigated until date for their isopentenol producing capabilities [42, 79]. The extensive studies on isopentenol producing enzymes led to the identification of Nudix hydrolases, such as *B. subtilis* NudF ($NudF_{Bs}$) and *E.coli* NudB ($NudB_{Ec}$) as the most efficient pyrophosphatases for the conversion of IPP and DMAPP in to their respective isopentenols. Further, it was identified that basal level expression of Nudix hydrolases was not enough to convert all the accumulated IPP and DMAPP into isopentenol efficiently. Therefore, for the optimal expression of Nudix hydrolases, the ribosomal-binding site (Shine-Dalgarno sequence or RBS) of NuB_{Ec} was optimized by using well-developed computational tools [44]. Optimization of RBS sequence improves the utilization of accumulated IPP and DMAPP and led to highest isoprenol titer ever achieved form a microbial host [44].

Initially, it was believed that the $NudF_{Bs}$ or $NudB_{Ec}$ directly transforms IPP and DMAPP into isoprenol and prenol, respectively, by removing pyrophosphate groups. Recently, it was found that pyrophosphatases remove only one phosphate group from prenyl pyrophosphates and convert them into prenyl monophosphates [75]. Conversion of prenyl monophosphates into their respective alcohol is catalyzed by endogenous phosphatases. Until now, the *E. coli* AphA enzyme has been recognized as the most effective phosphatase for the production of terpenoid alcohols from prenyl monophosphates [25, 75]. Therefore, identification of more efficient phosphatases from other microbial sources might be beneficial for improving isopentenol titers from engineered microbes. Alternatively, expression of AphA with either $NudF_{Bs}$ or $NudB_{Ec}$ might also be explored for improving the availability of prenyl monophosphates to AphA and thus increasing the yield of isopentenols.

As we have discussed, isoprene and prenol are synthesized from DMAPP. Therefore, its adequate supply is necessary for their high-yield production. We have also seen that the MVA pathway synthesizes IPP as the first prenyl precursor, and DXP pathway produces IPP and DMAPP in a ratio 5:1. In both the cases, the enzyme, IDI, maintains the balance between IPP and DMAPP inside microbial hosts. Overexpression of IDI from both the eukaryotic [54] and prokaryotic [80] sources has shown improved titers of DMAPP-derived hemiterpenoids. Another reason for the lesser availability of DMAPP is its quicker utilization by IspA for the biosynthesis of FPP. To counter this challenge, expression of IDI with $NudB_{Ec}$ has been explored to increase the yield of prenol in engineered *E. coli*. It has been seen that protein fusion product of IDI and $NudB_{Ec}$ has shown higher prenol titer than their co-expression [79]. Similarly, protein fusion products of IspS with IDI might also be helpful to increase isoprene titers in recombinant microbes.

4.3.2.2 Pathway Engineering for Monoterpenoid Production

The monoterpenes subclass of terpenoids consists of ten carbon atoms carrying cyclic and cyclic compounds. They are the major constituents of the essential oils of several aromatic plants and the resin of high-altitude plants. As discussed, GPP is the common precursor to all monoterpenoids and is synthesized by GPPS. However, microbes do not possess the GPPS enzyme, and the endogenous enzyme IspA quickly uptakes the GPP produced by it for further biosynthesis of FPP. Thus, to increase GPP flux inside microbial hosts, several variants of GPPS have been expressed in microbes. The GPPSs from *Abies grandis, Pinus abies,* and *Pinus taeda* are some of the widely utilized enzymes to improve GPP flux in microbial hosts [11]. The next step after improving flux toward GPP is to express heterologous monoterpene synthase enzymes, such as pinene synthase (PiS) and limonene synthase (LiS), of plant sources because microbial hosts do not carry these enzymes naturally. We have already discussed that the eukaryotic enzymes consist of a few domains that hinder their efficient expression in prokaryotes. To remove the unnecessary domains, such as membrane spanning domains and plastid transit domain, these enzymes need to be truncated. Truncated enzymes due to their reduced size do not form protein aggregates like their nontruncated versions. Thus, truncation also helps in soluble expression of heterologous proteins. In addition to truncation, codon optimization is also important for the high-level expression of heterologous monoterpene synthase enzymes. Both the truncated and codon optimized variants of GPPS and monoterpene synthase enzymes have shown higher terpenoid titers in recombinant hosts as compared to their nonoptimized versions [81–83].

Like Nudix hydrolases, RBS optimization of GPPS has also been explored for the efficient conversion of IPP and DMAPP into GPP [84]. The RBS optimized GPPS has shown improved yield of monoterpenoid-based metabolite indicating enhanced accumulation/availability of GPP for downstream monoterpenoid synthase enzyme. It has also been found that the GPPS and monoterpene synthase enzymes, such as PiS, can be inhibited by GPP via feedback regulation. Thus, the efficient conversion of GPP into monoterpenes is crucial to rescue host from GPP-induced feedback inhibition of monoterpenoid biosynthesis machinery. For this, protein fusion product of monoterpene synthase with GPPS can be an excellent alternative [83] because it would improve the availability of GPP for monoterpene synthase and hence might improve desired monoterpene titer. Alternatively, RBS optimization for GPPS can also be sought to improve its expression in prokaryotic hosts. It has also been found that the activity of monoterpene synthases, such as PiS, depends upon the presence of divalent metal ion Mn^{2+} [85]. However, the *E. coli* cytosol generally lacks Mn^{2-} and is rich in Mg^{2+}. Tashiro et al. [85] developed a mutant version of PiS (PiS_{mut}), which showed reduced dependency on Mn^{2+} and increased terpenoid titer in recombinant *E. coli* over its native version. Thus, the improved variant of GPP developed using directed mutagenesis might be more beneficial for the efficient utilization of GPP and for high-yield production of monoterpenoids.

4.3.2.3 Pathway Engineering for Sesquiterpenoid Production

The sesquiterpenes are the largest subclass of terpenoid super family, which contains both cyclic (mono-, bi- or tri-cyclic) and acyclic compounds with several biological activities including antifungal, antimalarial, and anti-inflammatory [86, 87]. In addition, they play a varied physiological role in plants, insects, and fungi [4]. The FPP is the common precursor for the biosynthesis of sesquiterpenoids in the living system. In microbial hosts, it is produced by IspA enzyme that utilizes IPP and DMAPP, whereas eukaryotic systems possess FPPS enzymes that utilize GPP and IPP for the production of FPP. To improve the FPP flux, overexpression of endogenous IspA was preferred over the expression of heterologous FPPS [37, 88, 89]. However, only the overexpression of IspA is not sufficient because IspA requires both IPP and DMAPP to biosynthesize FPP, and microbial cells produce more IPP and DMAPP. Therefore, for improving DMAPP supply, co-expression of IDI with IspA was utilized [90]. It was observed that overexpression of IDI with IspA improved farnesene titer in recombinant *E. coli* [90] indicating the dependency of IspA on IPP:DMAPP balance inside the microbial cells.

Like monoterpenes, most of the commercially important sesquiterpenes are heterologous to microbial hosts. Therefore, to convert FPP into desired sesquiterpenoids, heterologous sesquiterpenoid synthases, such as amorphadiene synthase (ADS), farnesene synthase (FaS), and bisabolene synthase (BiS), need to be incorporated in microbial hosts. Production of acyclic or monocyclic sesquiterpenoids, such as farnesene, farnesol, and bisabolene, is comparatively easy than the complex sesquiterpenoids amorphadiene. This is because the synthesis of simpler sesquitepenoids generally requires single catalyzed step after FPP, whereas biosynthesis of complex sesquitepenoids requires multisteps, which are catalyzed by cytochrome P450 enzymes (CYP450s or CYPs). Together with the terpenoid synthases, the CYPs control the regio- and stereochemistry of a variety of bioactive terpenoids [26]. However, expression of CYPs in microbial hosts is still a challenge and may put extra burden on microbial chassis. Therefore, researchers are currently focusing on the precursor molecules for complex terpenoids so that it can be converted into complex terpenoid via few chemical modifications.

4.3.2.4 Pathway Engineering for Higher Terpenoid Production

The biosynthesis of higher terpenoids, such as diterpene, triterpene, and carotenoids, also requires expression of heterologous enzymes. For example, to synthesize diterpenes, GGPP is the common precursor, which is biosynthesized by heterologous GGPPS enzyme. GGPP consumes FPP and IPP to biosynthesize GGPP. Similarly, heterologous squalene synthase (SqS) and phytoene synthase (PhS) are required for the triterpenes and carotenoid biosynthesis in microbial hosts, respectively. The SqS condenses two FPP molecules to convert them into squalene, which is itself a bioactive compound and precursor to triterpenoids, while PhS utilizes two GGPP molecules to produce phytoene. Thus, we can say that accumulation of FPP is crucial for higher terpenoid production, which is governed by the activity of IDI and IspA.

Among diterpenoids, production of taxadiene is widely studied due to its economical importance. Taxadiene is a precursor to potential chemotherapeutic agent, paclitaxel (Taxol), which is a complex terpenoid. As discussed, biosynthesis of complex terpenoid requires CYPs, which is challenge for microbial system; therefore, the expression of heterologous taxadiene synthase (TxS) is currently being utilized together with the expression of IspA and GGPPS for the production of taxadiene [91, 92]. The taxadiene can be further converted to Taxol via chemical routes. Similarly, to produce squalene from *B. subtilis*, co-expression of IspA and SqS was utilized together with other DXP pathway components [93]. Further, to improve the availability of FPP for SqS for efficient conversion of FPP into squalene, protein fusion product of IspA and SqS has also been successfully utilized in microbial hosts [94, 95].

Microbial carotenoid production was explored in *E. coli* in the early 90s by incorporating carotenoid biosynthesis from *Erwinia uredovora* [96]. During early days, the carotenoid production was explored without optimizing the upper terpenoid modules (MVA or DXP). Later, the discovery of DXP pathway and successful incorporation of MVA into *E. coli* microbial enable researchers to improve prenyl precursor production for carotenoid production [97, 98]. To increase carotenoid yield, overexpression of PhS or CrtB is important for the condensation of two GGPP molecules. The carotenoid-specific enzymes such as phytoene desaturase (CrtI) need to be introduced to convert phytoene into lycopene [99]. Further, the downstream enzymes like Lycopene β-cyclase (CrtY), β-carotene ketolase (CrtW), and β-carotene hydroxylase (CrtZ), which act downstream to CrtI can give rise to wide variety of carotenoids such as β-carotene, zeaxanthin, anthaxanthin, and astaxanthin [100].

4.4 Conclusion

There is no doubt that metabolic engineering approaches have contributed tremendously in the area of microbial terpenoid production. Despite several success stories [101–103], bottlenecks still persist for achieving the high-yield production from microbial routes. Though the DXP pathway is stoichiometrically more competent than MVA pathway [104], the terpenoid yields obtained from the overexpression of DXP pathway are far lesser than the same achieved through incorporating heterologous MVA pathway. Addressing some of the bottlenecks of DXP pathway enzyme has shown some improvement; however, the terpenoids titers are still lagging behind to that of MVA pathway expression. It indicates that the metabolic flux to DXP pathway in microbial hosts is subjected to multilevel regulations, which need to be elucidated by conducting more dedicated studies on each enzyme of the DXP pathway and on the channelization of the carbon flux inside the host.

References

1. Ajikumar PK, Tyo K, Carlsen S et al (2008) Terpenoids: opportunities for biosynthesis of natural product drugs using engineered microorganisms. Mol Pharm 5:167–190
2. Lange BM, Rujan T, Martin W, Croteau R (2000) Isoprenoid biosynthesis: the evolution of two ancient and distinct pathways across genomes. Proc Natl Acad Sci U S A 97:13172–13177. https://doi.org/10.1073/pnas.240454797
3. Paduch R, Kandefer-Szerszeń M, Trytek M, Fiedurek J (2007) Terpenes: substances useful in human healthcare. Arch Immunol Ther Exp 55:315–327. https://doi.org/10.1007/s00005-007-0039-1
4. George KW, Alonso-Gutierrez J, Keasling JD, Lee TS (2015) Isoprenoid drugs, biofuels, and chemicals-artemisinin, farnesene, and beyond. In: Advances in biochemical engineering/biotechnology, pp 355–389
5. Ludwiczuk A, Skalicka-Woźniak K, Georgiev MI (2017) Terpenoids. In: Badal S, Delgoda R (eds) Pharmacognosy. Elsevier, pp 233–266
6. Tippmann S, Chen Y, Siewers V, Nielsen J (2013) From flavors and pharmaceuticals to advanced biofuels: production of isoprenoids in *Saccharomyces cerevisiae*. Biotechnol J 8: 1435–1444. https://doi.org/10.1002/biot.201300028
7. Gupta P, Phulara SC (2015) Metabolic engineering for isoprenoid-based biofuel production. J Appl Microbiol 119:605–619. https://doi.org/10.1111/jam.12871
8. Gupta P, Phulara S (2021) Homologous and heterologous expression strategies in microbes. In: Gupta P, Phulara S (eds) Biotechnology of terpenoid production from microbial cell factories. Academic Press, US, pp 103–132
9. Zhou X, Zhu H, Liu L et al (2010) A review: recent advances and future prospects of taxol-producing endophytic fungi. Appl Microbiol Biotechnol 86:1707–1717. https://doi.org/10.1007/s00253-010-2546-y
10. Thomas P, Farjon A (2011) Taxus wallichiana. IUCN Red List Threat. Species e. T46171879A9730085
11. Phulara SC, Chaturvedi P, Gupta P (2016) Isoprenoid-based biofuels: homologous expression and heterologous expression in prokaryotes. Appl Environ Microbiol 82:5730–5740. https://doi.org/10.1128/AEM.01192-16
12. Kusari S, Singh S, Jayabaskaran C (2014) Rethinking production of Taxol W (paclitaxel) using endophyte. Trends Biotechnol 32:304–311
13. Guan Z, Xue D, Abdallah II et al (2015) Metabolic engineering of *Bacillus subtilis* for terpenoid production. Appl Microbiol Biotechnol 99:9395–9406. https://doi.org/10.1007/s00253-015-6950-1
14. Lin PC, Pakrasi HB (2019) Engineering cyanobacteria for production of terpenoids. Planta 249:145–154. https://doi.org/10.1007/s00425-018-3047-y
15. Vickers CE, Williams TC, Peng B, Cherry J (2017) Recent advances in synthetic biology for engineering isoprenoid production in yeast. Curr Opin Chem Biol 40:47–56. https://doi.org/10.1016/j.cbpa.2017.05.017
16. Ward VCA, Chatzivasileiou AO, Stephanopoulos G (2018) Metabolic engineering of *Escherichia coli* for the production of isoprenoids. FEMS Microbiol Lett 365:fny079. https://doi.org/10.1093/femsle/fny079
17. Bian G, Deng Z, Liu T (2017) Strategies for terpenoid overproduction and new terpenoid discovery. Curr Opin Biotechnol 48:234–241. https://doi.org/10.1016/j.copbio.2017.07.002

18. Schempp FM, Drummond L, Buchhaupt M, Schrader J (2018) Microbial cell factories for the production of terpenoid flavor and fragrance compounds. J Agric Food Chem 66:2247–2258. https://doi.org/10.1021/acs.jafc.7b00473
19. Breitmaier E (2006) Terpenes: flavors, fragrances, pharmaca, pheromones. Wiley, Weinheim
20. Wilding EI, Brown JR, Bryant AP et al (2000) Identification, evolution, and essentiality of the mevalonate pathway for isopentenyl diphosphate biosynthesis in gram-positive cocci. J Bacteriol 182:4319–4327
21. Rodríguez-Concepción M (2006) Early steps in isoprenoid biosynthesis: multilevel regulation of the supply of common precursors in plant cells. Phytochem Rev 5:1–15. https://doi.org/10.1007/s11101-005-3130-4
22. Schwender R, Seemann M, Lichtenthaler HK, Rohmer M (1996) Biosynthesis of isoprenoids (carotenoids, sterols, prenyl side-chains of chlorophylls and plastoquinone) via a novel pyruvate/glyceraldehyde 3-phosphate non-mevalonate pathway in the green alga *Scenedesmus obliquus*. Biochem J 316:73–80
23. Kanehisa M, Goto S (2000) KEGG: kyoto encyclopedia of genes and genomes. Nucleic Acids Res 28:27–30. https://doi.org/10.1093/nar/27.1.29
24. Kanehisa M, Goto S, Sato Y et al (2012) KEGG for integration and interpretation of large-scale molecular data sets. Nucleic Acids Res 40:109–114. https://doi.org/10.1093/nar/gkr988
25. Zada B, Wang C, Park J-B et al (2018) Metabolic engineering of *Escherichia coli* for production of mixed isoprenoid alcohols and their derivatives. Biotechnol Biofuels 11:210. https://doi.org/10.1186/s13068-018-1210-0
26. Daletos G, Stephanopoulos G (2020) Protein engineering strategies for microbial production of isoprenoids. Metab Eng Commun 11:e00129
27. Phulara SC, Chaurasia D, Diwan B et al (2018) In-situ isopentenol production from *Bacillus subtilis* through genetic and culture condition modulation. Process Biochem 72:47–54. https://doi.org/10.1016/j.procbio.2018.06.019
28. Banerjee A, Wu Y, Banerjee R et al (2013) Feedback inhibition of deoxy- D -xylulose-5-phosphate synthase regulates the methylerythritol 4-phosphate pathway. J Biol Chem 288:16926–16936. https://doi.org/10.1074/jbc.M113.464636
29. Withers ST, Gottlieb SS, Lieu B et al (2007) Identification of isopentenol biosynthetic genes from *Bacillus subtilis* by a screening method based on isoprenoid precursor toxicity. Appl Environ Microbiol 73:6277–6283. https://doi.org/10.1128/AEM.00861-07
30. Zhou K, Zou R, Stephanopoulos G, Too H-P (2012a) Metabolite profiling identified methylerythritol cyclodiphosphate efflux as a limiting step in microbial isoprenoid production. PLoS One 7:e47513. https://doi.org/10.1371/journal.pone.0047513
31. Bentley FK, Zurbriggen A, Melis A (2014) Heterologous expression of the mevalonic acid pathway in cyanobacteria enhances endogenous carbon partitioning to isoprene. Mol Plant 7:71–86. https://doi.org/10.1093/mp/sst134
32. Sasaki Y, Eng T, Herbert RA et al (2019) Engineering *Corynebacterium glutamicum* to produce the biogasoline isopentenol from plant biomass hydrolysates. Biotechnol Biofuels 12:1–15. https://doi.org/10.1186/s13068-019-1381-3
33. Martin VJJ, Pitera DJ, Withers ST et al (2003) Engineering a mevalonate pathway in *Escherichia coli* for production of terpenoids. Nat Biotechnol 21:796–802. https://doi.org/10.1038/nbt833
34. Keasling JD (2012) Synthetic biology and the development of tools for metabolic engineering. Metab Eng 14:189–195. https://doi.org/10.1016/j.ymben.2012.01.004
35. Wang C, Liwei M, Bin PJ et al (2018) Microbial platform for terpenoid production: *Escherichia coli* and Yeast. Front Microbiol 9:2460. https://doi.org/10.3389/fmicb.2018.02460
36. Wang C, Zada B, Wei G, Kim S-W (2017) Metabolic engineering and synthetic biology approaches driving isoprenoid production in *Escherichia coli*. Bioresour Technol 241:430–438. https://doi.org/10.1016/J.BIORTECH.2017.05.168

37. Peralta-Yahya PP, Ouellet M, Chan R et al (2011) Identification and microbial production of a terpene-based advanced biofuel. Nat Commun 2:483. https://doi.org/10.1038/ncomms1494
38. Harada H, Yu F, Okamoto S et al (2009) Efficient synthesis of functional isoprenoids from acetoacetate through metabolic pathway-engineered *Escherichia coli*. Appl Microbiol Biotechnol 81:915–925. https://doi.org/10.1007/s00253-008-1724-7
39. Okamura E, Tomita T, Sawa R et al (2010) Unprecedented acetoacetyl-coenzyme A synthesizing enzyme of the thiolase superfamily involved in the mevalonate pathway. Proc Natl Acad Sci U S A 107:11265–11270
40. Alonso-Gutierrez J, Chan R, Batth TS et al (2013) Metabolic engineering of *Escherichia coli* for limonene and perillyl alcohol production. Metab Eng 19:33–41. https://doi.org/10.1016/J.YMBEN.2013.05.004
41. Tsuruta H, Paddon CJ, Eng D et al (2009) High-level production of amorpha-4,11-diene, a precursor of the antimalarial agent artemisinin, in *Escherichia coli*. PLoS One 4:e4489. https://doi.org/10.1371/journal.pone.0004489
42. Zheng Y, Liu Q, Li L et al (2013) Metabolic engineering of *Escherichia coli* for high-specificity production of isoprenol and prenol as next generation of biofuels. Biotechnol Biofuels 6:57
43. Dueber JE, Wu GC, Malmirchegini GR et al (2009) Synthetic protein scaffolds provide modular control over metabolic flux. Nat Biotechnol 27:753–759. https://doi.org/10.1038/nbt.1557
44. George KW, Thompson MG, Kang A et al (2015) Metabolic engineering for the high-yield production of isoprenoid-based C 5 alcohols in E. coli. Sci Rep 5:11128. https://doi.org/10.1038/srep11128
45. Kazieva E, Yamamoto Y, Tajima Y et al (2017) Characterization of feedback-resistant mevalonate kinases from the methanogenic archaeons *Methanosaeta concilii* and *Methanocella paludicola*. Microbiol (United Kingdom) 163:1283–1291. https://doi.org/10.1099/mic.0.000510
46. Yoon S-H, Lee Y-M, Kim J-E et al (2006) Enhanced lycopene production in *Escherichia coli* engineered to synthesize isopentenyl diphosphate and dimethylallyl diphosphate from mevalonate. Biotechnol Bioeng 94:1025–1032. https://doi.org/10.1002/bit.20912
47. Yoon SH, Lee SH, Das A et al (2009) Combinatorial expression of bacterial whole mevalonate pathway for the production of β-carotene in *E. coli*. J Biotechnol 140:218–226. https://doi.org/10.1016/j.jbiotec.2009.01.008
48. Redding-Johanson AM, Batth TS, Chan R et al (2011) Targeted proteomics for metabolic pathway optimization: application to terpene production. Metab Eng 13:194–203. https://doi.org/10.1016/j.ymben.2010.12.005
49. Rohdich F, Hecht S, Gärtner K et al (2002) Studies on the nonmevalonate terpene biosynthetic pathway: metabolic role of IspH (LytB) protein. Proc Natl Acad Sci U S A 99:1158–1163. https://doi.org/10.1073/pnas.032658999
50. Henke NA, Frohwitter J, Peters-Wendisch P, Wendisch VF (2018) Carotenoid production by recombinant *Corynebacterium glutamicum*: strain construction, cultivation, extraction, and quantification of carotenoids and terpenes. In: Barreiro C, Barredo J-L (eds) Microbial carotenoids: methods and protocols, methods in molecular biology. Springer Science+Business Media, LLC, part of Springer Nature 2018, New York, pp 127–141
51. Beuttler H, Hoffmann J, Jeske M et al (2011) Biosynthesis of zeaxanthin in recombinant *Pseudomonas putida*. Appl Microbiol Biotechnol 89:1137–1147. https://doi.org/10.1007/s00253-010-2961-0
52. Mi J, Becher D, Lubuta P et al (2014) De novo production of the monoterpenoid geranic acid by metabolically engineered *Pseudomonas putida*. Microb Cell Factories 13:170. https://doi.org/10.1186/s12934-014-0170-8
53. Phelan RM, Sekurova ON, Keasling JD, Zotchev SB (2014) Engineering terpene biosynthesis in *Streptomyces* for production of the advanced biofuel precursor bisabolene. ACS Synth Biol 4:393–399. https://doi.org/10.1021/sb5002517

54. Gao X, Gao F, Liu D et al (2016) Engineering the methylerythritol phosphate pathway in cyanobacteria for photosynthetic isoprene production from CO2. Energy Environ Sci 9:1400–1411. https://doi.org/10.1039/c5ee03102h
55. Kang M-K, Eom J-H, Kim Y et al (2014) Biosynthesis of pinene from glucose using metabolically-engineered *Corynebacterium glutamicum*. Biotechnol Lett 36:2069–2077. https://doi.org/10.1007/s10529-014-1578-2
56. Kudoh K, Kawano Y, Hotta S et al (2014) Prerequisite for highly efficient isoprenoid production by cyanobacteria discovered through the over-expression of 1-deoxy-D-xylulose 5-phosphate synthase and carbon allocation analysis. J Biosci Bioeng 118:20–28. https://doi.org/10.1016/j.jbiosc.2013.12.018
57. Xue J, Ahring BK (2011) Enhancing isoprene production by genetic modification of the 1-deoxy-d-xylulose-5-phosphate pathway in *Bacillus subtilis*. Appl Environ Microbiol 77: 2399–2405. https://doi.org/10.1128/AEM.02341-10
58. Boghigian BA, Armando J, Salas D, Pfeifer BA (2012) Computational identification of gene over-expression targets for metabolic engineering of taxadiene production. Appl Microbiol Biotechnol 93:2063–2073. https://doi.org/10.1007/s00253-011-3725-1
59. Kudoh K, Kubota G, Fujii R et al (2017) Exploration of the 1-deoxy-D-xylulose 5-phosphate synthases suitable for the creation of a robust isoprenoid biosynthesis system. J Biosci Bioeng 123:300–307. https://doi.org/10.1016/j.jbiosc.2016.10.005
60. Zhou K, Zou R, Stephanopoulos G, Too H-P (2012b) Enhancing solubility of deoxyxylulose phosphate pathway enzymes for microbial isoprenoid production. Microb Cell Factories 11: 148. https://doi.org/10.1186/1475-2859-11-148
61. Han X, Ning W, Ma X et al (2020) Improving protein solubility and activity by introducing small peptide tags designed with machine learning models. Metab Eng Commun 11:e00138. https://doi.org/10.1016/j.mec.2020.e00138
62. Nguyen TKM, Ki MR, Son RG, Pack SP (2019) The NT11, a novel fusion tag for enhancing protein expression in *Escherichia coli*. Biotechnol Isoprenoids Adv Biochem Eng 103:2205–2216. https://doi.org/10.1007/s00253-018-09595-w
63. Han X, Wang X, Zhou K, Valencia A (2019) Develop machine learning-based regression predictive models for engineering protein solubility. Bioinformatics 35:4640–4646. https://doi.org/10.1093/bioinformatics/btz294
64. Zhao Y, Yang J, Qin B et al (2011) Biosynthesis of isoprene in *Escherichia coli* via methylerythritol phosphate (MEP) pathway. Appl Microbiol Biotechnol 90:1915–1922. https://doi.org/10.1007/s00253-011-3199-1
65. Soliman SSM, Tsao R, Raizada MN (2011) Chemical inhibitors suggest endophytic fungal paclitaxel is derived from both mevalonate and non-mevalonate-like pathways. J Nat Prod 74: 2497–2504. https://doi.org/10.1021/np200303v
66. Xue D, Abdallah II, de Haan IEM et al (2015) Enhanced C30 carotenoid production in *Bacillus subtilis* by systematic overexpression of MEP pathway genes. Appl Microbiol Biotechnol 99:5907–5915. https://doi.org/10.1007/s00253-015-6531-3
67. Kim S, Keasling JD (2001) Metabolic engineering of the nonmevalonate isopentenyl diphosphate synthesis pathway in *Escherichia coli* enhances lycopene production. Biotechnol Bioeng 72:408–415
68. Lv X, Xu H, Yu H (2013) Significantly enhanced production of isoprene by ordered coexpression of genes *dxs*, *dxr*, and *idi* in *Escherichia coli*. Appl Microbiol Biotechnol 97: 2357–2365. https://doi.org/10.1007/s00253-012-4485-2
69. Englund E, Shabestary K, Hudson EP, Lindberg P (2018) Systematic overexpression study to find target enzymes enhancing production of terpenes in *Synechocystis* PCC 6803, using isoprene as a model compound. Metab Eng 49:164–177. https://doi.org/10.1016/j.ymben.2018.07.004
70. Haiyuan L, Shu H, Liping X et al (2018) Overexpression of key enzymes of the 2-C-methyl-D-erythritol-4-phosphate (MEP) pathway for improving squalene production in *Escherichia coli*. African J Biotechnol 16:2307–2316. https://doi.org/10.5897/ajb2017.16235

71. Zhou K, Zou R, Zhang C et al (2013) Optimization of amorphadiene synthesis in *Bacillus subtilis* via transcriptional, translational, and media modulation. Biotechnol Bioeng 110:2556–2561. https://doi.org/10.1002/bit.24900
72. Frank A, Groll M (2017) The methylerythritol phosphate pathway to isoprenoids. Chem Rev 117:5675–5703. https://doi.org/10.1021/acs.chemrev.6b00537
73. Boghigian BA, Salas D, Ajikumar PK et al (2011) Analysis of heterologous taxadiene production in K- and B-derived *Escherichia coli*. Appl Microbiol Biotechnol 93:1651–1661. https://doi.org/10.1007/s00253-011-3528-4
74. Wang C, Zhou J, Jang H et al (2013) Engineered heterologous FPP synthases-mediated Z,E -FPP synthesis in *E. coli*. Metab Eng 18:53–59. https://doi.org/10.1016/j.ymben.2013.04.002
75. Kang A, George KW, Wang G et al (2016) Isopentenyl diphosphate (IPP)-bypass mevalonate pathways for isopentenol production. Metab Eng 34:25–35. https://doi.org/10.1016/j.ymben.2015.12.002
76. Yang J, Nie Q, Liu H et al (2016) A novel MVA-mediated pathway for isoprene production in engineered *E. coli*. BMC Biotechnol 16:1. https://doi.org/10.1186/s12896-016-0236-2
77. Ilmén M, Oja M, Huuskonen A et al (2015) Identification of novel isoprene synthases through genome mining and expression in *Escherichia coli*. Metab Eng 31:153–162. https://doi.org/10.1016/j.ymben.2015.08.001
78. Li M, Nian R, Xian M, Zhang H (2018) Metabolic engineering for the production of isoprene and isopentenol by *Escherichia coli*. Appl Microbiol Biotechnol 102:7725–7738. https://doi.org/10.1007/s00253-018-9200-5
79. Chou HH, Keasling JD (2012) Synthetic pathway for production of five-carbon alcohols from isopentenyl diphosphate. Appl Environ Microbiol 78:7849–7855. https://doi.org/10.1128/AEM.01175-12
80. Kim JH, Wang C, Jang HJ et al (2016) Isoprene production by *Escherichia coli* through the exogenous mevalonate pathway with reduced formation of fermentation byproducts. Microb Cell Factories 15:214. https://doi.org/10.1186/s12934-016-0612-6
81. Lange BM (2015) Biosynthesis and biotechnology of high-value p-menthane monoterpenes, including menthol, carvone, and limonene. In: Schrader J, Bohlmann J (eds) Biotechnology of isoprenoids. Advances in biochemical engineering/biotechnology. Springer International Publishing, Switzerland 2015 Biosynthesis, pp 319–353
82. Mendez-Perez D, Alonso-Gutierrez J, Hu Q et al (2017) Production of jet fuel precursor monoterpenoids from engineered *Escherichia coli*. Biotechnol Bioeng 114:1703–1712. https://doi.org/10.1002/bit.26296
83. Sarria S, Wong B, Martín HG et al (2014) Microbial synthesis of pinene. ACS Synth Biol 3:466–475. https://doi.org/10.1021/sb4001382
84. Zhou J, Wang C, Yang L et al (2015) Geranyl diphosphate synthase: an important regulation point in balancing a recombinant monoterpene pathway in Escherichia coli. Enzym Microb Technol 68:50–55. https://doi.org/10.1016/j.enzmictec.2014.10.005
85. Tashiro M, Kiyota H, Kawai-Noma S et al (2016) Bacterial production of pinene by a laboratory-evolved pinene-synthase. ACS Synth Biol 5:1011–1020. https://doi.org/10.1021/acssynbio.6b00140
86. Awouafack MD, Tane P, Kuete V, Eloff JN (2013) Sesquiterpenes from the medicinal plants of Africa. In: Kuete V (ed) Medicinal plant research in Africa: pharmacology and chemistry. Elsevier Inc., pp 33–103
87. Buckle J (2015) Basic plant taxonomy, basic essential oil chemistry, extraction, biosynthesis, and analysis. In: Clinical aromatherapy. Churchill Livingstone, pp 37–72
88. Wang C, Park J, Choi E, Kim S (2016) Farnesol production in *Escherichia coli* through the construction of a farnesol biosynthesis pathway- application of PgpB and YbjG phosphatases. Biotechnol J 11:1291–1297. https://doi.org/10.1002/biot.201600250.Submitted
89. Wang C, Yoon SH, Jang HJ et al (2011) Metabolic engineering of *Escherichia coli* for α-farnesene production. Metab Eng 13:648–655. https://doi.org/10.1016/j.ymben.2011.08.001

90. Zhu F, Zhong X, Hu M et al (2014) In vitro reconstitution of mevalonate pathway and targeted engineering of farnesene overproduction in *Escherichia coli*. Biotechnol Bioeng 111:1396–1405. https://doi.org/10.1002/bit.25198
91. Abdallah II, Pramastya H, van Merkerk R et al (2019) Metabolic engineering of *Bacillus subtilis* toward taxadiene biosynthesis as the first committed step for taxol production. Front Microbiol 10:218. https://doi.org/10.3389/fmicb.2019.00218
92. Kai G, Xu H, Zhou C et al (2011) Metabolic engineering tanshinone biosynthetic pathway in *Salvia miltiorrhiza* hairy root cultures. Metab Eng 13:319–327. https://doi.org/10.1016/j.ymben.2011.02.003
93. Song Y, Guan Z, Van Merkerk R et al (2020) Production of squalene in *Bacillus subtilis* by squalene synthase screening and metabolic engineering. J Agric Food Chem 68:4447–4455. https://doi.org/10.1021/acs.jafc.0c00375
94. Choi SY, Lee HJ, Choi J et al (2016) Photosynthetic conversion of CO2 to farnesyl diphosphate-derived phytochemicals (amorpha-4,11-diene and squalene) by engineered cyanobacteria. Biotechnol Biofuels 9:202. https://doi.org/10.1186/s13068-016-0617-8
95. Choi SY, Wang JY, Kwak HS et al (2017) Improvement of squalene production from CO2 in Synechococcus elongatus PCC 7942 by metabolic engineering and scalable production in a photobioreactor. ACS Synth Biol 6:1289–1295. https://doi.org/10.1021/acssynbio.7b00083
96. Misawa N, Nakagawa M, Kobayashi K et al (1990) Elucidation of the *Erwinia uredovora* carotenoid biosynthetic pathway by functional analysis of gene products expressed in *Escherichia coli*. J Bacteriol 172:6704–6712. https://doi.org/10.1128/jb.172.12.6704-6712.1990
97. Liu N, Liu B, Wang G et al (2020) Lycopene production from glucose, fatty acid and waste cooking oil by metabolically engineered *Escherichia coli*. Biochem Eng J 155:107488. https://doi.org/10.1016/j.bej.2020.107488
98. Wei Y, Mohsin A, Hong Q et al (2018) Enhanced production of biosynthesized lycopene via heterogenous MVA pathway based on chromosomal multiple position integration strategy plus plasmid systems in *Escherichia coli*. Bioresour Technol 250:382–389. https://doi.org/10.1016/J.BIORTECH.2017.11.035
99. Lee PC, Mijts BN, Schmidt-Dannert C (2004) Investigation of factors influencing production of the monocyclic carotenoid torulene in metabolically engineered *Escherichia coli*. Appl Microbiol Biotechnol 65:538–546. https://doi.org/10.1007/s00253-004-1619-1
100. Henke N, Heider S, Peters-Wendisch P, Wendisch V (2016) Production of the marine, carotenoid astaxanthin by metabolically engineered *Corynebacterium glutamicum*. Mar Drugs 14:124. https://doi.org/10.3390/md14070124
101. Marliere P, Anissimova M, Chayot R, Delcourt M (2013) Process for the production of isoprenol from mevalonate employing a diphosphomevalonate decarboxylase
102. Renninger NS, Mcphee DJ (2008) Fuel compositions comprising farnesane and farnesane derivatives and method. 2
103. Ubersax JA, Platt DM (2010) Genetically modified microbes producing isoprenoids
104. Dugar D, Stephanopoulos G (2011) Relative potential of biosynthetic pathways for biofuels and bio-based products. Nat Biotechnol 29:1074–1078. https://doi.org/10.1038/nbt.2055

Chapter 5
Modern Analytical Techniques for Extraction, Purification, and Structural Characterization of Microbial Bioactive Compounds

Pramod Rawat, Yashaswi Singh, Manisha Bisht, and Manoj Pal

Abstract Analytical techniques play a vital role in extraction, purification, and molecular characterization of bioactive molecules. The selection of appropriate analytical method depends mainly on the specific properties of the bioactive compound being isolated. Since these bioactive compounds derived from microorganisms found in different environmental conditions, ranging from moderate to extreme, therefore it is impossible to apply a single analytical method universally. Moreover, conventional analytical methods often fall short, especially when multiple nonessential compounds co-elute during the initial solvent extraction and chromatographic purification processes. Nevertheless, significant improvements and advancements are being made in existing analytical methods to enhance the speed and accuracy of the isolation process. Several advanced techniques, such as solid phase extraction (SPE), supercritical fluid extraction (SFE), liquid chromatography-mass spectrometry (LC–MS), single-crystal X-ray diffraction (SCXRD), and two-dimensional nuclear magnetic resonance (2D-NMR), are uncovering the way for future advancements in the characterization of bioactive compounds.

Keywords Bioactive molecules · Microorganisms · Solvent extraction · Chromatographic purification · Structural characterization

P. Rawat · Y. Singh
Department of Biotechnology, Graphic Era (Deemed to be University), Dehradun, Uttarakhand, India

M. Bisht
Department of Chemistry, LSM-Government (PG)-College, Pithoragarh, Uttarakhand, India

M. Pal (✉)
Department of Microbiology, Graphic Era (Deemed to be University), Dehradun, Uttarakhand, India
e-mail: manojpal.bt@geu.ac.in

R. Soni et al. (eds.), *Microbial Bioactive Compounds*,
https://doi.org/10.1007/978-3-031-40082-7_5

5.1 Introduction

Microbial habitats often present challenging conditions, including high temperature, high pressure, high salinity, and high pressure. In order to thrive in these harsh environments, microorganisms have evolved various adaptation mechanisms, one of which involves the synthesis of specific bioactive molecules [1, 2]. Bacteria, actinomycetes, fungi, and microalgae isolated from diverse environments are a rich source of various valuable bioactive compounds, such as antibiotics, food enzymes, industry-used enzymes, vitamins, biopesticides, biodegradable plastics, antifungal compounds, anticancer compounds, antioxidants, and immunomodulators [3]. Nonetheless, researchers face substantial challenges when it comes to isolating and characterizing these bioactive molecules from complex biological sources. These compounds often exist in minuscule quantities, buried within a sea of complex mixtures, necessitating sophisticated analytical techniques to unravel their secrets. Consequently, a diverse range of powerful tools and methodologies has been developed to facilitate their discovery, isolation, and characterization.

Initially for the extraction of bioactive molecules, microbial isolates are cultured in the laboratory using appropriate growth media and conditions. Generally, the culturing process starts on a small scale, and once optimized, it is scaled up in fermenters or bioreactors and photobioreactors. Following mass culturing, the focus shifts toward the separation and purification of bioactive molecules, which usually starts with different conventional and advanced extraction methods. Various extraction methods are frequently employed to separate bioactive molecules from microorganisms, including solid-phase extraction, liquid–liquid extraction, supercritical fluid extraction, microwave-assisted extraction, and enzymatic extraction, among others [4]. Each technique has its own advantages, limitations, and compatibility with microorganism sources. Moreover, the influence of extraction parameters, including solvent selection, extraction time, temperature, and methods for microbial cell disruption, should not be disregarded, as they significantly affect the efficiency and selectivity of extracting bioactive compounds.

However, the crude extract, which contains various nonessential components, cannot undergo further characterization until it is subjected to purification. To achieve purification of the bioactive compound, a range of chromatographic procedures are employed, including gas chromatography (GC), high-performance liquid chromatography (HPLC), and mass spectrometry (MS) interfaced chromatographic techniques, such as LC–MS, and GC–MS [5]. Subsequently, after chromatographic purification, molecular-level characterization can be carried out using advanced spectroscopic methods, such as tandem mass spectrometry (MS/MS), X-ray crystallography, nuclear magnetic resonance (NMR), and Fourier-transform infrared spectroscopy (FT-IR) [6].

However, the crude extract, which contains different nonessential components, cannot be further characterized until it is purified. To purify the bioactive compound, various chromatographic procedures are used, such as high-performance liquid chromatography (HPLC), gas chromatography (GC), and mass spectrometry

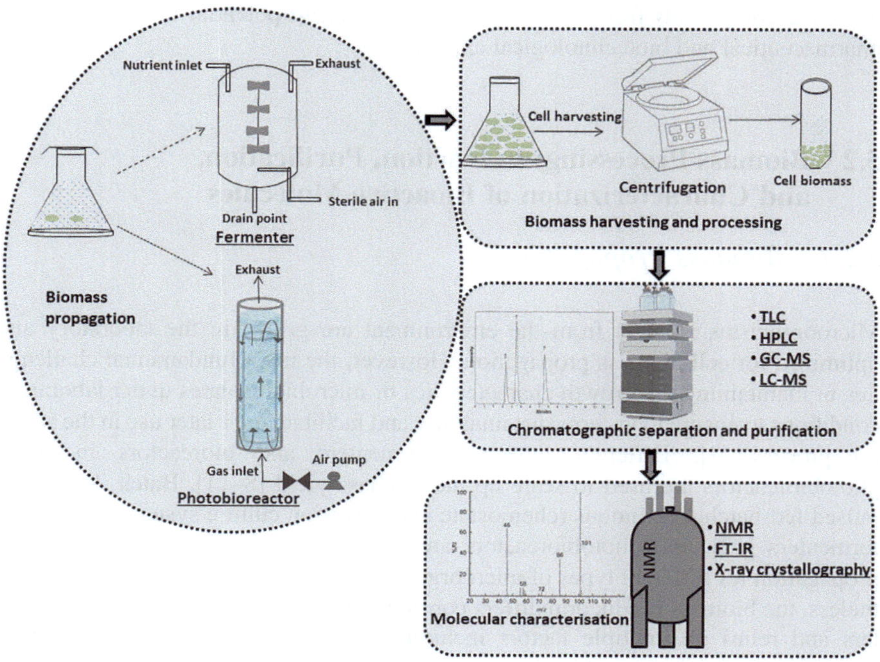

Fig. 5.1 General schematic approaches for extraction, purification, and characterization of bioactive molecules

(MS) interfaced chromatographic techniques, such as GC–MS and LC–MS. After chromatographic purification, molecular-level characterization can be performed using highly sophisticated spectroscopic methods, such as nuclear magnetic resonance (NMR), Fourier-transform infrared spectroscopy (FT-IR), tandem mass spectrometry (MS/MS), and X-ray crystallography.

In all these processes, the extraction of bioactive compounds from microorganisms is a critical step for isolating and studying their potential therapeutic properties. The choice of an appropriate extraction method greatly influences the yield, purity, and bioactivity of the isolated compounds. Therefore, it is essential to evaluate various extraction techniques to optimize the extraction process and obtain maximum recovery of bioactive compounds. The second most important step for the characterization of bioactive compounds is chromatographic purification, the credibility of which depends on the purity of the bioactive component for further characterization at the molecular level using different spectroscopic techniques. The choice of extraction, purification, and characterization methods depends on different factors, such as compound stability, target bioactivity, ease of scale-up, and downstream applications, which should be considered during method selection. General schematic bioprocess methods are indicated in Fig. 5.1. By optimizing the extraction process, researchers can enhance the discovery and development of novel bioactive compounds from microorganisms, leading to potential breakthroughs in pharmaceutical and biotechnological applications.

5.2 Biomass Processing; Extraction, Purification, and Characterization of Bioactive Molecules

5.2.1 Biomass Propagation

Microorganisms isolated from the environment are grown in the laboratory and optimized for cell biomass propagation. However, the main fundamental challenge lies in maintaining the growth characteristics of microbial isolates under laboratory conditions to ensure long-term sustainability and facilitate their later use in the scale-up process [7]. Different types of fermenters and bioreactors including photobioreactors are used to scale up the biomass yield [8–11]. Batch, fed-batch, pulsed fed-batch, continuous (chemostate and perfusion culture system), solid-state fermenters [12], and photobioreactors are commonly used for the cell biomass propagation for different types of microorganisms including microalgae [13]. Nonetheless, the biomass production differs considerably across various types of fermenters and relies on multiple factors including design, size, sensor-based control, regulation of nutrient supply, gas exchange, and mixing [14]. Yet, lot of improvements are required in bioreactor design to optimize the cost of cell biomass propagation and growth media compositions to grow viable but nonculturable cells (VBNC) [15].

5.2.2 Solvent Extraction

The extraction of bioactive compounds from microorganisms is a critical step in the process of isolating and studying their potential therapeutic properties. The choice of an appropriate extraction method greatly influences the yield, purity, and bioactivity of the isolated compounds. Therefore, it is essential to choose right extraction method to optimize the maximum bioactive compound recovery. A number of extraction methods from conventional [16–18] (soxhlet, maceration, and hydrodistillation) to emerging methods such as supercritical fluid [19, 20], subcritical fluid [21], microwave assisted [22], ultrasonic assisted [23, 24], and enzyme assisted [25] have been used for the extraction of bioactive molecules from microorganisms. However, most of the solvent extraction methods are more popularized for the extraction of plant-based bioactive compounds, and they have been less commonly utilized for isolating microbial bioactive compounds. The emerging advanced extraction methods including green extraction methods could be suitable methods over conventional extraction methods for the isolation of bioactive compounds.

5.2.3 Chromatographic Purification

Different microbial extracts (solvent extracts and fractional parts) can be further purified through various types of column chromatography (e.g., liquid chromatography and gas chromatography). The selection of appropriate solvent systems and stationary phases, tailored to the polarity of the bioactive fraction, allows for effective purification. Liquid chromatographic methods commonly used for the isolation of antibiotics and bioactive compounds [26] include normal phase [27] [28], reversed phase [19], ion exchange [29], size exclusion [30], and affinity chromatography [31]. The selection of the appropriate mode depends on the properties of the compounds of interest and the desired separation objectives. Among all the liquid chromatographic procedures, reverse phase is the most commonly employed procedure [26] because many bioactive compounds often possess varying degrees of hydrophobicity, making reversed phase chromatography an excellent choice for their isolation. The nonpolar stationary phase, such as C_{18}, interacts with the hydrophobic regions of the compounds, allowing for efficient separation [32]. Table 5.1 indicates the some selected chromatographic procedures employed for the isolation of bioactive compounds from microorganisms.

In the near future, the requirements of chromatographic procedures employed for the isolation of bioactive compounds from microorganisms are expected to evolve in response to advancements in technology and the growing demand for novel therapeutic agents. One key requirement will be the development of high-throughput and automated chromatographic systems that can efficiently handle large sample volumes and minimize manual intervention. Additionally, there will be a growing need for improved resolution and selectivity in separating complex mixtures of bioactive compounds. This will drive the development of advanced stationary phases, such as novel sorbents and hybrid materials, which can provide enhanced separation capabilities. Another important aspect will be the integration of chromatographic techniques with complementary analytical methods, such as mass spectrometry and nuclear magnetic resonance spectroscopy, to enable rapid compound identification and structural elucidation. Furthermore, there will be an increased emphasis on sustainability, pushing for the use of greener solvents, reduced energy consumption, and recycling of chromatographic materials. Overall, the future requirements of chromatographic procedures for isolating bioactive compounds from microorganisms will revolve around efficiency, selectivity, integration, and sustainability to meet the ever-expanding needs of drug discovery and natural product research.

Table 5.1 Selected chromatographic procedures used for the isolation and purification of bioactive compounds from microorganisms

S. No.	Chromatographic procedure	Microorganism	Bioactive compound isolated	References
1.	TLC	*Bacillus* sp.	Bacitracin	[33]
2.	SPE	*Bacillus lichenform*	Bacitracin	[34]
3.	Cation exchange, SPE	*Micrococcus luteus*	Bacitracin	[35]
4.	C_{18}-HPLC	*Bacillus* sp.	Bacitracin	[36]
5.	HPLC and supercritical fluid extraction (SFE)	*Penicillium expansum*, *Aspergillus fumigatus*, and *Streptomyces* sp.	Chaetogiobosin A, mycolutein, and luteoreticulin, 7,8-dihydro-7,8-epoxy-1-hydroxy-3-hydroxymethylxanthone-8-carboxylic acid methyl ester, and sydowinin B	[19]
6.	HPLC	*Nocardiopsis* sp., SCA21	4-bromophenol, and Bis (2-ethylhexyl) phthalate	[37]
7.	HPLC, LC–MS/MS	*Fusarium proliferatum* CECT 20569	Beauvericin (BEA)	[38]
8.	TLC, HPLC, and LC–MS/MS	*Streptomyces cavourensis* TN638	Cyclo-(Leu-Pro), Cyclo-(Val-Pro), Cyclo-(Phe-Pro), nonactin, monactin, dinactin, and trinactin	[39]
9.	GC–MS	*Streptomyces albidoflavus* 321.2	Dibutyl phthalate	[40]
10.	TLC and GC	*Streptomyces* sp., TN256 strain	N-[2-(1H-indol-3-yl)-2 oxo-ethyl] acetamide 'alkaloid' derivative; di-(2-ethylhexyl) phthalate, a phthalate derivative; 1-Nonadecene and Cyclo (L-Pro-L-Tyr) a diketopiperazine 'DKP' derivative	[41]
11.	HPLC	*Cladosporium* sp., F14	3-phenyl-2-propenoic acid, cyclo-(Phe-Pro), cyclo-(Val-Pro) 3-phenyl-2-propenoic acid, and bis (2-ethylhexyl)phthalate	[42]
12.	TLC, HPLC	*Aspergillus ostianus*	Circumdatins A and B and benzodiazepine alkaloids	[43]
13.	HPLC	*Chondrostereum* sp.	Hirsutane sesquiterpenoid	[44]
14.	HPLC	*Aspergillus* sp.	Aspergilone A & B	[45]

(continued)

Table 5.1 (continued)

S. No.	Chromatographic procedure	Microorganism	Bioactive compound isolated	References
15.	TLC	*S. chibaensis* AUBN1/7	Resistoflavine	[46]
16	TLC, HPLC	*Nocardiopsis alba* MSA10	Lipopeptide biosurfactant	[47]
17.	TLC, HPLC, and LC–MS	*Nocardiopsis* sp., GRG 2 (KT 235641)	1,4-diaza-2, 5-dioxo-3-isobutyl bicyclo[4.3.0] nonane (DDIBN)	
18.	TLC, HPLC, GC–MS, and LC–MS	*Streptomyces akiyoshiensis* GRG 6 (KY457710)	pyrrolo[1,2-a]pyrazine-1,4-dione, and hexahydro-3	[48]
19.	Sephadex G-25 gel column chromatography, and, IRC-50 ion-exchange resin, and TLC	*Streptomyces ahygroscopicus*	ε-poly-l-lysine (ε-PL)	[49]
20.	Ion exchange chromatography through DEAE Sepharose CL-6B column	*Streptomyces fradiae* NEAE-82	L-asparaginase	[50]
21.	Thin-layer chromatography (TLC)	*Nocardiopsis dassonvillei*	Tetrodotoxin	[51]
22.	Anion-exchange chromatography	*Pseudonocardia thermophila*	Thermoactive amidase	[52]
23.	Supercritical fluid extraction (SFE)	*Myxococcus xanthus* DK1622	Chloroxanthic acid A	[53]

5.2.4 Structural Characterization of Bioactive Molecules

The structural characterization of bioactive compounds isolated from microorganisms plays a vital role in understanding their therapeutic potential and mechanisms of action. Various types of advanced analytical techniques are used to determine the chemical structure, stereochemistry, and conformational properties [54]. Spectroscopic methods such as nuclear magnetic resonance (NMR) spectroscopy and mass spectrometry are commonly used to identify the composition and structure of compounds [55].

NMR spectroscopy provides valuable information about the arrangement of atoms in the molecule and helps in the determination of the compound's stereochemistry. Mass spectrometry, on the other hand, enables the measurement of the compound's molecular weight and fragmentation patterns, facilitating compound identification and providing insights into its structural features. Additionally, other techniques like X-ray crystallography and HR-TEM may be employed to visualize the three-dimensional structure of the bioactive compound, allowing for a more comprehensive understanding of its shape and spatial arrangement. Advanced NMR techniques are extensively employed in the structural characterization of bioactive compounds isolated from microorganisms. One such technique is multidimensional NMR spectroscopy, which involves the acquisition of multiple NMR spectra with different pulse sequences to correlate nuclear spins and establish connectivity between atoms [56]. Through techniques like COSY (correlation spectroscopy) [57], HMQC (heteronuclear multiple-quantum coherence) [58], and HMBC (heteronuclear multiple-bond correlation) [58], the interatomic relationships, and bond connectivity within the compound can be determined. Additionally, advanced NMR techniques such as NOESY (nuclear overhauser effect spectroscopy) provide valuable information about the spatial arrangement of atoms in the molecule, allowing for the determination of molecular conformation and stereochemistry [59]. The use of selective NMR experiments, such as selective TOCSY (total correlation spectroscopy) and selective HSQC (heteronuclear single-quantum coherence) [60], enables the identification and assignment of specific functional groups within the compound. Overall, advanced NMR techniques play a critical role in elucidating the structural features of bioactive compounds from microorganisms, helping in their characterization and understanding of their biological activities.

Another most advanced technique used to characterize bioactive molecules isolated from microorganism is X-ray crystallography. Similar to 2D-NMR techniques, it also allows researchers to determine the three-dimensional structure of these compounds at an atomic level, providing crucial insights into their chemical composition and spatial arrangement. By growing single crystals of the bioactive compound and subjecting them to X-ray diffraction, scientists can measure the angles and intensities of diffracted X-rays, which are then used to calculate the electron density distribution within the crystal [61]. This information enables the generation of an accurate molecular model, revealing the positions of individual atoms and their connectivity within the compound. X-ray crystallography helps in understanding the stereochemistry, molecular interactions, and overall conformation of the bioactive compound, aiding in the design of more effective drugs and therapeutic interventions. Furthermore, this technique contributes to the elucidation of structure–activity relationships [62], facilitating the optimization and development of novel pharmaceuticals derived from microorganisms (Table 5.2).

To determine the molecular mass of isolated bioactive compound, mass spectroscopy is used. Mass spectrometry utilizes various ionization processes, including electrospray ionization (ESI) [73], matrix-assisted laser desorption/ionization (MALDI) [82], and atmospheric pressure chemical ionization (APCI) [83], depending on the structural complexity and size of the bioactive molecule. Similarly, based on the desired resolution, accuracy, mass range, and other factors to achieve optimal results in their experiments, various types of mass analyzers are used such as quadrupole [84], time-of-flight (TOF) [85], Ion Trap, Orbitrap, Fourier Transform Ion Cyclotron Resonance (FT-ICR), and magnetic sector. Even, among these, quadrupole and TOF are most commonly used methods. However, further interfacing of chromatographic techniques to mass spectroscopic methods, such as LC–MS, and GC–MS, has facilitated the characterization process fast and more accurate.

Table 5.2 Different spectroscopic methods; NMR, MS, and X-ray used for characterization of bioactive molecules

S. No.	Spectroscopic method	Microorganism	Bioactive compound isolated	References
1.	NMR, mass, single-crystal X-ray diffraction (SCXRD)	*Nocardia* sp. ALAA 2000	Chrysophanol 8-methyl ether, asphodelin; 4,7′-bichrysophanol, and justicidin B, in addition to a novel bioactive compound ayamycin; 1,1-dichloro-4-ethyl-5-(4-nitro-phenyl)-hexan-2-one	[32]
2.	X-ray diffraction	*Penicillium vinaceum* (strain no. X17)	Quinazoline alkaloid ((-)-(1R,4R)-1,4-(2,3)-indolmethane-1-methyl-2,4-dihydro-1H-pyrazino-[2,1-b]-quinazoline-3,6-dione)	[63]
3.	Single crystal X-ray diffraction (SCXRD)	*Periconia* sp.	piperine (5-(3, 4-methylenedioxyphenyl)-1-piperidinopent-2, 4-dien-1-one)	[64]
4.	1D, 2D NMR, ESI HR-Mass, and X-ray crystallography	*Aspergillus* sp., ASCLA	Isoshamixanthone, epiisoshamixanthone, sterigmatocystin, arugosin C, norlichexanthone, diorcinol, ergosterol, and methyllinoleate	[65]
5.	NMR and X-ray diffraction analyses	*Aspergillus glaucus*	Aspergiolide A	[66]
6.	1D- and 2D NMR, HRESIMS, MS/MS, and electronic circular dichroism calculation and single-crystal X-ray diffraction	*Penicillium* sp., ZZ380	Penicipyrrodiether A and phenol A derivative	[67]
7.	1D NMR, HRESIMS, and X-ray crystallography	*Diaporthe* sp., GZU-1021	Diaporthichalasins A–C, and biatriosporin N	[68]
8.	1D NMR, 2D (COSY, HMQC, HMBC, NOESY) NMR, X-ray	*Nocardiopsis* sp.	Terretonin N	[69]
9.	NMR, HRESIMS, electronic circular dichroism (ECD) calculation, and X-ray diffraction	*Streptomyces* sp., ZZ446	Streptopyrazinones A – D	[70]
10.	MS, NMR, and X-ray crystallography	*Streptomyces* sp. SN194	Diterpenoids (chloroxaloterpin A and B)	[71]

(continued)

Table 5.2 (continued)

S. No.	Spectroscopic method	Microorganism	Bioactive compound isolated	References
11.	HR-ESI-MS, NMR, and single-crystal X-ray diffraction (SCXRD)	*Streptomyces anandii* H41-59	Anandins A and B	[72]
12.	ESIMS, 1D and 2D NMR data, and X-ray crystallography	*Aspergillus carbonarius*	Carbonarones A, and B	[73]
13.	HR-ESI-MS, X-ray diffraction, and NMR	*Chaetomium globosum*	Azaphilones	[74]
14.	NMR, HRESIMS, ECD, single-crystal X-ray diffraction (SCXRD)	*Streptomyces* sp. ZZ1956	Hygrocins K–U and Streptophenylpropanamide A	[75]
15.	ESIMS, 1D and 2D NMR data, and X-ray crystallography	*Micromonospora echinospora* SCSIO 04089	Angucyclinone derivatives and anthracene	[76]
16.	X-ray analysis	*Alternaria alternata*	Alternariol methyl ether (AME)	[77]
17.	HRESIMS, NMR and single-crystal X-ray diffraction (SCXRD)	*Penicillium* sp. SY2107	Mixed 16 metabolites	[78]
18.	Single-crystal X-ray diffraction (SCXRD)	*Emericella dentata* Nq45	Meleagrin, haenamindole, isorugulosuvine, secalonic acid D, ergosterol, and cerebroside A	[79]
19.	NMR, HRESIMS, electronic circular dichroism (ECD), ^{13}C NMR, and X-ray single-crystal diffraction (SCXRD)	*Penicillium* sp. ZZ380	Penicipyrroether A and Pyrrospirone J	[80]
20.	1D, 2D NMR and ECD	*Talaromyces scorteus* AS-242	Talascortenes A–G and 5α,9β dihydroxyisocupressic acid	[81]

LC–MS is a hybrid technique that combines liquid chromatography (LC) and mass spectrometry (MS) to separate and detect individual components within a complex mixture. In this method, a liquid mobile phase carries the sample through a stationary phase, separating the components based on their physicochemical properties. The eluted compounds are then introduced into the mass spectrometer, where they are ionized and analyzed based on their mass-to-charge ratio (m/z). LC–MS provides high sensitivity, selectivity, and the ability to handle complex mixtures. Liquid chromatography-mass spectrometry (LC–MS) and tandem mass spectrometry (MS/MS) are powerful analytical techniques used in the structure determination of bioactive compounds. MS/MS, also known as tandem mass spectrometry or MS2, is a technique that involves performing a second round of mass spectrometry on selected precursor ions obtained from the LC–MS analysis. In this process, the selected precursor ion is fragmented into smaller product ions using collision-induced dissociation (CID) or other fragmentation techniques. The resulting fragmentation patterns provide valuable structural information about the compound, including the arrangement of atoms and the presence of specific functional groups. Moreover, LC–MS and MS/MS can be combined with other techniques such as nuclear magnetic resonance (NMR) spectroscopy and high-resolution mass spectrometry to further enhance the structural determination of bioactive compounds. The combination of multiple analytical techniques increases confidence in the structural elucidation and can help researchers understand the chemical diversity and biological activities of natural bioactive compounds derived from microorganisms. In spite of significant contribution of these techniques in structural characterization of bioactive compounds, many challenges exist depending on purity, and structural complexity of biomolecules.

5.2.5 Challenges and Future Scope

A number of challenges exist from culturing of microorganisms to isolation, purification, and finally spectroscopic structural characterization of bioactive molecules. Once culture conditions are optimized, further extraction and purification remain major tasks. Chromatography techniques, such as high-performance liquid chromatography (HPLC), are commonly used for compound separation and purification. However, challenges can arise in the determination of the compound's structure:

(a) *Co-elution*: Sometimes, compounds with similar physicochemical properties can co-elute, making it difficult to differentiate and assign structures. In such cases, additional separation methods, such as preparative chromatography or orthogonal chromatographic techniques, may be employed to isolate individual compounds for further analysis.
(b) *Impurities and matrix effects*: Presence of impurities or complex matrices can interfere with the detection and identification of the target compound. Extensive sample preparation techniques, such as solid-phase extraction or sample derivatization, can be used to reduce interference and enhance the compound's detectability.

Mass spectrometry (MS) also faces challenges in mass determination, mainly due to different ionization efficiencies and fragmentation capabilities of bioactive compounds.

(a) *Ionization efficiency*: Different compounds exhibit different ionization efficiencies, which can affect the accuracy of mass spectral data. Careful optimization of ionization techniques, such as electrospray ionization (ESI) or matrix-assisted laser desorption/ionization (MALDI), is necessary to ensure efficient ionization and accurate mass determination.

(b) *Fragmentation pattern analysis*: Interpreting the fragmentation patterns obtained from MS analysis can be complex, particularly for large and structurally diverse compounds. The use of tandem mass spectrometry (MS/MS) or high-resolution MS can provide more detailed fragmentation data, aiding in structural elucidation.

NMR is a highly sophisticated tool used to elucidate atomic arrangement inside the molecules, but it also depends on:

(a) *Compound solubility*: Poor solubility of the compound in NMR solvents can impede data acquisition and spectral analysis. Optimization of solvents or the use of advanced NMR techniques, such as microscale NMR or diffusion-ordered spectroscopy (DOSY) [86], can overcome solubility issues.

(b) *Complex spectra*: In the case of structurally complex compounds, overlapping peaks and multiplicity can make spectral interpretation difficult. Advanced NMR techniques like 2D-NMR spectroscopy (e.g., COSY, HSQC, and HMBC) can be employed to resolve overlapping signals and provide additional structural information.

Again similar to NMR, single-crystal X-ray diffraction or crystallography is a powerful method for determining the 3D structure of bioactive compounds. However, it has its own challenges:

(a) *Obtaining suitable crystals:* Obtaining high-quality single crystals can be a significant challenge, especially for compounds with low crystallinity or limited availability. Techniques such as recrystallization, co-crystallization, or cryocrystallography can be employed to improve crystal quality or increase the chances of obtaining suitable crystals.

(b) *Radiation damage*: Exposure to X-rays during crystallographic data collection can lead to radiation damage to the crystal, resulting in poor data quality or structural changes. To mitigate this, low-temperature data collection, limited exposure time, and advanced data collection strategies like multicrystal or serial crystallography are employed.

5.3 Conclusion

Microorganisms are rich source of many value-added bioactive compounds, and their isolation, purification, and structural characterization always remain a challenge. However, various types of analytical techniques such as solvent extraction, chromatographic purification, and spectroscopic methods are used to characterize the bioactive molecules. By uncovering the chemical diversity present in microorganisms, these techniques open up avenues for bioprospecting and drug discovery, offering potential solutions to unmet medical needs and challenges in various industries. In summary, the isolation, purification, and characterization of bioactive compounds from microorganisms using analytical techniques enable researchers to harness the vast potential of these microorganisms as a source of valuable molecules. These techniques provide critical insights into the structural features and functional properties of bioactive compounds, paving the way for their further exploration and application in diverse fields.

References

1. Mahajan G, Balachandran L (2015) Biodiversity in production of antibiotics and other bioactive compounds. In: Mukherjee J (ed) Advances in biochemical engineering/biotechnology, vol 147. Springer, Berlin, pp 37–58. https://doi.org/10.1007/10_2014_268
2. Moopantakath J, Imchen M, Anju VT, Busi S, Dyavaiah M, Martínez-Espinosa RM, Kumavath R (2023) Bioactive molecules from haloarchaea: scope and prospects for industrial and therapeutic applications. Front Microbiol 14. https://doi.org/10.3389/fmicb.2023.1113540
3. Thompson TP, Gilmore BF (2023) Exploring halophilic environments as a source of new antibiotics. Crit Rev Microbiol 1–30. https://doi.org/10.1080/1040841X.2023.2197491
4. Shakoor R, Hussain N, Younas S, Bilal M (2023) Novel strategies for extraction, purification, processing, and stability improvement of bioactive molecules. J Basic Microbiol 63(3–4): 276–291. https://doi.org/10.1002/jobm.202200401
5. Fernandes C, Ribeiro R, Pinto M, Kijjoa A (2023) Absolute stereochemistry determination of bioactive marine-derived cyclopeptides by liquid chromatography methods: an update review (2018–2022). Molecules 28(2). https://doi.org/10.3390/molecules28020615
6. Gomes AR, Duarte AC, Rocha-Santos TAP (2016) Analytical techniques for discovery of bioactive compounds from marine fungi. In: Merillon J-M, Ramawat KG (eds) Fungal metabolites. Springer International Publishing, pp 1–20. https://doi.org/10.1007/978-3-319-19456-1_9-1
7. Bodor A, Bounedjoum N, Vincze GE, Erdeiné Kis Á, Laczi K, Bende G, Szilágyi Á, Kovács T, Perei K, Rákhely G (2020) Challenges of unculturable bacteria: environmental perspectives. Rev Environ Sci Biotechnol 19(1):1–22. https://doi.org/10.1007/s11157-020-09522-4
8. Godbole V, Pal MK, Gautam P (2021) A critical perspective on the scope of interdisciplinary approaches used in fourth-generation biofuel production. Algal Res 58:102436. https://doi.org/10.1016/j.algal.2021.102436
9. Koller M (2018) A review on established and emerging fermentation schemes for microbial production of polyhydroxyalkanoate (PHA) biopolyesters. Fermentation 4(2). https://doi.org/10.3390/fermentation4020030
10. Masojídek J, Torzillo G (2014) Mass cultivation of freshwater microalgae☆. In Reference module in earth systems and environmental sciences. Elsevier. https://doi.org/10.1016/b978-0-12-409548-9.09373-8
11. Terefe NS (2021) Recent developments in fermentation technology: toward the next revolution in food production. In Juliano P, Buckow R, Nguyen MH, Knoerzer K, and J. B. T.-F. E. I. A. the F. S. C. Sellahewa (eds) Food engineering innovations across the food supply chain. Academic, pp 89–106. https://doi.org/10.1016/B978-0-12-821292-9.00026-1
12. Venkata Mohan S, Rohit MV, Chiranjeevi P, Chandra R, Navaneeth B (2015) Heterotrophic microalgae cultivation to synergize biodiesel production with waste remediation: progress and perspectives. Bioresour Technol 184:169–178. https://doi.org/10.1016/j.biortech.2014.10.056
13. Rios Pinto LF, Ferreira GF, Tasic M (2020) Cultivation techniques. In: Galanakis CMBT-M (ed) Microalgae: cultivation, recovery of compounds and applications. Academic, pp 1–33. https://doi.org/10.1016/B978-0-12-821218-9.00001-3

14. Stanbury PF, Whitaker A, Hall SJ (2017). Design of a fermenter. In Stanbury PF, Whitaker A, & S. J. B. T.-P. of F. T. (Third E. Hall) (eds) Principles of fermentation technology. Butterworth-Heinemann, pp 401–485. https://doi.org/10.1016/b978-0-08-099953-1.00007-7
15. Oliver JD (2010) Recent findings on the viable but nonculturable state in pathogenic bacteria. FEMS Microbiol Rev 34(4):415–425. https://doi.org/10.1111/j.1574-6976.2009.00200.x
16. Ardiles P, Cerezal-Mezquita P, Salinas-Fuentes F, Órdenes D, Renato G, Ruiz-Domínguez MC (2020) Biochemical composition and phycoerythrin extraction from red microalgae: a comparative study using green extraction technologies. Processes 8(12):1–16. https://doi.org/10.3390/pr8121628
17. Saadouli I, El Euch IZ, Trabelsi E, Mosbah A, Redissi A, Ferjani R, Fhoula I, Cherif A, Sabatier JM, Sewald N, Ouzari HI (2020) Isolation, characterization and chemical synthesis of large spectrum antimicrobial cyclic dipeptide (L-leu-l-pro) from streptomyces misionensisv16r3y1 bacteria extracts. a novel1H NMR metabolomic approach. Antibiotics 9(5). https://doi.org/10.3390/antibiotics9050270
18. Singh LS, Sharma H, Talukdar NC (2014b) Production of potent antimicrobial agent by actinomycete, Streptomyces sannanensis strain SU118 isolated from phoomdi in Loktak Lake of Manipur, India. BMC Microbiol 14(1):278. https://doi.org/10.1186/s12866-014-0278-3
19. Cocks S, Wrigley SK, Ine^s Chicarelli-Robinson M, Smith RM (1995) High-performance liquid chromatography comparison of supercritical-fluid extraction and solvent extraction of microbial fermentation products. J Chromatogr A 697(1):115–122. https://doi.org/10.1016/0021-9673(94)00817-S
20. Molino A, Larocca V, Di Sanzo G, Martino M, Casella P, Marino T, Karatza D, Musmarra D (2019) Extraction of bioactive compounds using supercritical carbon dioxide. Molecules 24(4). https://doi.org/10.3390/molecules24040782
21. Ho TC, Chun B-S (2019) Extraction of bioactive compounds from Pseuderanthemum palatiferum (Nees) Radlk. Using subcritical water and conventional solvents: a comparison study. J Food Sci 84(5):1201–1207. https://doi.org/10.1111/1750-3841.14501
22. Pasquet V, Chérouvrier J-R, Farhat F, Thiéry V, Piot J-M, Bérard J-B, Kaas R, Serive B, Patrice T, Cadoret J-P, Picot L (2011) Study on the microalgal pigments extraction process: performance of microwave assisted extraction. Process Biochem 46(1):59–67. https://doi.org/10.1016/j.procbio.2010.07.009
23. Al Khawli F, Martí-Quijal FJ, Pallarés N, Barba FJ, Ferrer E (2021) Ultrasound extraction mediated recovery of nutrients and antioxidant bioactive compounds from phaeodactylum tricornutum microalgae. Appl Sci (Switzerland) 11(4):1–19. https://doi.org/10.3390/app11041701
24. Lavilla I, Bendicho C (2017). Chapter 11 - Fundamentals of ultrasound-assisted extraction. In H Dominguez González & M. J. B. T.-W. E. of BC González Muñoz (eds) Water extraction of bioactive compounds. Elsevier, pp 291–316. https://doi.org/10.1016/B978-0-12-809380-1.00011-5
25. Puri M, Sharma D, Barrow CJ (2012) Enzyme-assisted extraction of bioactives from plants. Trends Biotechnol 30(1):37–44. https://doi.org/10.1016/j.tibtech.2011.06.014
26. Duarte K, Rocha-Santos TAP, Freitas AC, Duarte AC (2012) Analytical techniques for discovery of bioactive compounds from marine fungi. TrAC - Trends Anal Chem 34:97–110. https://doi.org/10.1016/j.trac.2011.10.014
27. Parrot D, Legrave N, Intertaglia L, Rouaud I, Legembre P, Grube M, Suzuki MT, Tomasi S (2016) Cyaneodimycin, a bioactive compound isolated from the culture of Streptomyces cyaneofuscatus associated with Lichina confinis. Eur J Org Chem 2016(23):3977–3982. https://doi.org/10.1002/ejoc.201600252
28. Nguyen Van M, Phat N, Linh D (2021) Purification of bioactive compound from endophytes Bacillus sp. RD26 of Phyllanthus amarus Schum. et Thonn. Pharmacophore 12:29–36. https://doi.org/10.51847/4DEIRdiC4c
29. Lu R, Fasano S, Madayiputhiya N, Morin NP, Nataro J, Fasano A (2009) Isolation, identification, and characterization of small bioactive peptides from Lactobacillus GG conditional media

that exert both anti-gram-negative and Gram-positive bactericidal activity. J Pediatr Gastroenterol Nutr 49(1):23–30. https://doi.org/10.1097/MPG.0b013e3181924d1e
30. Jeong D, Kim DH, Kang IB, Kim H, Song KY, Kim HS, Seo KH (2017) Characterization and antibacterial activity of a novel exopolysaccharide produced by Lactobacillus kefiranofaciens DN1 isolated from kefir. Food Control 78:436–442. https://doi.org/10.1016/j.foodcont.2017.02.033
31. Duan M, Bai J, Yang J, Shi P, Bian L (2021) Screening and identification of bioactive components resistant to metallo-beta-lactamase from Schisandra chinensis (Turcz.) Baill. by metalloenzyme-immobilized affinity chromatography. J Chromatogr B Anal Technol Biomed Life Sci 1165:122524. https://doi.org/10.1016/j.jchromb.2021.122524
32. El-Gendy MMA, Hawas UW, Jaspars M (2008) Novel bioactive metabolites from a marine derived bacterium nocardia sp. ALAA 2000. J Antibiot 61(6):379–386. https://doi.org/10.1038/ja.2008.53
33. Bossuyt R, Van Renterghem R, Waes G (1976) Identification of antibiotic residues in milk by thin-layer chromatography. J Chromatogr A 124(1):37–42. https://doi.org/10.1016/S0021-9673(00)87834-6
34. Gallagher JB, Love PW, Knotts LL, Collaborators (1982) High pressure liquid chromatographic determination of bacitracin in premix feeds and finished feeds: collaborative study. J Assoc Off Anal Chem 65(5):1178–1185. https://doi.org/10.1093/jaoac/65.5.1178
35. Webster GK (1997) Liquid chromatographic analysis of bacitracin methylene disalicylate in feed. J AOAC Int 80(4):732–736. https://doi.org/10.1093/jaoac/80.4.732
36. Capitán-Vallvey LF, Navas N, Titos A, Checa R (2001) Determination of the antibiotic zinc bacitracin in animal food by high-performance liquid chromatography with ultraviolet detection. Chromatographia 54(1):15–20. https://doi.org/10.1007/BF02491826
37. Siddharth S, Rai VR (2019) Isolation and characterization of bioactive compounds with antibacterial, antioxidant and enzyme inhibitory activities from marine-derived rare actinobacteria, Nocardiopsis sp. SCA21. Microb Pathog 137:103775. https://doi.org/10.1016/j.micpath.2019.103775
38. Meca G, Sospedra I, Soriano JM, Ritieni A, Moretti A, Mañes J (2010) Antibacterial effect of the bioactive compound beauvericin produced by Fusarium proliferatum on solid medium of wheat. Toxicon 56(3):349–354. https://doi.org/10.1016/j.toxicon.2010.03.022
39. Kaaniche F, Hamed A, Elleuch L, Chakchouk-Mtibaa A, Smaoui S, Karray-Rebai I, Koubaa I, Arcile G, Allouche N, Mellouli L (2020) Purification and characterization of seven bioactive compounds from the newly isolated Streptomyces cavourensis TN638 strain via solid-state fermentation. Microb Pathog 142:104106. https://doi.org/10.1016/j.micpath.2020.104106
40. Roy RN, Laskar S, Sen SK (2006) Dibutyl phthalate, the bioactive compound produced by Streptomyces albidoflavus 321.2. Microbiol Res 161(2):121–126. https://doi.org/10.1016/j.micres.2005.06.007
41. Smaoui S, Mathieu F, Elleuch L, Coppel Y, Merlina G, Karray-Rebai I, Mellouli L (2012) Taxonomy, purification and chemical characterization of four bioactive compounds from new Streptomyces sp. TN256 strain. World J Microbiol Biotechnol 28(3):793–804. https://doi.org/10.1007/s11274-011-0872-6
42. Qi S-H, Xu Y, Xiong H-R, Qian P-Y, Zhang S (2009) Antifouling and antibacterial compounds from a marine fungus Cladosporium sp. F14. World J Microbiol Biotechnol 25(3):399–406. https://doi.org/10.1007/s11274-008-9904-2
43. Ookura R, Kito K, Ooi T, Namikoshi M, Kusumi T (2008) Structure revision of circumdatins A and B, benzodiazepine alkaloids produced by marine fungus Aspergillus ostianus, by X-ray crystallography. J Org Chem 73(11):4245–4247. https://doi.org/10.1021/jo800348d
44. Li H-J, Lan W-J, Lam C-K, Yang F, Zhu X-F (2011) Hirsutane sesquiterpenoids from the marine-derived fungus chondrostereum sp. Chem Biodivers 8(2):317–324. https://doi.org/10.1002/cbdv.201000036
45. Shao C-L, Wang C-Y, Wei M-Y, Gu Y-C, She Z-G, Qian P-Y, Lin Y-C (2011) Aspergilones A and B, two benzylazaphilones with an unprecedented carbon skeleton from the

gorgonian-derived fungus Aspergillus sp. Bioorg Med Chem Lett 21(2):690–693. https://doi.org/10.1016/j.bmcl.2010.12.005
46. Gorajana A, Venkatesan M, Vinjamuri S, Kurada BVVSN, Peela S, Jangam P, Poluri E, Zeeck A (2007) Resistoflavine, cytotoxic compound from a marine actinomycete, Streptomyces chibaensis AUBN1/7. Microbiol Res 162(4):322–327. https://doi.org/10.1016/j.micres.2006.01.012
47. Gandhimathi R, Seghal Kiran G, Hema TA, Selvin J, Rajeetha Raviji T, Shanmughapriya S (2009) Production and characterization of lipopeptide biosurfactant by a sponge-associated marine actinomycetes Nocardiopsis alba MSA10. Bioprocess Biosyst Eng 32(6):825–835. https://doi.org/10.1007/s00449-009-0309-x
48. Nadar Rajivgandhi G, Ramachandran G, Li J-L, Yin L, Manoharan N, Rajesh Kannan M, Antony Joseph Velanganni A, Alharbi NS, Kadaikunnan S, Khaled JM, Li W-J (2020) Molecular identification and structural detection of anti-cancer compound from marine Streptomyces akiyoshiensis GRG 6 (KY457710) against MCF-7 breast cancer cells. J King Saud Univ Sci 32(8):3463–3469. https://doi.org/10.1016/j.jksus.2020.10.008
49. Chen J, Liu H, Xia Z, Zhao X, Wu Y, An M (2019) Purification and structural analysis of the effective anti-TMV compound ε-poly-L-lysine produced by Streptomyces ahygroscopicus. Molecules 24(6). https://doi.org/10.3390/molecules24061156
50. El-Naggar NE-A, Deraz SF, Soliman HM, El-Deeb NM, El-Ewasy SM (2016) Purification, characterization, cytotoxicity and anticancer activities of L-asparaginase, anti-colon cancer protein, from the newly isolated alkaliphilic Streptomyces fradiae NEAE-82. Sci Rep 6(1): 32926. https://doi.org/10.1038/srep32926
51. Wu Z, Xie L, Xia G, Zhang J, Nie Y, Hu J, Wang S, Zhang R (2005) A new tetrodotoxin-producing actinomycete, Nocardiopsis dassonvillei, isolated from the ovaries of puffer fish Fugu rubripes. Toxicon 45(7):851–859. https://doi.org/10.1016/j.toxicon.2005.02.005
52. Egorova K, Trauthwein H, Verseck S, Antranikian G (2004) Purification and properties of an enantioselective and thermoactive amidase from the thermophilic actinomycete Pseudonocardia thermophila. Appl Microbiol Biotechnol 65(1):38–45. https://doi.org/10.1007/s00253-004-1607-5
53. Bader CD, Neuber M, Panter F, Krug D, Müller R (2020) Supercritical fluid extraction enhances discovery of secondary metabolites from myxobacteria. Anal Chem 92(23):15403–15411. https://doi.org/10.1021/acs.analchem.0c02995
54. Hassan S, Meenatchi R, Pachillu K, Bansal S, Brindangnanam P, Arockiaraj J, Kiran GS, Selvin J (2022) Identification and characterization of the novel bioactive compounds from microalgae and cyanobacteria for pharmaceutical and nutraceutical applications. J Basic Microbiol 62(9): 999–1029. https://doi.org/10.1002/jobm.202100477
55. Singh KS, Majik MS, Tilvi S (2014a) Chapter 6 - Vibrational spectroscopy for structural characterization of bioactive compounds. In: Rocha-Santos T, Duarte ACBT-CAC (eds) Analysis of marine samples in search of bioactive compounds, vol 65. Elsevier, pp 115–148. https://doi.org/10.1016/B978-0-444-63359-0.00006-9
56. Gomes AR, Duarte AC, Rocha-Santos TAP (2017) Analytical techniques for discovery of bioactive compounds from marine fungi. In: Mérillon J-M, Ramawat KG (eds) Fungal metabolites. Springer International Publishing, pp 415–434. https://doi.org/10.1007/978-3-319-25001-4_9
57. Numan M, Shah M, Asaf S, Ur Rehman N, Al-Harrasi A (2022) Bioactive compounds from endophytic bacteria Bacillus subtilis strain EP1 with their antibacterial activities. Metabolites 12(12). https://doi.org/10.3390/metabo12121228
58. Shaala LA, Youssef DTA (2015) Identification and bioactivity of compounds from the fungus Penicillium sp. CYE-87 isolated from a marine tunicate. Mar Drugs 13(4):1698–1709. https://doi.org/10.3390/md13041698
59. Betancur LA, Forero AM, Vinchira-Villarraga DM, Cárdenas JD, Romero-Otero A, Chagas FO, Pupo MT, Castellanos L, Ramos FA (2020) NMR-based metabolic profiling to follow the

production of anti-phytopathogenic compounds in the culture of the marine strain Streptomyces sp. PNM-9. Microbiol Res 239:126507. https://doi.org/10.1016/j.micres.2020.126507

60. Sharma D, Singh VP, Singh RK, Joshi CS, Sharma V (2020) Isolation and characterization of bioactive compounds from natural resources: metabolomics and molecular approaches. In Srivastava AK, Kannaujiya VK, Singh RK, & D. B. T.-E. D. as a S. for A. M. Singh (eds) Evolutionary diversity as a source for anticancer molecules. Academic, pp 77–101. https://doi.org/10.1016/B978-0-12-821710-8.00004-7
61. Deschamps JR (2010) X-ray crystallography of chemical compounds. Life Sci 86(15–16): 585–589. https://doi.org/10.1016/j.lfs.2009.02.028
62. Tojo S, Kohno T, Tanaka T, Kamioka S, Ota Y, Ishii T, Kamimoto K, Asano S, Isobe Y (2014) Crystal structures and structure-activity relationships of imidazothiazole derivatives as IDO1 inhibitors. ACS Med Chem Lett 5(10):1119–1123. https://doi.org/10.1021/ml500247w
63. Zheng CJ, Li L, Zou JP, Han T, Qin LP (2012) Identification of a quinazoline alkaloid produced by Penicillium vinaceum, an endophytic fungus from Crocus sativus. Pharm Biol 50(2): 129–133. https://doi.org/10.3109/13880209.2011.569726
64. Verma VC, Lobkovsky E, Gange AC, Singh SK, Prakash S (2011) Piperine production by endophytic fungus Periconia sp. Isolated from Piper longum L. J Antibiot 64(6):427–431. https://doi.org/10.1038/ja.2011.27
65. Kamel RA, Abdel-Razek AS, Hamed A, Ibrahim RR, Stammler HG, Frese M, Sewald N, Shaaban M (2020) Isoshamixanthone: a new pyrano xanthone from endophytic Aspergillus sp. ASCLA and absolute configuration of epiisoshamixanthone. Nat Prod Res 34(8): 1080–1090. https://doi.org/10.1080/14786419.2018.1548458
66. Du L, Zhu T, Fang Y, Liu H, Gu Q, Zhu W (2007) Aspergiolide A, a novel anthraquinone derivative with naphtho[1,2,3-de]chromene-2,7-dione skeleton isolated from a marine-derived fungus Aspergillus glaucus. Tetrahedron 63(5):1085–1088. https://doi.org/10.1016/j.tet.2006.11.074
67. Song T, Chen M, Ge ZW, Chai W, Li XC, Zhang Z, Lian XY (2018) Bioactive penicipyrrodiether A, an adduct of GKK1032 analogue and phenol A derivative, from a marine-sourced fungus Penicillium sp. ZZ380. J Org Chem 83(21):13395–13401. https://doi.org/10.1021/acs.joc.8b02172
68. Liu Y, Ruan Q, Jiang S, Qu Y, Chen J, Zhao M, Yang B, Liu Y, Zhao Z, Cui H (2019) Cytochalasins and polyketides from the fungus Diaporthe sp. GZU-1021 and their anti-inflammatory activity. Fitoterapia 137:104187. https://doi.org/10.1016/j.fitote.2019.104187
69. Hamed A, Abdel-Razek AS, Frese M, Stammler HG, El-Haddad AF, Ibrahim TMA, Sewald N, Shaaban M (2018) Terretonin N: a new meroterpenoid from Nocardiopsis sp. Molecules 23(2). https://doi.org/10.3390/molecules23020299
70. Chen M, Chai W, Zhu R, Song T, Zhang Z, Lian X-Y (2018) Streptopyrazinones A–D, rare metabolites from marine-derived Streptomyces sp. ZZ446. Tetrahedron 74(16):2100–2106. https://doi.org/10.1016/j.tet.2018.03.028
71. Bi Y, Yu Z (2016) Diterpenoids from Streptomyces sp. SN194 and their antifungal activity against botrytis cinerea. J Agric Food Chem 64(45):8525–8529. https://doi.org/10.1021/acs.jafc.6b03645
72. Zhang YM, Liu BL, Zheng XH, Huang XJ, Li HY, Zhang Y, Zhang TT, Sun DY, Lin BR, Zhou GX (2017) Anandins A and B, two rare steroidal alkaloids from a marine Streptomyces anandii H41-59. Mar Drugs 15(11). https://doi.org/10.3390/md15110355
73. Zhang Y, Zhu T, Fang Y, Liu H, Gu Q, Zhu W (2007) Carbonarones A and B, new bioactive γ-Pyrone and α-Pyridone derivatives from the marine-derived fungus Aspergillus carbonarius. J Antibiot 60(2):153–157. https://doi.org/10.1038/ja.2007.15
74. Borges WS, Mancilla G, Guimarães DO, Durán-Patrón R, Collado IG, Pupo MT (2011) Azaphilones from the endophyte Chaetomium globosum. J Nat Prod 74(5):1182–1187. https://doi.org/10.1021/np200110f

75. Yi W, Newaz AW, Yong K, Ma M, Lian XY, Zhang Z (2022) New hygrocins K–U and streptophenylpropanamide A and bioactive compounds from the marine-associated streptomyces sp. ZZ1956. Antibiotics 11(11). https://doi.org/10.3390/antibiotics11111455
76. Fang Z, Jiang X, Zhang Q, Zhang L, Zhang W, Yang C, Zhang H, Zhu Y, Zhang C (2020) S-bridged thioether and structure-diversified angucyclinone derivatives from the South China sea-derived micromonospora echinospora SCSIO 04089. J Nat Prod 83(10):3122–3130. https://doi.org/10.1021/acs.jnatprod.0c00719
77. Palanichamy P, Kannan S, Murugan D, Alagusundaram P, Marudhamuthu M (2019) Purification, crystallization and anticancer activity evaluation of the compound alternariol methyl ether from endophytic fungi Alternaria alternata. J Appl Microbiol 127(5):1468–1478. https://doi.org/10.1111/jam.14410
78. Kaleem S, Qin L, Yi W, Lian XY, Zhang Z (2020) Bioactive metabolites from the mariana trench sediment-derived fungus Penicillium sp. SY2107. Mar Drugs 18(5). https://doi.org/10.3390/md18050258
79. Hamed A, Abdel-Razek AS, Araby M, Abu-Elghait M, El-Hosari DG, Frese M, Soliman HSM, Stammler HG, Sewald N, Shaaban M (2021) Meleagrin from marine fungus Emericella dentata Nq45: crystal structure and diverse biological activity studies. Nat Prod Res 35(21):3830–3838. https://doi.org/10.1080/14786419.2020.1741583
80. Song T, Tang M, Ge H, Chen M, Lian X, Zhang Z (2019) Novel bioactive penicipyrroether a and pyrrospirone J from the marine-derived penicillium sp. zz380. Mar Drugs 17(5). https://doi.org/10.3390/md17050292
81. Meng LH, Li XM, Zhang FZ, Wang YN, Wang BG (2020) Talascortenes A-G, highly oxygenated diterpenoid acids from the sea-anemone-derived endozoic fungus talaromyces scorteus AS-242. J Nat Prod 83(8):2528–2536. https://doi.org/10.1021/acs.jnatprod.0c00628
82. Lindner B (2000) Matrix-assisted laser desorption/ionization time-of-flight mass spectrometry of lipopolysaccharides. In: Holst O (ed) Methods in molecular biology (Clifton, N.J.), vol 145. Humana Press, pp 311–325. https://doi.org/10.1385/1-59259-052-7:311
83. Jerz G, Elnakady YA, Braun A, Jäckel K, Sasse F, Al Ghamdi AA, Omar MOM, Winterhalter P (2014) Preparative mass-spectrometry profiling of bioactive metabolites in Saudi-Arabian propolis fractionated by high-speed countercurrent chromatography and off-line atmospheric pressure chemical ionization mass-spectrometry injection. J Chromatogr A 1347:17–29. https://doi.org/10.1016/j.chroma.2014.04.068
84. Darwesh OM, Mahmoud RH, Abdo SM, Marrez DA (2022) Isolation of Haematococcus lacustris as source of novel anti-multi-antibiotic resistant microbes agents; fractionation and identification of bioactive compounds. Biotechnol Rep 35:e00753. https://doi.org/10.1016/j.btre.2022.e00753
85. Li Y, Zhou G, Xing S, Tu P, Li X (2015) Identification of echinacoside metabolites produced by human intestinal bacteria using ultraperformance liquid chromatography–quadrupole time-of-flight mass spectrometry. J Agric Food Chem 63(30):6764–6771. https://doi.org/10.1021/acs.jafc.5b02881
86. Beretta G, Artali R, Caneva E, Orlandini S, Centini M, Facino RM (2009) Quinoline alkaloids in honey: Further analytical (HPLC-DAD-ESI-MS, multidimensional diffusion-ordered NMR spectroscopy), theoretical and chemometric studies. J Pharm Biomed Anal 50(3):432–439. https://doi.org/10.1016/j.jpba.2009.05.029

Chapter 6
Application of Alternative Technologies for the Recovery of Bioactive Compounds from Microbial Sources

Susana Ochoa and J. Felipe Osorio-Tobón

Abstract Indiscriminate chemical compound usage, such as agrochemicals, pesticides, fungicides, antibiotics, drugs, and other synthetic products, such as dyes, polymers, and heavy metals, have devastated soils, waters, and even living beings themselves. In this way, the development process for providing bioactive compounds for human, animal, and environmental health is one of the most urgent needs to enhance the balance between human exploitation and nature. These needs focus on obtaining healthy and safe products obtained in sustainable processes. Therefore, the search for natural compounds that can be used as nutrients, natural pesticides, and antibacterial or anticancer agents is increased.

Keywords Bioactive · Metabolites · Antimicrobial · Ultrasound-assisted extraction

6.1 Introduction

Plants, animals, and microorganisms are valuable sources of natural products with bioactivity [1, 2]. The natural products market was valued at USD 189 billion in 2021. This market is projected to reach USD 300 Billion by 2030 [3]. Natural compound research from plants is a well-known field. However, further research regarding natural compounds obtained from microorganisms is necessary.

Microorganisms produce primary metabolites such as amino acids, carbohydrates, proteins, and enzymes [3, 4]. Moreover, microorganisms can produce secondary metabolites with potential use for conservation or protection. These compounds are recognized by their bioactive properties, such as antimicrobial, antioxidant, anticarcinogenic, antiparasitic, and anti-inflammatory [2]. Therefore, the production and recovery of these compounds allow for obtaining high-value products. As can be observed in Table 6.1, microorganisms such as Archaea, bacteria, fungi, yeast, algae, and even some parasites produce bioactive compounds.

S. Ochoa · J. F. Osorio-Tobón (✉)
Faculty of Health Sciences, University Institution Colegio Mayor de Antioquia (COLMAYOR), Medellín, Antioquia, Colombia
e-mail: juan.tobon@colmayor.edu.co

R. Soni et al. (eds.), *Microbial Bioactive Compounds*,
https://doi.org/10.1007/978-3-031-40082-7_6

Table 6.1 Bioactive compounds derived from different microbial sources

Source	Organism	Bioactive compound (s)	Bioactive properties	References
Archaea	*Haloferax larsenii HA1*	Halocin	Antibacterial protein and cell protector	Kumar and Tiwari [5]
	Halorubrum sp. *SH1*	Bacterioruberin	Antioxidant	de la Vega et al. [6]
Bacteria	***Streptomyces***			
	Streptomyces anulatus NEAE-94	Unsaturated and saturated fatty acids, alkenes, fatty acid esters, alkanes, and triterpenes	Antimicrobial activity against *Staphylococcus aureus*	El-Naggar et al. [7]
	Streptomyces globisporus BU2018	Exopolysaccharides	Antioxidant	Abdel-Aziz et al. [8]
	Streptomyces tunisialbus sp	Fatty acids and menaquinones	Antimicrobial activity against gram-positive, and gram-negative bacteria, yeast, and filamentous fungi	Ayed et al. [9]
	Streptomyces sp. BO7	Biphenyls	Antibacterial, antioxidant, and anticancer	Taechowisan et al. [10]
	Bacillus			
	Bacillus licheniformis	Bacitracin	Antimicrobial	Ali et al. [11]
	Bacillus subtilis	Fengycin	Antifungal	Wu et al. [12]
	Pseudomonas			
	Pseudomonas cedrina	Biomass extract rich in diketopiperazines	Anticancer	Sánchez-Tafolla et al. [13]
	Lactobacillus			
	Lactobacillus coryniformis NA-3	Exopolysaccharides (α-rhamnose, α-mannose, α-galactose, and α-glucose)	Antioxidant and anti-biofilm	Xu et al. [14]
Fungus	*Aspergillus* fumigatus MF029	Chaetominine, sphingofungin, emodin, chaetominine, sphingofungin, and trypacidin	Antitubercular activity	Song et al. [15]
	Aspergillus fumigatus	Biomass extract rich in phenolic compounds (rutin, quercetin, caffeic acid, kaempferol, and ellagic acid)	Antibiofilm, antiproliferative, antioxidant, and antimutagenic	Kaur et al. [16]

(continued)

Table 6.1 (continued)

Source	Organism	Bioactive compound (s)	Bioactive properties	References
	Fusarium redolens	Biomass extract rich in chrysophanol and fumaric acid	Antimicrobial	Nazir et al. [17]
Yeast	*Metschnikowia yeast genus*	Alkaloids, antibiotics, and long-chain fatty acids	Antifungal	Fernandez-San Millan et al. [18]
Microalgae	*Nannochloropsis gaditana*	Omega-3 eicosapentaenoic acid	Prevention of cardiovascular diseases	Martínez et al. [19]
	Spirulina (Arthrospira platensis)	Phycocyanin	Anti-inflammatory antioxidant, antiviral, immunity-boosting, and anticancer	Lauceri et al. [20]
	Nannochloropsis oculata and *Porphyridium purpureum*	Biomass extract rich in anticancer and antioxidant activities	Anticancer	Garcia-Parra et al. [21]

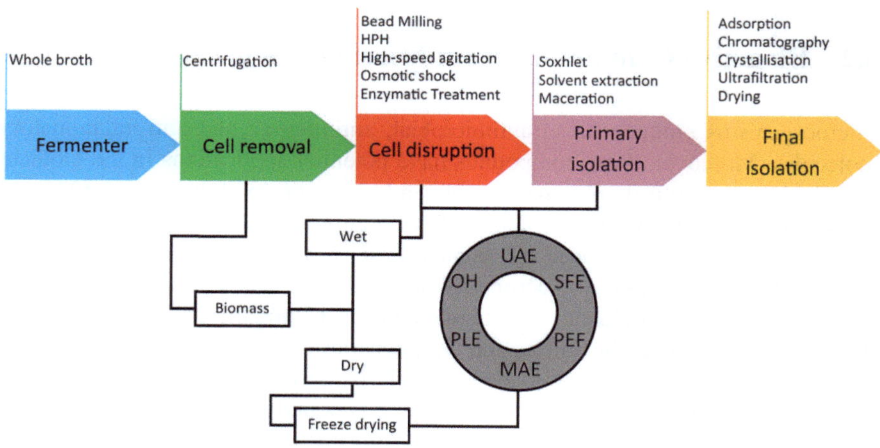

Fig. 6.1 Steps involved in the recovery of bioactive compounds from microorganisms

The production of the compounds depends on the environmental or culture media conditions. Moreover, most of these bioactive compounds could have high demand in the medical, pharmacological, and food industries [22].

The recovery of bioactive compounds from microbial sources comprises several steps in a downstream process. A further selection of recovery and purification steps will depend on whether the product is inside or outside the cell. Figure 6.1 represents the main steps involved in the recovery of bioactive compounds from microorganisms. Cell disruption is an initial step that is fundamental in the recovery of compounds, and thus, the disruption method choice is crucial. In this context, the cell wall composition also influences the disruption method performance. For example, some microalgae with high cellulose, glucose, and mannose contents have more rigid cell walls. Moreover, cells in the stationary phase or growth in rich nutrient media can have strong cell walls, which influences the selection of the disruption method and its parameters [23].

Among conventional disruption methods, mechanical and nonmechanical methods such as bead milling, high-pressure homogenization (HPH), osmotic shock, and enzymatic treatments are used for bioactive compound recovery. In the past decades, many alternative technologies have been explored for bioactive extraction from microbial sources, such as ultrasound-assisted extraction (UAE), supercritical fluid extraction (SFE), pulsed electric field extraction (PEF), microwave-assisted extraction (MAE), and others (Fig. 6.1). These alternative approaches are considered environmentally friendly and enhance extraction yields. In this context, the extraction with no previous cell disruption represents an excellent alternative to less energy-consuming process development. Alternative extraction techniques such as UAE, PEF, and MAE are mechanical methods that apply mechanical forces by waves or electric currents to break the cellular membrane.

6.2 Bioactive Compounds from Microbial Sources

Microorganisms produce novel antimicrobial, antitumoral, and anti-inflammatory molecules. Moreover, these compounds have potential applications in the biotechnological, nutraceutical, pharmaceutical, and environmental industries [24]. As shown in Table 6.1, bioactive compounds with antimicrobial, antifungal, and antioxidant activities are produced by Archea, Prokaryotic, and Eukaryotic domains. Moreover, the production of macromolecules such as amino acids, proteins, lipids, and carbohydrates is influenced by temperature, pH, humidity, aeration, and substrate. In this context, the interaction between these parameters and niches like oceans, mangroves, and caverns, even specific parts of plants or animal represent new opportunities to find novel bioactive compounds [25]. However, further research is necessary to identify the easy-to-cultivate and most productive microorganisms to scale up the production on a large scale for their subsequent application. Next, mainly microbial sources of bioactive compounds are described.

6.2.1 Main Microbial Sources of Bioactive Compounds

6.2.1.1 Archea Bioactive Compounds

Archaea compounds are produced under extreme conditions such as salt saturation, high temperature, and elevated UV radiation [26]. *Archea* produces bioactive compounds such as exopolysaccharides, carotenoids, and proteins. These compounds have potential applications in biomedical, pharmaceutical, cosmetic, environmental, and industrial fields. Gómez-Villegas et al. [27] reported strains of *Haloarchaea* as a potential source of compounds. *Haloferax larsenii* HA1 [28] produces halocins and sulfolobicins [29]. Halocins and sulfolobicins are proteinic compounds that could be used as food preservative because it causes cellular deformation and release of cell contents leading to cell death. Carotenoids are produced by *Haloarcula japónica*, *Halobacterium salinarum*, and *Halococcus morrhuae*. Moreover, extremozymes are produced by *Pyrococcus furiosus*, *Thermococcus littoralis*, and *Thermus aquaticus* [30]. On the other hand, some pigments, such as carotenoids, bacteriorhodopsin, and bacterioruberin, are produced by *Halobacterium Salinarum*. These compounds help cells to adapt to hypersaline conditions by acting as a water barrier, allowing ions and oxygen molecules to pass through the cell membrane. Therefore, these compounds can be used as antioxidants and photoprotective in food and cosmetics.

6.2.1.2 Bacteria Bioactive Compounds

Actinomycetes are one of the most reported genera that produce bioactive compounds. For example, *Streptomyces* spp. can synthesize microbial compounds such as vinaceuline, bafilomycin, and antimycin [24]. Moreover, antioxidant compounds such as violacein and prodigiosin are produced by *Streptomyces rubrireticuli* and *S. longisporus ruber* [31]. These molecules or their derivatives are known for their antimalarial, antibacterial, and anticancer activities. Other strains of the genera *Bacillus, Pseudomonas, Myxobacteria, Cyanobacteria* [1], and *Lactobacillus* [32] can produce other bioactive molecules. For example, the antimicrobial compounds bacitracin and bacilysin are produced by *B. liquenoformes* and *B. subtilis*, respectively [33]. On the other hand, *Pseudomonas* spp. (*P. aeruginosa, P. fluorescent,* and *Pseudomonas chlororaphisin*) produce antimicrobial compounds such as pyocyanin [3].

Lactic acid bacteria (LAB), such as *Lactococcus* and *Pediococcus*, have been reported to produce bacteriocins. Bacteriocins are known as immunomodulators with antimicrobial activity [32]. Phenolic compounds with antioxidant properties such as chlorogenic acid and gallic acid are produced by several cyanobacterial species [31]. On the other hand, fabclavines, xenocoumacins, xenorhabdins, and PAX peptides are antiparasitic compounds identified in *Xenorhabdus* and *Photorhabdus* [34]. Strains of *Photorhabdus luminescens* and *Xenorhabdus nematophila* showed anti-trypanosomal activity and potential use to develop novel drugs against Chagas disease [35]. Prokaryotes are an excellent choice for bioactive compound production due to their metabolic versatility and easy handling in the laboratory. Moreover, synthetic biology tools or heterologous systems could enhance bioactive compound production [31, 36].

6.2.1.3 Fungi and Yeast Bioactive Compounds

Fungi are eukaryotic organisms known to habit almost all ecological niches of the Earth, especially where there are organic sources and are in a state of decomposition. Many bioactive compounds are generated after mycelial growth and can affect direct sporulation. For example, *Aspergillus nidulans* and *Fusarium gramineaum* produce linoleic acid and zearalenone. Moreover, *Alternaria alternata* produces melanins as a proactive compound from UV rays. In addition, *Aspergillus terreus* has been recognized to produce lovastatin, which has metabolic activity. The most reported groups of secondary metabolites in this domain are polyketides, no ribosomal peptides, and terpenes [3]. Some endophytic fungi isolated from plants possess antimicrobial, antioxidant, and cytotoxic activities [37]. For example, compounds with antiprotozoal, antibacterial, and antiviral activity have been found in endophytic fungi such as *Colletotrichum, Diaporthe, Fusarium, Trichoderma, Penicillium*, and *Xylariagenera* [25]. Moreover, yeasts and other nonfilamentous eukaryotic microorganisms produce metabolites with antifungal such as piperidine and protoemetine (alkaloids), p-coumaroyl quinic acid (phenylpropanoid), which are produced by *Metschnikowia pulcherrim*a [18]. Moreover, the yeast used in fermented beverage production (wine, beer) produces alcohols such as tyrosol, which is responsible for the flavor in fermented beverages and is recognized for its antioxidant and cardioprotective properties [38]. Yeast can also be used as a model cell for genetic engineering assays to produce compounds derivate from plants when the expression in a complex system is required [39].

6.2.1.4 Microalgae Bioactive Compounds

Microalgae are found in oceans, fresh and wastewater, and extreme environments. Microalgae are an excellent source of metabolites such as fatty acids, carbohydrates, proteins, vitamins, and bioactive compounds [19]. Antimicrobial and anticancer compounds have been found in microalgae [21]. For example, phenolic compounds and hydroxycinnamic acids such as gallic acid, chlorogenic acid, ferulic acid, and caffeic acid have been found in *Chlorella vulgaris, Haematococcus pluvialis, Diacronema lutheri, Phaeodactylum tricornutum, Tetraselmissuecica, Ankistrodesmussp., Spirogyrasp., Euglena cantabrica, Caespitella pascheri, and Porphyridiumpurpureum* [31]. Metabolites such as exopolysaccharides with immunomodulatory, anti-inflammatory, antiviral, antifungal, and antibacterial capacities are produced by *Porphyridium* sp., *Arthospira* sp., and *Chlorella* sp. [21]. On the other hand, some specific compounds, such as polyunsaturated aldehydes with anticancer activity, are found in marine diatoms [40]. Therefore, microalgae could be an excellent source of novel bioactive compounds with multiple applications.

6.3 Production of Bioactive Compounds

During the past few decades, the bioactive compounds market from microbial sources has been growing due to its impact on the agriculture, food, and pharmaceutical industries. For example, the agroindustry uses bioactive compounds for pest control and plant growth promotion. Therefore, to take advantage of all properties offered by microorganisms, it is necessary to develop a sustainable process that produces bioactive compounds at low cost and high quality and effectiveness. Next, an overview of the main compounds produced by microbial sources is presented. Almost all bioactive compounds from microorganisms are related to their antimicrobial activity. Currently, there is a concern regarding the increase in antimicrobial resistance. As mentioned before, some antimicrobial molecules can disrupt the cell membrane. For example, membrane synthesis is inhibited by lipopeptides and polymyxin produced by *Bacillus* sp. and *Paenibacillus polymyxa* [33], respectively. In that context, antibiotics such as streptomycin, gentamicin, and tetracycline are produced by *Streptomyces griseus*, *Micromonospora purpurea*, and *Streptomyces aureofasciencs* [24]. These antibiotics inhibit protein synthesis in cells. Many of these compounds are recovered from marine microorganisms, which grow in extreme temperatures, under osmotic stress [41]. On the other hand, compounds such as bacteriocins isolated from the gastrointestinal tract are recognized for their immunomodulatory properties and antimicrobial capacity. Moreover, bacteriocins are used as a food preservative [32].

In the case of microalgae, the antimicrobial activity is related to the overproduction of fatty acids that can reduce the ability to breathe and cause cell death [24]. On the other hand, antifungal compounds such as glycolipids produced by *Bacillus licheniformis* can inhibit *Aspergillus niger* [2]. Antioxidant compounds such as polyphenols, carotenoids, or exopolysaccharides are produced by *Aspergillus* spp. *and Artrospira* sp., among others. These compounds can scavenge free radicals and are recognized for their photo-protective properties. Exopolysaccharides are high-molecular-weight carbohydrate polymers with radical scavenging activities, metal chelation activity, and lipid peroxidation inhibition [31]. These compounds are one of the most exploited bioactive substances due to their antiaging capacity. Other bioactive properties, such as anticholinesterase, antituberculosis, and antimalarial activity, have been shown in microorganisms. Further research is necessary to explore the microbial capacity to obtain bioactive compounds. Combined with the develop alternative extraction processes can allow the obtaining of pure and safe molecules with application in the medical field, pharmaceuticals, food, and environment industries.

6.3.1 Conventional Extraction Processes

Conventional extraction methods such as maceration, Soxhlet, solvent extraction, and hot reflux extraction have been used for compound recovery from microbial sources. Although they consume large quantities of solvents and employ longer extraction times, these processes are recognized for their simplicity and low-cost implementation [42]. Generally, they are used as a reference to compare with alternative technologies. Maceration is a straightforward extraction process carried out at ambient temperature under agitation. Bioactive compounds from *Pleurotus ostreatus* were recovered by maceration after 90 min at 25 °C, 150 rpm, using water and ethanol as extraction solvents [43]. Different fractions rich in proteins and phenolic compounds were recovered depending on the solvent proportion. For example, a mixture with 95% ethanol enhances protein extraction, while 50% ethanol increases the content of phenolic compounds. In a similar approach, Daud et al. [44] recovered red pigments from the fungus *Monascus purpureus* at 30 °C but during 16 h under agitation (180 rpm). The red pigment solubility depends on the solvent polarity, where the best solvent was 60% ethanol, which allows a maximum yield of 207 AU/g dry fermented solids. On the other hand, hot water extraction is used for obtaining polysaccharides from mushrooms. For example, polysaccharides were obtained from *Ganoderma resinaceum* by hot reflux extraction at 100 °C for 8 h [45].

Recently, Soxhlet extraction of oil from the microalgae *Spirogyra* [46] and *Chlorella pyrenoidosa* [47] was studied using extraction times ranging between 1 and 4 h at boiling temperatures depending on the solvents (n-Hexane and 2-Methyltetrahydrofuran). Oil extraction from *Spirogyra* required previous drying and milling pretreatments to enhance the extraction process due to a finer algae size causing better contact with the solvent. Moreover, usually, the more dried, the higher yield. In the drying/dehydration processes, freeze-drying is one of the most used methods. Low temperatures employed by freeze-drying keep the integrity of the compounds. However, freeze-drying is energy consuming, which could limit its application at the industrial scale. In this context, the direct extraction from wet biomass is getting more attention in the scientific community. For example, wet biomass was used as raw material in the oil recovery from *Chlorella pyrenoidosa* by Soxhlet [47]. Although lipid extraction is enhanced, total fatty acids content presented a reduction. Therefore, cell pretreatment influences extraction performance in conventional extraction processes as well as in alternative techniques. Thus, further research is necessary to establish the optimal process conditions. As mentioned, high temperatures and long extraction times are characteristics of conventional extraction methods. Therefore, energy consumption and thermally compound degradation are drawbacks that must be overcome. However, these techniques will continue to be used to compare alternative or novel technologies.

6.3.2 *Alternative Extraction Technologies*

Over the past decades, the recovery of bioactive compounds using alternative technologies has been focused on by researchers in many fields, such as food, chemical, or biotechnology. Generally, alternative extraction methods are more feasible than conventional extraction methods. The use of solvents generally recognized as safe (GRAS), the shorter extraction times, and the higher extraction yields are the main advantages of these methods. Among the most popular alternative extraction methods, ultrasound, microwaves, supercritical fluids, pressurized fluids, and electric fields are the most used technologies for bioactive extraction from microbial sources. Next, the concepts and applications of alternative extraction methods will be presented.

6.3.2.1 Ultrasound-Assisted Extraction (UAE)

Ultrasound is one of the most used extraction technologies employed in the bioactive compound recovery from fruits, vegetables, herbs, spices, seeds, and microorganisms. In UAE, cell disruption is caused by an acoustic phenomenon known as cavitation. In cavitation, the ultrasound waves generate rarefaction and compression cycles, creating gas bubbles in the cytoplasm. Once the bubbles have reached a maximum size, they collapse and release large amounts of energy (5000 K and 2000 atm) [48]. The cell wall is disrupted due to mechanical effects, the solvent has more intimate contact with the target compounds, and the extraction rates are enhanced [49]. Extraction parameters such as ultrasound power, frequency, temperature, solvent, type of device, and extraction time influence the extraction process. As shown in Table 6.2, temperatures ranging from 25 °C and 70 °C are used for compound recovery. Ultrasonic power from 100 W to 1000 W and short extraction times are used (e.g., minutes). Regarding extraction solvents, GRAS solvents such as water, ethanol, buffers, and deep eutectic solvents (DESs) are used. DESs are recognized as environmentally friendly, inexpensive, and chemically stable [50].

Depending on the microorganism, the effect on extraction parameters can be different. Generally, an increase in the ultrasound power, temperature, and extraction time increases the recovery of the compounds. The increase in the ultrasonic power enhances the cavitation, the cell structure is disrupted faster, and the solvent penetrates more efficiently [51]. The increase in temperature enhances the solubility of the compounds. Moreover, the denaturation of the membrane can be promoted by temperature [52]. However, excessive temperature or ultrasonic power may trigger the degradation of the compounds through the failures of the chemical structure or the generation of ROS [53]. Regarding frequency, cell disruption is enhanced by the acceleration caused by higher frequencies. The higher the frequency, the smaller the cavity sizes and the faster the bubbles collapse [48].

Table 6.2 UAE applications of bioactive compounds from microbial sources

Species	Compound (s) recovered	Extraction conditions	References
Morchella importuna	Polysaccharides	62 °C, 600 W, 31 min, choline chloride/oxalic acid (DESs)	Pan et al. [50]
Dictyosphaerium sp.	Polysaccharides	50 °C, 500 W, 50 min, water	Chen et al. [51]
Saccharomyces cerevisiae	Polysaccharides	70 °C, 1000 W, 8 h, 0.2 M sodium hydroxide	Eom et al. [58]
Arthrospira Platensis	Lutein/zeaxanthin	60–70 °C, 10 min, methanol	Sam et al. [59]
Chlorella vulgaris and *Porphyridium purpureum*	Carotenoids	70% power, ethanol (60%)	Vintila et al. [60]
Nannochloropsis gaditana	Omega-3 long chain-polyunsaturated fatty acids	50 °C,100 W, 30 min, ethanol	Castejón and Marko [61]
Diaporthe schini	Antioxidant compounds	25 °C, 400 W, pulsed mode (0.93), 15 min, ethanol	da Rosa et al. [62]
Saccharomyces cerevisiae, saccharomyces boulardii, Metschnikowia fructicola andTorulaspora delbrueckii	Mannoproteins	80% amplitude, 4 min, 0.1 M phosphate buffer, pH 6.5	Snyman et al. [63]
Grifola frondosa	Polysaccharides	65 °C, 4.5 h, water	Ji et al. [64]
Agrocybe cylindracea	Dietary fiber	Ultrasonic-assisted enzymatic method, the α-amylase concentration of 1.50%, protamex concentration of 1.20%, 150 W	Jia et al. [65]
Haematococcus pluvialis	Astaxanthin	25 °C, 80% amplitude, pulsed mode (3 min off and 12 min on), $(NH_4)_2SO_4$ salt solution/2-propanol	Khoo et al. [66]
Porphyridium cruentum and *Porphyridium purpureum*	Proteins, carbohydrates, lipids, fatty acids and phycoerythrin	30 °C, 100 W, 13–15 min, 50 mM Na-phosphate buffer (*P. cruentum*), and water (*P. purpureum*)	Ardiles et al. [67]

6.3.2.2 Supercritical Fluid Extraction (SFE)

SFE uses substances at temperatures and pressures above their critical point. These substances are known as supercritical fluids (SCF). Above the critical point, the fluids can diffuse as gas and has liquid solvation power [54]. For SFE, generally, before extraction, the microbial biomass is freeze-dried and disrupted. For instance, a ball mill is used to enhance the extraction of intracellular compounds from *Scenedesmus almeriensis* [55] and *Nannochloropsis* sp. [56]. CO_2 is the most used

Table 6.3 SFE applications of bioactive compounds from microbial sources

Species	Compound (s) recovered	Extraction conditions	References
Scenedesmus almeriensis	Lutein	65 °C, 550 bar, 14.48 g/min CO_2	Mehariya et al. [55]
Nannochloropsis sp.	Omega-3 fatty acids	75 °C, 550 bar, 14.48 g/min CO_2	Leone et al. [56]
Aurantiochytrium sp.	Omega-3 fatty acids and phenolic compounds	80 °C, 300 bar, 12 g/min CO_2	De Melo et al. [68]
Diaporthe schini	Antioxidant compounds	40 °C, 250 bar, 4 g/min CO_2, biomass:Ethanol, 1:1.5 (w/v)	da Rosa et al. [69]
Usnea subfloridana	Usnic acid	85 °C, 150 bar, 2 mL/min CO_2	Boitsova et al. [70]
Schizochytrium sp.	Docosahexaenoic acid (DHA)	77 °C, 465 bar, 5 mL/min CO_2, 1.25 mL/min ethanol	Rodríguez-España et al. [73]
Inonotus obliquus	Triterpenoids	50 °C, 350 bar, 3 mL/min CO_2	Huynh et al. [74]
Coccomyxa onubensis	Lutein and phenolic compounds	70 °C, 400 bar, 2 mL/min CO_2, 2.30 mL/min ethanol	Ruiz-Domínguez et al. [75]
Haematococcus pluvialis	Astaxanthin	50 °C, 500 bar, 2 L/min CO_2	Espinosa Álvarez et al. [76]
Chlorella vulgaris	Phenolic compounds	60 °C, 250 bar, 40 g/min CO_2 (ethanol 10% w/w)	Georgiopoulou et al. [77]

solvent in SFE, and it is recognized as safe (GRAS), inexpensive, has low toxicity, readily available, and has an easily accessible critical point (31 °C and 73.8 bar) [57]. Temperatures between 40 °C and 85 °C and pressures between 250 and 550 bar are suitable for bioactive compound extraction (Table 6.3). The selectivity of the CO_2 is modified by changing the temperature and pressure. For example, the solvent density increases as the temperature increases, and the solvent density increases as the pressure increases. Thus, the solubility of the intracellular compounds is enhanced by increasing the pressure at a constant temperature. This behavior has been observed in the recovery of omega-3 fatty acids and phenolic compounds from *Nannochloropsis* sp. [56] and *Aurantiochytrium* sp. [68].

Although CO_2 is the most common SFC used for bioactive compound extraction from microbial sources by SFE, it only allows the extraction of nonpolar compounds as lipids. Thus, CO_2 is used mainly for lipid or fatty acid extraction from microbial sources such as microalgae, as shown in Table 6.3. On the other hand, for polar compound extraction (e.g., phenolic compounds), a co-solvent such as ethanol is necessary. For instance, ethanol is used as a co-solvent for bioactive compound extraction from fungi [69], microalgae [55], and lichen [70]. SFE allows obtaining higher purity extracts while solvent recycling is possible.

Table 6.4 MAE applications of bioactive compounds from microbial sources

Species	Compound (s) recovered	Extraction conditions	References
Nannochloropsis oceanica	Proteins	40 °C, 700 W, 30 min, choline acetate	Motlagh et al. [72]
Rhizopus oryzae	Chitosan	300 W, 22 min, 1 N NaOH	Sebastian et al. [81]
Haematococcus pluvialis	Astaxanthin	75 °C, 700 W, 7 min dimethyl sulfoxide	Aslanbay Guler et al. [82]
Auxenochlorella Protothecoides	Lipids	2.8 kW, 200 μs pulse, cell suspension	Zhang et al. [83]
Chlorella vulgaris and Botryococcus braunii	Lipids	400 W, 40 s, cell suspension	Rokicka et al. [84]
Kappaphycus alvarezii	β-Carotene, chlorophyll, antioxidants	45 °C, 170 W, 12.5–14.5 min 80% methanol	Baskararaj et al. [85]
Psylocibe cubensis	Psilocin and psilocybin	50 °C, 600 W, 5 min, 60% methanol	Polo-Castellano et al. [86]
Lactococcus lactis	Menaquinones	50 °C, 600 W, 5 min, ethanol	Lee et al. [87]
Porphyridium cruentum and *Porphyridium purpureum*	Proteins, carbohydrates, lipids, fatty acids, and phycoerythrin	200 W, 60 s, 50 mM Na-phosphate buffer/ water (54:46 v/v)	Ardiles et al. [67]

6.3.2.3 Microwave-Assisted Extraction (MAE)

MAE is an alternative technology recognized by the shorter extraction time and the use of GRAS solvents, which increase extraction yields and preserve the integrity of the extracts. MAE has several applications, mainly in the food industry, regarding the extraction of bioactive compounds. In the compound recovery from microorganisms, microwaves are applied to a cell suspension prepared with an organic solvent. The cell suspension can be prepared using wet or dried biomass. Among organic solvents, dielectric or polar solvents such as water or ethanol are preferred. Microwaves with frequencies ranging between 300 MHz and 300 GHz cause fast boiling of the intracellular liquid, which increases the internal pressure and the size expansion of the cells, producing cell disruption [71]. However, although microwaves can cause cell disruption, previous cell disruption (e.g., high-pressure or bead milling) of microorganisms such as microalgae is recommended before MAE [72]. This pretreatment increases the cell wall disruption and enhances the extraction yields. Microwave power, temperature, solvent, extraction time, and matrix are the main parameters that influence MAE. Interaction between solvent and compounds is fundamental because the target compound should be highly soluble, and the solvent must have a high dielectric constant. Solvents such as water, methanol, and ethanol can absorb high amounts of microwave energy, and as shown in Table 6.4, these solvents are used for compound recovery from microorganisms. Temperature and extraction time are related. Higher temperatures and longer extraction times allow an increase in extraction yields [78]. For example, as can be observed in Table 6.4, the recovery of compounds from microorganisms is performed using temperatures and extraction times up to 75 °C and 30 min. However, exposure to high temperatures during prolonged times triggers compound degradation.

6.3.2.4 Pulsed Electric Field (PEF) Extraction

PEF is a nonthermal technology with growing interest in biotechnology industries for cell disruption due to its many advantages. PEF is an environmentally friendly process with shorter extraction times that increases extraction yields, avoiding triggering the degradation of the bioactive compounds [79]. In PEF, the sample is placed in the treatment chamber where a uniform and strong electric field is applied. The pass of short high-voltage electric pulses causes an electropermeabilization of the cell membranes without altering the bioactive compounds [80]. This permeabilization allows the recovery of the compounds from the microorganisms, minimizing the formation of cell debris with further simplification of the downstream operations.

As can be observed in Table 6.5, electrical impulses ranging between 15 kV/cm and 40 kV/cm are enough to allow the recovery of the compounds from the microorganisms. Although the effects of the electric field strength depend on the matrix characteristics, this range generates the irreversible permeabilization of microbial cells [88]. During membrane permeabilization, many transmembrane pores are formed, which enhances solvent penetration and further extraction of the bioactive compounds. For example, in the extraction of carotenoids from *Xanthophyllomyces dendrorhous* after PEF at 20 kV/cm for 135 μs, 80% of permeabilization was obtained, which increases the extraction yield up to 70% of total carotenoids contained in the yeast suspension. Moreover, extraction parameters such as extraction time, pulse width, conductivity, and pulse frequency can also influence permeabilization and PEF efficiency. For example, the increase from 25 kV/cm to 40 kV/cm in the electric field in the extraction of lipids from *Chlorella* cells increases lipid extraction [89]. However, when the electric field strength reaches a threshold value, the lipid extraction yield decreases due to the release of other compounds and the generation of large amounts of cell debris. This technology could represent many advantages at the industrial scale due to its low-energy consumption and easy incorporation into the processing line [90].

6.3.2.5 Other Alternative Technologies

Other alternative technologies are used for bioactive compound recovery from vegetal sources but with less intensive application in microbial sources. For example, pressurized liquid extraction (PLE) uses solvents above their boiling point but

Table 6.5 PEF applications of bioactive compounds from microbial sources

Species	Compound (s) recovered	Extraction conditions	References
Chlorella vulgaris	Water-soluble proteins, carbohydrates, and lipids	25 °C, 20 kV/cm, 100 kJ/kg_{SUSP}, 5 μs of pulse width, water (1 h), ethyl acetate (3 h)	Carullo et al. [80]
Chlorella	Lipids	35 kV/cm, the conductivity of 400 μS/cm, water,30 min	Zhang et al. [89]
Chlorella pyrenoidosa	Lipids	25 °C, 20 kV/cm, 6 μs of pulse width, chloroform/methanol	Han et al. [101]
Nannochloropsis oculata	Carbohydrates, proteins, and pigments	30 °C, 40 kV/cm, 10 μs of pulse width, water,30 min	Zhang et al. [102]
Saitozyma podzolica	Lipids	20 °C, 15 kV/cm, 1 μs of pulse width, ethanol and hexane,	Gorte et al. [103]
Xanthophyllomyces dendrorhous	Carotenoids, astaxanthin	25 °C, 20 kV/cm, 3 μs of pulse width, ethanol	Aguilar-Machado et al. [104]

below their critical point, applying high pressures. The high pressure allows deeper penetration of the solvent, and the temperature reduces the solvent viscosity, enhancing the extraction of the compounds [78]. Currently, PLE is mainly applied to contaminant detection in several areas. However, PLE can also be used for compound recovery from microorganisms after drying and cell disruption. Unsaturated fatty acids and carotenoids have been recovered from oleaginous yeasts [91] and microalgae [92–94] using temperatures ranging from 80 °C to 150 °C and pressures of 100 bar. An alternative to PLE is continuous pressurized solvent extraction (CPSE), which uses lower temperatures and pressures than PLE, keeping similar or even higher yields. For example, carotenoids and phycobiliproteins from *Cyanobium* sp. LEGE 06113 by CPSE at 70 °C and 1.5 mL/min (ethanol) have been recovered using CPSE [95].

Ultrahigh pressure extraction (UHPE) is similar to PLE but uses higher pressures (up to 8000 bar). This variation allows performing the extraction process without previous cell disruption due to higher pressure can break down the cell membrane. For example, UHPE (one cycle at 6000 bar at 50 °C) enhanced the extraction of carotenoids from *Haematococcus pluvialis* and *Porphyridium cruentum* microalgae compared with PLE [96]. However, the performance also depends on the microorganisms and type of compound. For example, although UHPE (one cycle at 1000 bar and 50 °C) was not superior to PLE in the carotenoid extraction from *Nannochloropsis oceanica*, UHPE increased the extraction of polyunsaturated fatty acids [93].

Ohmic heating (OH) is based on the Joule effect, where an electric current flows through resistive materials such as the cell wall [97]. Heating in OH is faster and more homogeneous than traditional thermal treatments. Moreover, OH causes cell wall breakdown, enhancing the mass transfer of intracellular compounds. For example, the ethanolic extracts obtained from *Cyanobium* sp. by OH (70 °C,

5 min, and 20 kHz) showed high antioxidant capacity [95]. Moreover, yields and antioxidant activity obtained by OH were better than the extraction by homogenization. Recently, OH has been applied in bioactive compound recovery from microalgae. In this context, nutrients from *Coelastrella* sp. LFR1 [98] were recovered by OH at 217 V/cm and 100 °C, showing higher performance for cell disruption. This combination allows the yield increase of chlorophyll and proteins in microalgae biomass. OH is also used for the recovery of bioactive compounds from *Spirulina platensis*. This photosynthetic cyanobacterium is recognized for producing antioxidant, antiviral, anti-cancer, and anti-inflammatory compounds. Ferreira-Santos et al. [99, 100] reported the feasibility of OH in the recovery of intracellular compounds using temperatures between 30 °C and 50 °C, 4 V/cm, and 20 kHz of frequency. OH is a technology with higher extraction yields, lower energy consumption, and shorter extraction times than conventional extraction processes with potential use at an industrial scale (Table 6.5).

6.4 Conclusions and Future Perspectives

Microorganisms can be an excellent source of valuable compounds with applications in several industries. Antibiofilm, antiproliferative, antioxidant, antimicrobial, anti-inflammatory, and antimutagenic activities are found in bioactive compounds recovered from microbial sources. Depending on the localization of the compounds, different pretreatments and extraction techniques can be explored. Pretreatment and extraction techniques are applied depending on the localization of the compounds. Pretreatment as drying or cell disruption is necessary for intracellular compounds. Some alternative extraction techniques like UAE, MAE, and OH allow simultaneous cell disruption and extraction. Alternative extraction technologies have higher yields and shorter extraction times than conventional processes. However, the initial cost and the lack of scaling-up criteria still are the main shortcomings. Therefore, large-scale systems development and further research regarding process optimization are necessary. To the extent that industrial-scale equipment and economically viable processes are developed, these alternative technologies could be more extensively used.

References

1. Abdel-Razek AS, El-Naggar ME, Allam A et al (2020) Microbial natural products in drug discovery. PRO 8:1–19. https://doi.org/10.3390/PR8040470
2. Pham JV, Yilma MA, Feliz A et al (2019) A review of the microbial production of bioactive natural products and biologics. Front Microbiol 10:1–27. https://doi.org/10.3389/fmicb.2019.01404
3. Omokhefe Bruce S (2022) Secondary metabolites from natural products. In: Secondary metabolites - trends and reviews. IntechOpen, p 310

4. Steele AD, Teijaro CN, Yang D, Shen B (2019) Leveraging a large microbial strain collection for natural product discovery. J Biol Chem 294:16567–16576. https://doi.org/10.1074/jbc.REV119.006514
5. Kumar V, Tiwari SK (2017a) Activity-guided separation and characterization of new halocin HA3 from fermented broth of Haloferax larsenii HA3. Extremophiles 21:609–621. https://doi.org/10.1007/s00792-017-0930-6
6. de la Vega M, Sayago A, Ariza J et al (2016) Characterization of a bacterioruberin-producing Haloarchaea isolated from the marshlands of the Odiel river in the southwest of Spain. Biotechnol Prog 32:592–600. https://doi.org/10.1002/btpr.2248
7. El-Naggar NEA, El-Bindary AAA, Abdel-Mogib M, Nour NS (2017) In vitro activity, extraction, separation and structure elucidation of antibiotic produced by Streptomyces anulatus NEAE-94 active against multidrug-resistant Staphylococcus aureus. Biotechnol Biotechnol Equip 31:418–430. https://doi.org/10.1080/13102818.2016.1276412
8. Abdel-Aziz SH, Awady MEE, Nasr-Eldin MA et al (2019) Production and assessment of antioxidant activity of exopolysaccharide from marine streptomyces *globisporus* BU2018. Egypt J Bot 59:645–655. https://doi.org/10.21608/ejbo.2019.6847.1274
9. Ayed A, Slama N, Mankai H et al (2018) Streptomyces tunisialbus sp. nov., a novel Streptomyces species with antimicrobial activity. Antonie van Leeuwenhoek. Int J Gen Mol Microbiol 111:1571–1581. https://doi.org/10.1007/s10482-018-1046-4
10. Taechowisan T, Chaisaeng S, Phutdhawong WS (2017) Antibacterial, antioxidant and anticancer activities of biphenyls from Streptomyces sp. BO-07: an endophyte in Boesenbergia rotunda (L.) Mansf A. Food Agric Immunol 28:1330–1346. https://doi.org/10.1080/09540105.2017.1339669
11. Ali S, Nelofer R, Andleeb S et al (2018) Biosynthesis and optimization of bacitracin by mutant bacillus licheniformis using submerged fermentation. Indian J Biotechnol 17:251–260
12. Wu JJ, Chou HP, Huang JW, Deng WL (2021) Genomic and biochemical characterization of antifungal compounds produced by Bacillus subtilis PMB102 against Alternaria brassicicola. Microbiol Res 251:126815. https://doi.org/10.1016/j.micres.2021.126815
13. Sánchez-Tafolla L, Padrón JM, Mendoza G et al (2019) Antiproliferative activity of biomass extract from Pseudomonas cedrina. Electron J Biotechnol 40:40–44. https://doi.org/10.1016/j.ejbt.2019.03.010
14. Xu X, Peng Q, Zhang Y et al (2020) A novel exopolysaccharide produced by lactobacillus coryniformis NA-3 exhibits antioxidant and biofilm-inhibiting properties in vitro. Food Nutr Res 64:10.29219/fnr.v64.3744
15. Song Z, Liu Y, Gao J et al (2021) Antitubercular metabolites from the marine-derived fungus strain aspergillus fumigatus MF029. Nat Prod Res 35:2647–2654. https://doi.org/10.1080/14786419.2019.1660331
16. Kaur N, Arora DS, Kalia N, Kaur M (2020) Antibiofilm, antiproliferative, antioxidant and antimutagenic activities of an endophytic fungus aspergillus fumigatus from Moringa oleifera. Mol Biol Rep 47:2901–2911. https://doi.org/10.1007/s11033-020-05394-7
17. Nazir A, Hafeez S, Habeeb AR (2022) Bioactive potentials of endophyte (fusarium redolens) isolated from Olea europaea. Arch Microbiol 204:219. https://doi.org/10.1007/s00203-022-02826-9
18. Fernandez-San Millan A, Gamir J, Farran I et al (2022) Identification of new antifungal metabolites produced by the yeast Metschnikowia pulcherrima involved in the biocontrol of postharvest plant pathogenic fungi. Postharvest Biol Technol 192:15. https://doi.org/10.1016/j.postharvbio.2022.111995
19. Martínez PM, Ortiz-Martínez VM, Segado SS et al (2022) Deep eutectic solvents for the extraction of fatty acids from microalgae biomass: recovery of omega-3 eicosapentaenoic acid. Sep Purif Technol 300:8. https://doi.org/10.1016/j.seppur.2022.121842
20. Lauceri R, Cavone C, Zittelli GC et al (2023) High purity grade Phycocyanin recovery by decupling cell lysis from the pigment extraction: an innovative approach. Food Bioprocess Technol 16:111–121. https://doi.org/10.1007/s11947-022-02926-w

21. Garcia-Parra J, Fuentes-Grünewald C, Gonzalez D (2022) Therapeutic potential of microalgae-derived bioactive metabolites is influenced by different large-scale culture strategies. Mar Drugs 20:627. https://doi.org/10.3390/md20100627
22. Singh R, Kumar M, Mittal A, Mehta PK (2017) Microbial metabolites in nutrition, healthcare and agriculture. 3 Biotech 7:14. https://doi.org/10.1007/s13205-016-0586-4
23. Alhattab M, Kermanshahi-Pour A, Brooks MSL (2019) Microalgae disruption techniques for product recovery: influence of cell wall composition. J Appl Phycol 31:61–88. https://doi.org/10.1007/s10811-018-1560-9
24. Rani A, Saini KC, Bast F et al (2021a) A review on microbial products and their perspective application as antimicrobial agents. Biomol Ther 11(12):1860. https://doi.org/10.3390/biom11121860
25. Cadamuro RD, da Silveira Bastos IMA, da Silva IT et al (2021) Bioactive compounds from mangrove endophytic fungus and their uses for microorganism control. J Fungi 7(6):455. https://doi.org/10.3390/jof7060455
26. Pfeifer K, Ergal İ, Koller M et al (2021) Archaea biotechnology. Biotechnol Adv 47:107668. https://doi.org/10.1016/J.BIOTECHADV.2020.107668
27. Gómez-Villegas P, Vigara J, Vila M et al (2020) Antioxidant, antimicrobial, and bioactive potential of two new haloarchaeal strains isolated from odiel salterns (Southwest Spain). Biology (Basel) 9:1–20. https://doi.org/10.3390/biology9090298
28. Kumar V, Tiwari SK (2017b) Halocin HA1: an archaeocin produced by the haloarchaeon Haloferax larsenii HA1. Process Biochem 61:202–208. https://doi.org/10.1016/J.PROCBIO.2017.06.010
29. Besse A, Peduzzi J, Rebuffat S, Carré-Mlouka A (2015) Antimicrobial peptides and proteins in the face of extremes: lessons from archaeocins. Biochimie 118:344–355. https://doi.org/10.1016/J.BIOCHI.2015.06.004
30. Grivard A, Goubet I, de Souza Duarte Filho LM et al (2022) Archaea carotenoids: natural pigments with unexplored innovative potential. Mar Drugs 20(8):524
31. Peyrat LA, Tsafantakis N, Georgousaki K et al (2019) Terrestrial microorganisms: cell factories of bioactive molecules with skin protecting applications. Molecules 24(9):1836. https://doi.org/10.3390/molecules24091836
32. Hernández-González JC, Martínez-Tapia A, Lazcano-Hernández G et al (2021) Bacteriocins from lactic acid bacteria. A powerful alternative as antimicrobials, probiotics, and immunomodulators in veterinary medicine. Anim 11:17. https://doi.org/10.3390/ani11040979
33. Tran C, Cock IE, Chen X, Feng Y (2022) Antimicrobial bacillus: metabolites and their mode of action. Antibiotics 11:25. https://doi.org/10.3390/antibiotics11010088
34. Gulsen SH, Tileklioglu E, Bode E et al (2022) Antiprotozoal activity of different Xenorhabdus and Photorhabdus bacterial secondary metabolites and identification of bioactive compounds using the easyPACId approach. Sci Rep 12:13. https://doi.org/10.1038/s41598-022-13722-z
35. Antonello AM, Sartori T, Silva MB et al (2019) Anti-Trypanosoma activity of bioactive metabolites from Photorhabdus luminescens and Xenorhabdus nematophila. Exp Parasitol 204:7. https://doi.org/10.1016/j.exppara.2019.107724
36. Arendt P, Pollier J, Callewaert N, Goossens A (2016) Synthetic biology for production of natural and new-to-nature terpenoids in photosynthetic organisms. Plant J 87:16–37. https://doi.org/10.1111/tpj.13138
37. Merlin JN, Christhudas IVSN, Kumar PP, Agastian P (2013) Optimization of growth and bioactive metabolite production: fusarium solani. Asian J Pharm Clin Res 6:98–103
38. Mas A, Guillamon JM, Torija MJ et al (2014) Bioactive compounds derived from the yeast metabolism of aromatic amino acids during alcoholic fermentation. Biomed Res Int 2014:7. https://doi.org/10.1155/2014/898045
39. Chrzanowski G (2020) Saccharomyces cerevisiae—an interesting producer of bioactive plant polyphenolic metabolites. Int J Mol Sci 21:1–18. https://doi.org/10.3390/ijms21197343
40. Mohamed SS, Abdelhamid SA, Ali RH (2021) Isolation and identification of marine microbial products. J Genet Eng Biotechnol 19:10. https://doi.org/10.1186/s43141-021-00259-3

41. Rani A, Saini KC, Bast F et al (2021b) Microorganisms: a potential source of bioactive molecules for antioxidant applications. Molecules 26(4):1142. https://doi.org/10.3390/molecules26041142
42. da Costa WA, Bezerra FWF, de Oliveira MS et al (2019) Supercritical CO2 extraction and transesterification of the residual oil from industrial palm kernel cake with supercritical methanol. J Supercrit Fluids 147:179–187. https://doi.org/10.1016/J.SUPFLU.2018.10.012
43. Duarte Trujillo AS, Jiménez Forero JA, Pineda Insuasti JA et al (2020) Extracción de sustancias bioactivas de *Pleurotus ostreatus* (Pleurotaceae) por maceración dinámica. Acta Biológica Colomb 25:61–74. https://doi.org/10.15446/abc.v25n1.72409
44. Daud NFS, Said FM, Chisti Y, Yasin NHM (2021) Recovery of red pigments from Monascus purpureus FTC 5357 by extraction of fermented solids: operational conditions and kinetics. Brazilian Arch Biol Technol 64:1–11. https://doi.org/10.1590/1678-4324-2021200182
45. Bleha R, Třešňáková L, Sushytskyi L et al (2022) Polysaccharides from Basidiocarps of the polypore fungus Ganoderma resinaceum: isolation and structure. Polymers (Basel) 14(2):255. https://doi.org/10.3390/polym14020255
46. Aravind S, Barik D, Ragupathi P, Vignesh G (2020) Investigation on algae oil extraction from algae spirogyra by Soxhlet extraction method. Mater Today Proc 43:308–313. https://doi.org/10.1016/j.matpr.2020.11.668
47. de Jesus SS, Ferreira GF, Moreira LS et al (2019) Comparison of several methods for effective lipid extraction from wet microalgae using green solvents. Renew Energy 143:130–141. https://doi.org/10.1016/j.renene.2019.04.168
48. Liu Y, Liu X, Cui Y, Yuan W (2022) Ultrasound for microalgal cell disruption and product extraction: a review. Ultrason Sonochem 87:106054. https://doi.org/10.1016/j.ultsonch.2022.106054
49. Zheng S, Zhang G, Wang HJ et al (2021) Progress in ultrasound-assisted extraction of the value-added products from microorganisms. World J Microbiol Biotechnol 37:1–14. https://doi.org/10.1007/s11274-021-03037-y
50. Pan X, Xu L, Meng J et al (2022) Ultrasound-assisted deep eutectic solvents extraction of polysaccharides from Morchella importuna: optimization, physicochemical properties, and bioactivities. Front Nutr 9:1–11. https://doi.org/10.3389/fnut.2022.912014
51. Chen C, Zhao Z, Ma S et al (2020) Optimization of ultrasonic-assisted extraction, refinement and characterization of water-soluble polysaccharide from Dictyosphaerium sp. and evaluation of antioxidant activity in vitro. J Food Meas Charact 14:963–977. https://doi.org/10.1007/s11694-019-00346-7
52. Dahmen-Ben Moussa I, Masmoudi MA, Choura S et al (2021) Extraction optimization using response surface methodology and evaluation of the antioxidant and antimicrobial potential of polyphenols in *Scenedesmus* sp. and *Chlorella* sp. Biomass Convers Biorefin 13:7185. https://doi.org/10.1007/s13399-021-01850-x
53. Soquetta MB, Terra L d M, Bastos CP (2018) Green technologies for the extraction of bioactive compounds in fruits and vegetables. CYTA J Food 16:400–412. https://doi.org/10.1080/19476337.2017.1411978
54. Wang W, Rao L, Wu X et al (2021) Supercritical carbon dioxide applications in food processing. Food Eng Rev 13:570–591. https://doi.org/10.1007/s12393-020-09270-9
55. Mehariya S, Iovine A, Di Sanzo G et al (2019) Supercritical fluid extraction of lutein from scenedesmus almeriensis. Molecules 24:1–15. https://doi.org/10.3390/molecules24071324
56. Leone GP, Balducchi R, Mehariya S et al (2019) Selective extraction of ω-3 fatty acids from nannochloropsis sp. using supercritical CO2 extraction. Molecules 24(13):2406. https://doi.org/10.3390/molecules24132406
57. Sarkar S, Gayen K, Bhowmick TK (2022) Green extraction of biomolecules from algae using subcritical and supercritical fluids. Biomass Convers Biorefinery. https://doi.org/10.1007/s13399-022-02309-3

58. Eom SJ, Park JT, Kang MC et al (2022) Use of ultrasound treatment to extract mannan polysaccharide from Saccharomyces cerevisiae. J Food Process Eng 45:1–6. https://doi.org/10.1111/jfpe.14105
59. Sam KJ, Nair MS, Velmurugan S et al (2022) Extraction of lutein/zeaxanthin from Arthrospira Platensis and optimisation of the saponification process using the response surface methodology. Indian Chem Eng:1–11. https://doi.org/10.1080/00194506.2022.2101146
60. Vintila ACN, Vlaicu A, Radu E et al (2022) Evaluation of ultrasound assisted extraction of bioactive compounds from microalgae. J Food Meas Charact 16:2518–2526. https://doi.org/10.1007/s11694-022-01347-9
61. Castejón N, Marko D (2022) Fatty acid composition and cytotoxic activity of lipid extracts from Nannochloropsis gaditana produced by green technologies. Molecules 27(12):3710. https://doi.org/10.3390/molecules27123710
62. da Rosa BV, Sauzem G d S, Kuhn RC (2021) Obtaining antioxidant compounds from the endophytic fungus Diaporthe schini using heat- and ultrasound-assisted extraction. Brazilian J Chem Eng 38:189–195. https://doi.org/10.1007/s43153-021-00089-3
63. Snyman C, Nguela JM, Sieczkowski N et al (2021) Optimised extraction and preliminary characterisation of mannoproteins from non-saccharomyces wine yeasts. Foods 10:1–17. https://doi.org/10.3390/foods10050924
64. Ji H-y, Yu J, Chen X-y, Liu A-j (2019) Extraction, optimization and bioactivities of alcohol-soluble polysaccharide from *Grifola frondosa*. J Food Meas Charact 13:1645–1651. https://doi.org/10.1007/s11694-019-00081-z
65. Jia F, Liu X, Gong Z et al (2020) Extraction, modification, and property characterization of dietary fiber from Agrocybe cylindracea. Food Sci Nutr 8:6131–6143. https://doi.org/10.1002/fsn3.1905
66. Khoo KS, Chew KW, Yew GY et al (2020) Integrated ultrasound-assisted liquid biphasic flotation for efficient extraction of astaxanthin from Haematococcus pluvialis. Ultrason Sonochem 67:1–9. https://doi.org/10.1016/j.ultsonch.2020.105052
67. Ardiles P, Cerezal-Mezquita P, Salinas-Fuentes F et al (2020) Biochemical composition and phycoerythrin extraction from red microalgae: a comparative study using green extraction technologies. PRO 8:1–16. https://doi.org/10.3390/pr8121628
68. De Melo MMR, Sapatinha M, Pinheiro J et al (2020) Supercritical CO2 extraction of *Aurantiochytrium* sp. biomass for the enhanced recovery of omega-3 fatty acids and phenolic compounds. J CO2 Util 38:24–31. https://doi.org/10.1016/j.jcou.2020.01.014
69. da Rosa BV, Kuhn KR, Ugalde GA et al (2020) Antioxidant compounds extracted from Diaporthe schini using supercritical CO2 plus cosolvent. Bioprocess Biosyst Eng 43:133–141. https://doi.org/10.1007/s00449-019-02211-9
70. Boitsova TA, Brovko OS, Ivakhnov AD, Zhil'tsov DV (2020) Optimizing supercritical fluid extraction of Usnic acid from the lichen species Usnea subfloridana. Russ J Phys Chem B 14: 1135–1141. https://doi.org/10.1134/S1990793120070040
71. Zainuddin MF, Fai CK, Ariff AB et al (2021) Current pretreatment/cell disruption and extraction methods used to improve intracellular lipid recovery from oleaginous yeasts. Microorganisms 9:1–28. https://doi.org/10.3390/microorganisms9020251
72. Motlagh SR, Elgharbawy AA, Khezri R et al (2021) Ionic liquid-based microwave-assisted extraction of protein from *Nannochloropsis* sp. biomass. Biomass Convers Biorefin 13:8327. https://doi.org/10.1007/s13399-021-01778-2
73. Rodríguez-España M, Mendoza-Sánchez LG, Magallón-Servín P et al (2021) Supercritical fluid extraction of lipids rich in DHA from Schizochytrium sp. J Supercrit Fluids 179:1–7. https://doi.org/10.1016/j.supflu.2021.105391
74. Huynh N, Beltrame G, Tarvainen M et al (2022) Supercritical CO2 extraction of triterpenoids from Chaga sterile conk of Inonotus obliquus. Molecules 27(6):1880. https://doi.org/10.3390/molecules27061880

75. Ruiz-Domínguez MC, Medina E, Salinas F et al (2022) Methodological optimization of supercritical fluid extraction of valuable bioactive compounds from the acidophilic microalga *Coccomyxa onubensis*. Antioxidants 11(7):1248. https://doi.org/10.3390/antiox11071248
76. Espinosa Álvarez C, Vardanega R, Salinas-Fuentes F et al (2020) Effect of CO2 flow rate on the extraction of Astaxanthin and fatty acids from Haematococcus pluvialis using supercritical fluid technology. Molecules 25(24):6044. https://doi.org/10.3390/molecules25246044
77. Georgiopoulou I, Tzima S, Louli V, Magoulas K (2022) Supercritical CO2 extraction of high-added value compounds from Chlorella vulgaris: experimental design, modelling and optimization. Molecules 27(18):5884. https://doi.org/10.3390/molecules27185884
78. Carpentieri S, Soltanipour F, Ferrari G et al (2021) Emerging green techniques for the extraction of antioxidants from Agri-food by-products as promising ingredients for the food industry. Antioxidants 10:42
79. Azmi AAB, Sankaran R, Show PL et al (2020) Current application of electrical pre-treatment for enhanced microalgal biomolecules extraction. Bioresour Technol 302:122874. https://doi.org/10.1016/j.biortech.2020.122874
80. Carullo D, Abera BD, Scognamiglio M et al (2022) Application of pulsed electric fields and high-pressure. Foods 11:1–16
81. Sebastian J, Rouissi T, Brar SK et al (2019) Microwave-assisted extraction of chitosan from Rhizopus oryzae NRRL 1526 biomass. Carbohydr Polym 219:431–440. https://doi.org/10.1016/j.carbpol.2019.05.047
82. Aslanbay Guler B, Saglam-Metiner P, Deniz I et al (2022) Aligned with sustainable development goals: microwave extraction of astaxanthin from wet algae and selective cytotoxic effect of the extract on lung cancer cells. Prep Biochem Biotechnol 53(5):565–571. https://doi.org/10.1080/10826068.2022.2116455
83. Zhang Y, Soldatov S, Papachristou I et al (2022) Pulsed microwave pretreatment of fresh microalgae for enhanced lipid extraction. Energy 248:123555. https://doi.org/10.1016/j.energy.2022.123555
84. Rokicka M, Zieliński M, Dudek M, Dębowski M (2021) Effects of ultrasonic and microwave pretreatment on lipid extraction of microalgae and methane production from the residual extracted biomass. Bioenergy Res 14:752–760. https://doi.org/10.1007/s12155-020-10202-y
85. Baskararaj S, Theivendren P, Palanisamy P et al (2019) Optimization of bioactive compounds extraction assisted by microwave parameters from Kappaphycus alvarezii using RSM and ANFIS modeling. J Food Meas Charact 13:2773–2789. https://doi.org/10.1007/s11694-019-00198-1
86. Polo-Castellano C, Álvarez J, Palma M et al (2022) Optimization through a Box–behnken experimental design of the microwave-assisted extraction of the psychoactive compounds in hallucinogenic fungi (*Psylocibe cubensis*). J Fungi 8:4–15. https://doi.org/10.3390/jof8060598
87. Lee SY, Hu X, Stuckey DC (2022) Optimised "green solvent" extraction of long-chain menaquinones (vitamin K2) from wet Lactococcus lactis biomass. Sep Purif Technol 287:120560. https://doi.org/10.1016/j.seppur.2022.120560
88. Nowosad K, Sujka M, Pankiewicz U, Kowalski R (2021) The application of PEF technology in food processing and human nutrition. J Food Sci Technol 58:397–411. https://doi.org/10.1007/S13197-020-04512-4/TABLES/2
89. Zhang R, Gu X, Xu G, Fu X (2021) Improving the lipid extraction yield from chlorella based on the controllable electroporation of cell membrane by pulsed electric field. Bioresour Technol 330:1–9. https://doi.org/10.1016/j.biortech.2021.124933
90. Aguilar-Machado D, Montañez J, Raso J, Martínez JM (2022) Other applications of pulsed electric fields technology for the food industry. Food engineering services. Springer, pp 439–466. https://doi.org/10.1007/978-3-030-70586-2_15
91. Li Q, Kamal R, Chu Y et al (2020) Automated pressurized liquid extraction of microbial lipids from oleaginous yeasts. Appl Biochem Biotechnol 192:283–295. https://doi.org/10.1007/s12010-020-03331-9

92. Derwenskus F, Metz F, Gille A et al (2019) Pressurized extraction of unsaturated fatty acids and carotenoids from wet Chlorella vulgaris and *Phaeodactylum tricornutum* biomass using subcritical liquids. GCB Bioenergy 11:335–344. https://doi.org/10.1111/gcbb.12563
93. Gallego R, Bueno M, Chourio AM et al (2021) Use of high and ultra-high pressure based-processes for the effective recovery of bioactive compounds from Nannochloropsis oceanica microalgae. J Supercrit Fluids 167:105039. https://doi.org/10.1016/j.supflu.2020.105039
94. He Y, Huang Z, Zhong C et al (2019) Pressurized liquid extraction with ethanol as a green and efficient technology to lipid extraction of Isochrysis biomass. Bioresour Technol 293:122049. https://doi.org/10.1016/j.biortech.2019.122049
95. Pagels F, Pereira RN, Amaro HM et al (2021) Continuous pressurized extraction versus electric fields-assisted extraction of cyanobacterial pigments. J Biotechnol 334:35–42. https://doi.org/10.1016/j.jbiotec.2021.05.004
96. Bueno M, Gallego R, Chourio AM et al (2020) Green ultra-high pressure extraction of bioactive compounds from Haematococcus pluvialis and Porphyridium cruentum microalgae. Innov Food Sci Emerg Technol 66:1–10. https://doi.org/10.1016/j.ifset.2020.102532
97. Torgbo S, Sukatta U, Kamonpatana P, Sukyai P (2022) Ohmic heating extraction and characterization of rambutan (Nephelium lappaceum L.) peel extract with enhanced antioxidant and antifungal activity as a bioactive and functional ingredient in white bread preparation. Food Chem 382:132332. https://doi.org/10.1016/J.FOODCHEM.2022.132332
98. Al-Hilphy AR, Al-Musafer AM, Gavahian M (2020) Pilot-scale ohmic heating-assisted extraction of wheat bran bioactive compounds: effects of the extract on corn oil stability. Food Res Int 137:109649. https://doi.org/10.1016/J.FOODRES.2020.109649
99. Ferreira-Santos P, Nunes R, De Biasio F et al (2020) Influence of thermal and electrical effects of ohmic heating on C-phycocyanin properties and biocompounds recovery from Spirulina platensis. Lwt 128:109491. https://doi.org/10.1016/j.lwt.2020.109491
100. Ferreira-Santos P, Miranda SM, Belo I et al (2021) Sequential multi-stage extraction of biocompounds from Spirulina platensis: combined effect of ohmic heating and enzymatic treatment. Innov Food Sci Emerg Technol 71:1–12. https://doi.org/10.1016/j.ifset.2021.102707
101. Han SF, Jin W, Yang Q et al (2019) Application of pulse electric field pretreatment for enhancing lipid extraction from Chlorella pyrenoidosa grown in wastewater. Renew Energy 133:233–239. https://doi.org/10.1016/j.renene.2018.10.034
102. Zhang R, Marchal L, Lebovka N et al (2020) Two-step procedure for selective recovery of bio-molecules from microalga Nannochloropsis oculata assisted by high voltage electrical discharges. Bioresour Technol 302:1–8. https://doi.org/10.1016/j.biortech.2020.122893
103. Gorte O, Nazarova N, Papachristou I et al (2020) Pulsed electric field treatment promotes lipid extraction on fresh oleaginous yeast Saitozyma podzolica DSM 27192. Front Bioeng Biotechnol 8:1–14. https://doi.org/10.3389/fbioe.2020.575379
104. Aguilar-Machado D, Delso C, Martinez JM et al (2020) Enzymatic processes triggered by PEF for Astaxanthin extraction from Xanthophyllomyces dendrorhous. Front Bioeng Biotechnol 8: 1–12. https://doi.org/10.3389/fbioe.2020.00857

Chapter 7
Emerging Technologies for the Recovery of Microbial Bioactive Compounds

Pragati Srivastava and Hemant Dasila

Abstract For agriculture, industry, and society as a whole, microbially derived bioactive compounds are extremely important. They are frequently used as active components in food additives, medicine, and agriculture. Archaea, bacteria, fungus, protozoa, algae, and viruses are among the few examples of microorganisms. A vast array of distinctive compounds produced by the diversity of microorganisms has emerged as a viable source for cutting-edge biotechnology. Microorganisms are simple to grow and enable a more effective generation of natural bioactive compounds than do plants. In contrast to synthetic bioactive chemicals, most microbial ones are noncytotoxic and nonmutagenic. There is a huge variety of microorganisms, but very few of them have been cultivated and looked at for the generation of secondary metabolites. The pharmaceutical and neutraceutical industries are very interested in the phenols, flavonoids, steroids, and alkaloids that have been discovered in microalgae, bacteria, yeast, and actinomycetes. More extensive research is required in order to better understand and make the most use of these microbial bioctives because their mode of action has not yet been fully clarified. Taking advantage of nature's rich biodiversity, this could also result in the development of new medicines and applications.

7.1 Introduction

The diversity of microscopic microorganisms, including bacteria, protozoa, fungi, archaea, algae, and viruses, provides a range of special bioactive substances of pharmacological value that must be made commercially available for the benefit of mankind. A numeric value of around 23,000 secondary metabolites extracted from

P. Srivastava (✉)
Department of Microbiology, G. B. Pant University of Agriculture & Technology, Pantnagar, India

H. Dasila
Department of Microbiology, Akal College of Basic Sciences, Sirmaur, Himachal Pradesh, India

R. Soni et al. (eds.), *Microbial Bioactive Compounds*,
https://doi.org/10.1007/978-3-031-40082-7_7

microbes are been discovered so far. Among them 42% is contributed by both actinomycetes and fungi and rest of the 16% is bagged by eubacteria [1]. Bioactive compounds comprise of growth hormones, antitumor agents, pigments, antibiotics, etc., which comprehensively participates in the pharmaceuticals industries [2]. Scientists have recently focused their attention on cutting-edge techniques that allow the extraction of biologically active compounds without their degradation. Examples of these techniques include microwave-assisted extraction, enzyme-assisted extraction, supercritical fluid extraction, pressurized liquid extraction, and ultrasound-assisted extraction [3]. These efficient and emerging extraction methods are better than traditional methods in terms of performance, maximum yield, short processing time, and its ecologically sound approach. Earlier traditional methods of extraction include liquid–liquid extraction (LLE), solid–liquid extraction (SLE), and extraction in a Soxhlet system. The major stumbling block in the use of traditional method of extraction is the requirement of large amount of hazardous solvents and lengthy extraction durations [4].

Plant-based bioactive compounds, such as medicinal plants herbs shrubs, fruits, vegetables, cereals, and so forth, have historically provided the industrial market for these substances. However, due to climatic and regional variation, this method has certain drawbacks, due to which it affects the chemical content of plants and the challenge of ensuring the quality of agricultural products [5]. The industrial production of bioactive substances hence requires robust and sustainable manufacturing employing contemporary biotechnological technologies. Since the introduction of cell culture technologies, metabolic engineering, and synthetic biology, the biotechnological generation of bioactive substances has been intensively researched [6]. Large-scale fermentors are used for the optimum production of bioactive compounds. Instead of using wild-type organisms, this approach typically uses target transgenic organism with desired properties under precisely regulated process conditions. Microbes have been drawing more interest among these production hosts due to their unique characteristics in comparison to other organisms or tissues, such as rapid growth, ease of cultivation, and simplicity of genetic manipulation.

The best strategy for enhanced recovery of bioactive compounds will be selection of appropriate extraction technique, solvent type, and microbial class when using microorganisms. Although traditional organic solvent extraction methods are widely used, accessible, and enable quick extraction of bioactive compounds, they should be phased out gradually because they use a lot of solvents and run the risk of thermal denaturation or transformation of compounds of interest [7].

With regard to enhance production of desired bioactive compounds with less or no harmful impact on the target bioactives, novel extraction techniques, such as ultrasonic, enzyme, and microwave aided extraction alone or in combination, have illustrated significant advantages over conventional approaches:

7.2 Microwave-Assisted Extraction (MAE)

In particular, MAE has been used expansively for the extraction of bioactive compounds from plant materials [8, 9]. Two optimum frequencies are being utilized 915 MHz and 2450 MHz for ignition in both industrial and residential settings. Microwave has an electromagnetic property. According to Kaderides et al. [10], the heating effect produced by the microwave is the key mechanism of MAE, which enables higher temperature for extraction and a faster mass transfer rate. Microwave have the tendency to penetrate into the material into certain depths and cause interaction with the polar constituent in it that result in direct heating or bulk heating inside the solvent body and the sample matrix [11]. Figure 7.1 depicts the closed type microwave system.

Microwave-assisted extraction is functional only within a closed or open loop, so as the closed or open system's pressure rising above or remaining below atmospheric pressure, respectively. The schematic designs of a closed system with homogenous radiation are shown in Fig. 7.1 along with those of an open system with focused radiation. If temperature and pressure control systems are present in the equipment, the closed MAE system may also perform extraction under controlled temperature and pressure in a sealed vessel with uniform microwave heating. Due to the extraction solvent's boiling point being raised by the increased pressure in the closed vessel, it can reach higher working temperatures than the open system [12]. Despite the extreme pressure and temperature in the reactor that lead to efficient and fast extraction with less solvent consumption, they also increase the safety risks and equipment control requirement. An advance green process through microwave-assisted extraction of bioactive metabolites from *Arthrospira Platensis* (cyanobacteria) and evaluation of its bioactivity was conducted. Numerous microbiota-derived compounds have been synthesized by microorganisms and one of the important one is bioactive peptides. These peptides are known for their regulating cell cycle and cell signaling. It also plays an important role in maintaining hypertension, hyperglycemia, and damaged proteins. There are other bioactive compounds that are secreted by microorganisms which have same potential in maintaining other important biochemical aspects [13]. Extraction of these bioactive compounds is necessary, and some of the important methods include.

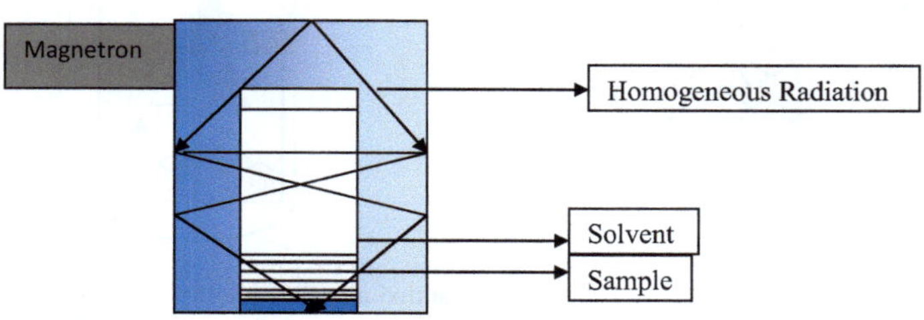

Fig. 7.1 Closed-type microwave system

7.3 Ultrasound-Assisted Extraction Method

Ultrasound extraction method of bioactive compounds is being performed at both industrial and laboratory scale. Ultrasonic wave causes cavitations in particle that arises due to collision, disturbance, and disruption of atoms present in the particle. Cavitations cause pores in the material results in mass transfer rate and solvent into biomass [14]. The big advantage of using ultrasound mediated extraction is that it can be set up with different configuration depending upon the nature of extraction. Ultrasound extraction method can be used with a variety of solvents like ethanol, water, acetone, methanol, ethyl acetate, and ethanol, but critical things are to be carried out by ultrasound-mediated extraction at lower temperature, which is necessary for maintaining the integrity of thermosensitive compounds [15]. Keeping this in mind, the extraction of gallic and ergatsterol form *Agaricus bisporus* was done with the cavitations method in which ethanol was being used as solvents [16]. Figure 7.2 demonstrates the principle of UAE.

7.4 Enzyme-Mediated Extraction

Enzyme-mediated extraction is useful for extracting phyto-chemicals from their respective associated cell wall. Presence of hemicelluloses, cellulose, and lignin in higher concentration makes difficult for conventional-based extraction method for extracting bioactive compounds [17]. Enzyme-mediated extraction provides an

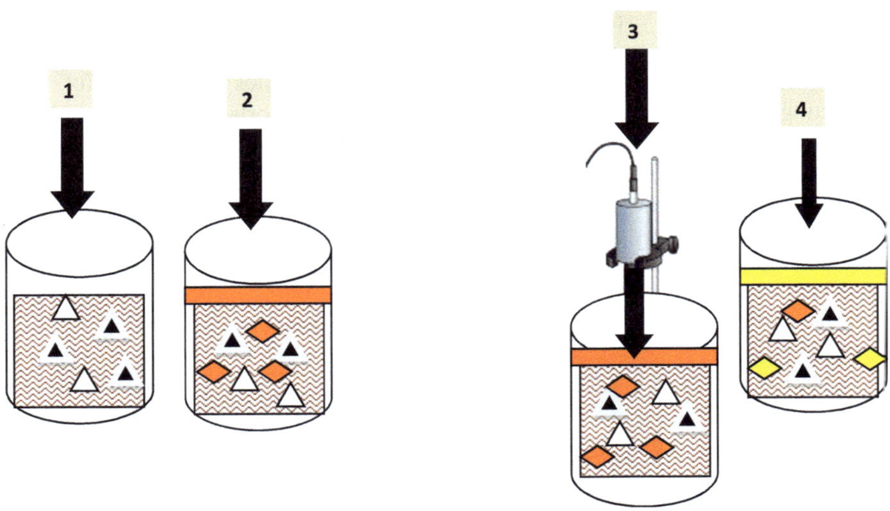

Fig. 7.2 Principle of the UAE. (1) The sample is added into the cylindrical vessel. (2) Addition of solvent for extraction. (3) The sample is brought into touch with the sonotrode. Sample sonication takes place. (4) The sample is prepared for cleanup

Table 7.1 List of bioactive compounds extracted from microorganisms, plants, and algae via enzyme-mediated extraction method

S. No.	Source	Bioactive compound	Enzyme	Reference
1.	*C. annuum baydgi*	Carotenoids and capsaicinoids	Cellulase, hemicellulase, pectinase	Salgado-Roman et al. [18]
2.	*Codium fragile*	Uronic acid, sugars, sulfates, and proteins	Cellulase, beta-glucanase, Ultraflo, Neutrase	Kulshreshtha et al. [19]
3.	*Fucus distichus*	Fucoidans	Alginate lyase	Nguyen et al. [20]
4.	*Helianthus annuus L.*	Carotenoids	Viscozyme	Ricarte et al. [21]
5.	*Haematococcus pluvialis*	Astaxanthin	Flavourzyme	Poojary et al. [22]

alternative method to it by employing enzymes like alpha-amylase, cellulose, hemicellulose, and pectinase, which are known to be involved in the digestion of cell wall [11]. Enzyme-mediated extraction offers several advantages like nature friendly and lower consumption of energy as compared to other techniques. Reduction of toxic solvents and efficient extraction of volatile and thermal compounds is also one of its bid advantage over conventional methods (Table 7.1). Extracted volatile compounds can be used in providing flavors, fragrnance, and pigments [17]. However, one disadvantage of using enzyme-mediated extraction is using larger volume of substances as it can be highly expensive [23].

7.5 Pressurized Liquid Extraction

According to Nieto et al. [24], PLE is an extraction method that inculcates expelling of analytes from solid matrixes by giving high temperatures (T_{extr}) and pressures (P_{extr}), typically up to 200 °C and over 200 bar, respectively, without reaching the critical point using liquid solvents [25]. These optimal conditions provide improved matrix kinetics by increasing solubility and mass transfer rates, which in turn increase solvent diffusivity [26]. The solvent and the sample are simultaneously administered into the extraction cell. The extraction cell comprises an oven chamber and a pressure valve, which altogether contributes in attaining suitable temperature and pressure in order to extract the compound present in the sample. Then, the extracted compound is cooled and collected in a moving steel chamber. Li et al. [27] depicted the automated pressurized liquid extraction (APLE) method for lipid extraction from dried cells of the oleaginous yeast species *Rhodosporidium toruloides* and *Cryptococcus curvatus*.

7.6 Application

Plant-derived polyphenols such as stilbenes and curcuminoid comprising one or more hydroxyl group acquired from phenylalanine or tyrosine aromatic amino acids are well popularized in food and cosmetic industries. Properties such as antioxidant and anti-inflammatory make them suitable for the prevention of heart disruption and cancer. Genetically engineered bacterium such as *E.coli and* yeast *Saccharomyces cerevisiae* are used for the production of polyphenols commercially because of their ease of cultivation in laboratory under modulated fermentation condition. At present, the use of novel extraction strategies for the extraction of polyphenol bioactives from microbes is in trend, and many new complex structures and biosynthetic metabolic pathways have been elucidated [28, 29].

Table 7.2 lists the biologically active compounds retrained from microbes.

Proteins are considered as essential component in maintaining good health by its consumption, and amino acids are the basic constituents' building blocks for its activity. Leucine, isoleucine, valine, threonine, lysine, methionine, phenyl- alanine, and tryptophan are among the eight amino acids essential required by the human body. All amino acids are commercially produced except for glycine, methionine, and aspartate. *C. glutamicum is* the most prominent strain for the production of amino acids and generally regarded as safe due to its resistant capacity against phage infection. Also *E.coli* is used for the industrial production of amino acids because it is stable at high fermentation temperatures [39, 40]. Also, vitamins which are

Table 7.2 The biologically active compounds retrained from microbes

Bioactive Compounds	Examples	References
1. Polysaccharides	Alginate, cellulose, fucoidan, laminarin, agar, carragenan, furcellaran, mannan, porphyrin, xylan, amylase, amylo-pectin, pectin, xylan, cellulose	Ghosh et al. [30]
2. Lipids/fatty acids	PUFAs, omega3 fatty acids: eicosapentaenoic acid, decosahexanoic acid; omega 6 fatty acids: γ linolenic acid and arachidonic acid	Priyadarshani and Rath [31]
3. Poly-phenols	Polyphenols, with flavonoids, stilbenes, and curcuminoids	Dudnik et al. [32]
4. Pigments	Phycocyanin, phycoerythrin, carotenoids, catotenes: γ car-otene and β carotene, lycopene, xanthophylls, chlorophyll, etc.	García-López et al. [33]
5. Antioxidants	Tocopherol, mycosporine-like amino acids	Young and Lowe [34]
6. Proteins and amino acids	Spirulina	Wan et al. [35]
7. Minerals	Ca, Mg, Zn, CO, Na, I, B	Hou [36]
8. Hormones	Auxins, gibberellins, ethylene, cytokinins	Mazzoli et al. [37]
9. Vitamins	B_{12}, K, C, E, D, A	Watanabe and Bito [38]

majorly required in the human body that too in very minute quantities are not synthesized inside the human body so required uptake externally. Vitamin B_{12} cobalamine, vitamin B_2 riboflavin, vitamin C ascorbic acid, and β carotene are produced industrially via chemical transformation reaction [41, 42] .

7.7 Conclusion

The capacity to produce bioactive compounds by microbes via different novel extraction techniques is discussed in this review, along with their significance as cutting-edge sources of naturally occurring bioactive compounds. Microorganisms have simple growth requirements and are easily cultivated under lab condition to produce desired bioactive compound with enhanced productivity via modulating the temperature, pressure, volume, and other necessary constituents. Compared to synthetic bioactive compounds such as antioxidants, the majority of microbial antioxidants are nonmutagenic and noncytotoxic. There is an immense heterogeneity in the microbial population, but its potential of producing bioactive compounds for many microbes are yet to be discovered that may have high biotechnological and pharmaceutical significance in the present era.

The pharmaceutical and nutraceutical sectors are becoming increasingly interested in the production of flavonoids, phenols, alkaloids, vitamins, and steroids using microorganisms including microalgae, bacteria, yeast, actinomycetes, and mushrooms. A few fungi-derived substances with promising antioxidant activity include isopestacin, astaxanthin, pestacin, and polysaccharides. These substances are employed as functional ingredients in food, nutraceutical products, cosmetics, and pharmaceuticals. In-depth investigation and research are required to increase the productivity of popularizing bioactive compounds from fungus.

In particular, *Saccharomyces cerevisiae* is the primary species used in the generation of bioactive substances. Long back yeast is considered as a good food supplement in food industries and is well employed for the production of beverages. Torularhodin, peroxiredoxins, and thioredoxins carotenoids that are gaining attention for its multifunctional role as antioxidant, anticancer, and antimicrobial property opens new possibilities for the discovery of new drugs having high economic importance. More thorough research is required to advance knowledge and maximizes the usage of these microbial bio-active because their mode of action has not been fully investigated. The creation of novel medications and applications that make use of nature's rich biodiversity may result from this as well.

References

1. de Castro I, Mendo S, Caetano T (2020) Antibiotics from haloarchaea: what can we learn from comparative genomics? Mar Biotechnol 22(2):308–316

2. Demain AL (2014) Importance of microbial natural products and the need to revitalize their discovery. J Ind Microbiol Biotechnol 41(2):185–201
3. Michalak I, Chojnacka K (2015) Algae as production systems of bioactive compounds. Eng Life Sci 15(2):160–176
4. Kadam SU, Tiwari BK, O'Donnell CP (2015) Extraction, structure and biofunctional activities of laminarin from brown algae. Int J Food Sci Technol 50(1):24–31
5. Zhu L, Huang Y, Zhang Y, Xu C, Lu J, Wang Y (2017) The growing season impacts the accumulation and composition of flavonoids in grape skins in two-crop-a-year viticulture. J Food Sci Technol 54:2861–2870
6. Pandey RP, Parajuli P, Koffas MA, Sohng JK (2016) Microbial production of natural and non-natural flavonoids: pathway engineering, directed evolution and systems/synthetic biology. Biotechnol Adv 34(5):634–662
7. Łubek-Nguyen A, Ziemichód W, Olech M (2022) Application of enzyme-assisted extraction for the recovery of natural bioactive compounds for nutraceutical and pharmaceutical applications. Appl Sci 12(7):3232
8. Pimentel-Moral S, Borras-Linares I, Lozano-Sanchez J, ArraezRoman D, Martınez-Ferez A, Segura-Carretero A (2018) Microwave-assisted extraction for Hibiscus sabdariffa bioactive compounds. J Pharm Biomed Anal 156:313–322. https://doi.org/10.1016/j.jpba.2018.04.050
9. Rodsamran P, Sothornvit R (2019) Extraction of phenolic compounds from lime peel waste using ultrasonic-assisted and microwave-assisted extractions. Food Biosci 28:66–73
10. Kaderides K, Papaoikonomou L, Serafim M, Goula AM (2019) Microwave-assisted extraction of phenolics from pomegranate peels: optimization, kinetics, and comparison with ultrasounds extraction. Chem Eng Process: Process Intensif 137:1–11
11. Azmir J, Sarker MZ, Rahman M, Khan MS, Awang M, Ferdosh S, Jahurul MHA, Ghafoor K, Norulaini NAN, Omar AKM (2013) Techniques for extraction of bioactive compounds from plant materials: a review. J Food Eng 117(4):426–436. https://doi.org/10.1016/j.jfoodeng.2013.01.014
12. Vinatoru M, Mason TJ, Calinescu I (2017) Ultrasonically assisted extraction (UAE) and microwave assisted extraction (MAE) of functional compounds from plant materials. TrAC Trends Anal Chem 97:159–178
13. Laroute V, Yasaro C, Narin W, Mazzoli R, Pessione E, Cocaign-Bousquet M et al (2016) GABA production in Lactococcus lactis is enhanced by arginine and co-addition of malate. Front Microbiol 7:1050. https://doi.org/10.3389/fmicb.2016.01050
14. Gonzalez M, Barrios S, Budelli E, Pérez N, Lema P, Heinzen H (2020) Ultrasound assisted extraction of bioactive compounds in fresh and freeze-dried Vitis vinifera cv Tannat grape pomace. Food Bioproduct Process 124:378–386. https://doi.org/10.1016/j.fbp.2020.09.012
15. Roohinejad S, Nikmaram N, Brahim M, Koubaa M, Khelfa A, Greiner R (2017) Potential of novel technologies for aqueous extraction of plant bioactives. In: Water extraction of bioactive compounds: from plants to drug development. Elsevier, pp 399–419. https://doi.org/10.1016/B978-0-12-809380-1.00016-4
16. Heleno SA, Diz P, Prieto MA, Barros L, Rodrigues A, Barreiro MF, Ferreira IC (2016) Optimization of ultrasound-assisted extraction to obtain mycosterols from Agaricus bisporus L. by response surface methodology and comparison with conventional Soxhlet extraction. Food Chem 197(B):1054–1063. https://doi.org/10.1016/j.foodchem.2015.11.108
17. Nadar SS, Rao P, Rathod VK (2018) Enzyme assisted extraction of biomolecules as an approach to novel extraction technology: a review. Food Res Int Elsevier Ltd 108:309–330. https://doi.org/10.1016/j.foodres.2018.03.006
18. Salgado-Roman M, Botello-Álvarez E, Rico-Martínez R, Jiménez-Islas H, Cárdenas-Manríquez M, Navarrete-Bolaños JL (2008) Enzymatic treatment to improve extraction of capsaicinoids and carotenoids from chili (Capsicum annuum) fruits. J Agric Food Chem 56(21):10012–10018. https://doi.org/10.1021/jf801823m
19. Kulshreshtha G, Burlot AS, Marty C, Critchley A, Hafting J, Bedoux G, Bourgougnon N, Prithiviraj B (2015) Enzyme-assisted extraction of bioactive material from Chondrus crispus

and Codium fragile and its effect on herpes simplex virus (HSV-1). Mar Drugs 13(1):558–580. https://doi.org/10.3390/md13010558
20. Nguyen TT, Mikkelsen MD, Tran V, Trang V, Rhein-Knudsen N, Holck J, Rasin AB, Cao H, Van T, Meyer AS (2020) Enzyme-assisted fucoidan extraction from brown macroalgae Fucus distichus subsp. evanescens and Saccharina latissima. Mar Drugs 18(6):296. https://doi.org/10.3390/md18060296
21. Ricarte GN, Coelho M, Marrucho IM, Ribeiro BD (2020) Enzyme-assisted extraction of carotenoids and phenolic compounds from sunflower wastes using green solvents. 3 Biotech 10(9):405. https://doi.org/10.1007/s13205-020-02393-0
22. Poojary MM, Orlien V, Passamonti P, Olsen K (2017) Enzyme-assisted extraction enhancing the umami taste amino acids recovery from several cultivated mushrooms. Food Chem 234: 236–244. https://doi.org/10.1016/j.foodchem.2017.04.157
23. Franco D, Munekata P, Agregan R, Bermudez R, Lopez-Pedrouso M, Pateiro M, Lorenzo JM (2020) Application of pulsed electric fields for obtaining antioxidant extracts from fish residues. Antioxidants 9(2):90. https://doi.org/10.3390/antiox9020090
24. Nieto A, Borrull F, Pocurull E, Marcé RM (2010) Pressurized liquid extraction: a useful technique to extract pharmaceuticals and personal-care products from sewage sludge. TrAC Trends Anal Chem 29(7):752–764
25. Perez-Vazquez A, Carpena M, Barciela P, Cassani L, Simal-Gandara J, Prieto MA (2023) Pressurized liquid extraction for the recovery of bioactive compounds from seaweeds for food industry application: a review. Antioxidants 12(3):612
26. Mena-García A, Ruiz-Matute AI, Soria AC, Sanz ML (2019) Green techniques for extraction of bioactive carbohydrates. TrAC Trends Anal Chem 119:115612
27. Li Q, Kamal R, Chu Y, Wang Q, Yu X, Huang Q (2020) Automated pressurized liquid extraction of microbial lipids from oleaginous yeasts. Appl Biochem Biotechnol 192:283–295. https://doi.org/10.1007/s12010-020-03331-9
28. Fang Z, Jones JA, Zhou J, Koffas MA (2018) Engineering Escherichia coli co-cultures for production of curcuminoids from glucose. Biotechnol J 13(5):1700576
29. Jones JA, Vernacchio VR, Collins SM, Shirke AN, Xiu Y, Englaender JA et al (2017) Complete biosynthesis of anthocyanins using E. coli polycultures. MBio 8(3):e00621–e00617
30. Ghosh S, Sarkar T, Pati S, Kari ZA, Edinur HA, Chakraborty R (2022) Novel bioactive compounds from marine sources as a tool for functional food development. Front Mar Sci 9:76
31. Priyadarshani I, Rath B (2012) Commercial and industrial applications of micro algae–a review. J Algal Biomass Util 3(4):89–100
32. Dudnik A, Gaspar P, Neves AR, Forster J (2018) Engineering of microbial cell factories for the production of plant polyphenols with health-beneficial properties. Curr Pharm Des 24(19): 2208–2225
33. García-López DA, Olguín EJ, González-Portela RE, Sánchez-Galván G, De Philippis R, Lovitt RW et al (2020) A novel two-phase bioprocess for the production of Arthrospira (spirulina) maxima LJGR1 at pilot plant scale during different seasons and for phycocyanin induction under controlled conditions. Bioresour Technol 298:122548
34. Young AJ, Lowe GL (2018) Carotenoids—antioxidant properties. Antioxidants 7(2):28
35. Wan D, Wu Q, Kuča K (2021) Spirulina. In: Nutraceuticals. Academic Press, pp 959–974
36. Hou T (2022) Bioactive compounds in mineral bioavailability: activities, structures, and mechanisms. Front Nut 9:1050670
37. Mazzoli R, Riedel K, Pessione E (2017) Bioactive compounds from microbes. Front Microbiol 8:392
38. Watanabe F, Bito T (2018) Vitamin B12 sources and microbial interaction. Exp Biol Med 243(2):148–158
39. Wu J, Liu Y, Zhao S, Sun J, Jin Z, Zhang D (2019) Application of dynamic regulation to increase L-phenylalanine production in Escherichia coli. J Microbiol Biotechnol 29:923–932
40. Wang J, Ma W, Zhou J, Wang X (2023) Microbial chassis design and engineering for production of amino acids used in food industry. Syst Microbiol Biomanuf 3(1):28–48

41. Acevedo-Rocha CG, Gronenberg LS, Mack M, Commichau FM, Genee HJ (2019) Microbial cell factories for the sustainable manufacturing of B vitamins. Curr Opin Biotechnol 56:18–29
42. Arun KB, Anoopkumar AN, Sindhu R, Binod P, Aneesh EM, Madhavan A, Awasthi MK (2023) Synthetic biology for sustainable food ingredients production: recent trends. Syst Microbiol Biomanuf 3(1):137–149

Chapter 8
Nanocarriers: Potential Vehicles for Managed Delivery of Bioactive Compounds in Therapeutics

Ashfaq Ahmad Shah and Amit Gupta

Abstract In recent years, the medical industry has used nanomaterials extensively in large part due to the fast advancement of nanotechnology. They are ideally suited for the target-specific and precisely controlled delivery of micro- and macromolecules in disease therapy due to their unique physical and chemical properties, which include minimized size, customized surface features, robust interactions with ligands, high carrier ability, and ease of adhering with both hydrophilic and hydrophobic substances. They have also shown outstanding promise pertaining to clinical applications with the goal of fine-tuning bioavailability, bio efficacy, and pharmacokinetics. The primary challenges in therapeutics include absorption, post-administration stability, and bioavailability of drugs and other bioactive compounds. Some crucial medications have limited gastrointestinal absorption and permeability in their bioactive state, get sometimes inactivated by pH and temperature changes, and produce disastrous off-target and unwanted side effects. Certain studies have also found that active efflux systems impact the assimilation of some currently integrated substances by causing structural changes across the gut wall. Furthermore, gut bacteria and/or enzymes degrade the fragile components of active chemicals into a range of metabolites, each with a distinct bioactivity from the original chemical molecule. By virtue of nanocarrier-mediated dispersion, their solubilization potential improved, absorption pathways altered, and metabolic breakdown by gut bacteria and enzymes substantially decreased. Combining nanobiotechnology with existing therapeutic procedures has proven beneficial in bringing novel and previously rejected bioactive compounds to the market to treat an extensive range of illnesses and disorders. As a result, we anticipate that nanotechnology will play a bigger role in disease diagnosis and treatment in years to come, perhaps assisting in the resolution of obstacles in present medical procedures. This chapter provides a thorough examination of the techniques and applications of nanoengineered delivery systems, as well as the pharmacokinetic features and drug-delivery mechanisms of

A. A. Shah · A. Gupta (✉)
Department of Microbiology, Graphic Era (Deemed to be) University, Dehradun, Uttarakhand, India
e-mail: dr.amitgupta.bt@geu.ac.in

R. Soni et al. (eds.), *Microbial Bioactive Compounds*,
https://doi.org/10.1007/978-3-031-40082-7_8

these nanocarriers. The potential downsides, problems, future improvements, and applications of nanocarriers in clinical care are addressed as well.

Keywords Nanocarriers · Controlled release · Nanoparticles · Polymeric micelles · Nanocapsules · Dendrimers · Quantum dots

8.1 Introduction

Despite the fact that drugs, medicines, and novel bioactive molecules of synthetic or natural origin are beneficial to human health in diseased conditions, the main issues related to these disease-modifying agents of synthetic or natural origin are related to their post-administration instability, off-target, and unwanted side effects. Any disease-modifying agent's therapeutic outcome is determined by how well its pharmacokinetic profile improves after therapy. In the present era, nanotechnology is addressing all of these challenges by integrating both biological and physical sciences through the application of nanostructures and nanophases in a wide range of fields of research, including nanomedicine [1, 2]. Nanomedicine and nano-delivery platforms are a new but fast-expanding field in which tiny materials are used as diagnostic instruments or to administer therapeutic medications to specified domains in a regulated way. Nanotechnology has proven to offer numerous advantages in the treatment of chronic human diseases since it allows for the precise delivery of medications to designated locations. Medication delivery, chemotherapy, biosensors, and tissue engineering are all using nanoparticles right now in biomedicine. It also covers the use of nano-dimensional materials in live cells, e.g., nano-robots, nano-sensors for diagnostic and sensory applications, and actuation of materials in living cells. Nanomaterials with dimensions that range from one to one hundred nanometers are influencing the cutting-edge fields of nanomedicine, from biological sensors to microfluidics, microarray screening to tissue engineering, and drug encapsulation to drug delivery [2, 3]. In nanotechnology, to produce nanomedicines, a curative agent at the nanoscale level has been utilized. Nanoparticles typically refer to small nanospheres made of atomic or molecularly produced materials [3, 4]. Because of nanoscale structural dimensions, they may travel more easily compared to sized materials, across the human body and can easily permeate through the tissue system thereby allowing for facile drug absorption by cells to bring out a significant activity at the desired area. The structural, mechanical, chemical, magnetic, biological, and electrical characteristics of nanoscale particles are all different. Due to their capacity to encapsulate medications or bind bioactive molecules to nanostructures and then administer them to particular tissues more precisely and in a regulated manner, nanomedicines have garnered a lot of scrutiny recently. Furthermore, nanostructures also facilitate the transport of water-insoluble medications to their target area, as well as reducing drug decomposition in the gastrointestinal tract. Because the nanostructures have standard absorptive endocytosis absorption mechanisms as well as enhanced oral bioavailability, they are substantially more rapidly absorbed and assimilated by body cells. Nanostructures last a long time in the circulatory system, allowing combination drugs to be administered at precisely the right amount. As a result, they have fewer negative effects and cause fewer plasma fluctuations [4, 5]. The positive zeta potential, as well as hydrophobicity of these nanoscopic particles, aids their absorption from the gastrointestinal tract. Other mechanisms found to be supportive in enhancing the absorption of bioactive molecules include electrostatic communication between positively charged nano-vehicle surfaces and negatively charged mucin, amplified transcytosis, and receptor-mediated endocytosis, interaction with junction proteins to modulate tight junctions, the microfold cells mediated phagocytosis of nanoparticles, and chylomicron-aided enterocytes uptake intervened by lipases for lipid-based nanocarriers. Also, there are many processes by which these nanosystems release their protected bioactive components once, inside the body e.g., desorption of adsorbed/surface-bound various components, matrix erosion, enzymatic degradation, matrix diffusion, dissolution, or a combination of either of the processes [5, 6].

The creation of an assortment of nanosized holders in the size range of 10 to 100 nm is a potential direction for nanotechnology in therapeutics. The particle size, surface properties, and shape of the nanoparticles all have an impact on how well-loaded molecules are absorbed across the gastrointestinal mucosa. Polymer nanocarriers (nanocapsules, polymeric micelles, etc.), molecular complexes (cyclodextrins inclusion complexes and ohytosomes), lipid-containing nanocarriers (nanostructured lipid capsules (NLC), solid-lipid nanocapsules (SLN), lipid nanospheres, micro-and nanoemulsions, micelles) are all examples of nanosystems for enhanced, controlled, and site-specific delivery of bioactive compounds [6, 7]. Because of the hydrophobicity as well as hydrophilicity inside the polymeric system, polymer-based nanocarriers may accept a wide range of medicinal compounds. Natural or manufactured biodegradable polymers are commonly used as the carrier material in polymer-based nanocarriers. Natural ones either polysaccharides or proteins are recommended since they have a lower level of toxicity. Plant-based polysaccharides such as pectin, gum arabic, alginate, starch, and its derivatives, cellulose and its derivatives, and animal-based polysaccharides such as xanthan gum, chitosan, etc. are used to formulate polymeric nanoparticles. Polyglycolic acid, polylactic acid, poly-cyanoacrylate alkyl esters, polyvinyl alcohol, polylactic-glycolic acid, etc. are examples of synthetic polymers. Polysaccharide nanoparticles having distinct characteristics are considered ideal carriers for hydrophilic drug delivery. Polysaccharide nanoparticles are natural biomaterials that are biodegradable, safe, nontoxic, and stable under various conditions. Polysaccharides are plentiful in nature and need little processing costs. In polymeric nanoparticles, bioactive medicines and plant secondary metabolites are embedded, dissolved, or adsorbed [7, 8]. The nanocarriers release drug molecules ensuring that none of the molecules are released until they reach systemic circulation, bypassing different physiological obstacles that may obstruct drug metabolism. After reaching the apical membrane of intestinal epithelial cells, the majority of the loaded nanoparticles take entry into enterocytes via transcellular transport. Small particles, e.g., 100 to 400 nm, are ingested by enterocytes through clathrin- and caveolae-mediated endocytosis apart from taken up by specialized Peyer's patches (M cells) and GALT follicles in the gastrointestinal tract. The endolysosomal breakdown is prevented by covering nanoparticles with cationic chitosan. To guarantee improved medication absorption, nanocarrier micelles modify membrane permeability and mucoadhesion inside the GI tract [8, 9].

8.2 Nanomaterials as Nanocarriers

Owing to the brutal microenvironment of sick tissues of the living system, insufficient affordability of the necessary dosage, adverse reactions, low therapeutic indices, and nonspecific targeting, many stalemates have been seen for traditional site-specific drug delivery systems [10]. But nanotechnology offers a smart solution by devising a nano-sized system, in which encapsulated drug is provided through a nano-vehicle to the targeted site with reasonable biocompatibility. It is being protected from degradation in a hostile physiological environment with low toxicity to healthy cells. In order to qualify as an ideal nanocarrier, it should possess certain ideal characteristics like suitable preparation and purification methods, nonreactivity with other drugs, requisite mechanical strength, stability, particle size, shape, high drug payload, encapsulation efficiency, enhanced in vivo residence time, good biocompatibility, and low toxicity [8, 11]. Nanomaterials being used as nanocarriers could be of natural or artificial origin. Some of the representative nanocarriers of natural origin are:

8.2.1 Chitosan

Due to excellent mucoadhesive properties, chitosan-containing nanomaterials are frequently used as site-specific drug delivery vehicles for various types of epithelia, such as buccal [2], nasal [3], eye [4], and others [5, 6]. Chitosan, alginate, and pectin-based nanocarriers have been screened for the oral administration of the drug for the oral cavity [7], whereas carboxymethyl chitosan is used for intra-nasal carbamazepine (CBZ) release, for the increased amount of the medication in the brain to reduce the systemic drug exposure [8].

8.2.2 Cellulose

Exploiting the presence of a number of hydroxyl groups protruding out of the cellulose structure to form the hydrogen bonds in the cellulose nanocrystals and the bioactive drug, showed promising results in controlled release of the repaglinide (an anti-hyperglycemic—RPG) [9–12]. Four derivatives of cellulose such as methylcellulose, cationic hydroxyethyl cellulose, hydroxypropyl methylcellulose, and sodium carboxymethylcellulose have been screened for the release of the controlled drug into the nasal mucosa, and none of them showed detrimental impacts on either tissues or cells [11–13].

8.2.3 *Liposomes*

Liposomes are basically vesicles of globular form which are made up of steroids and phospholipids. One of the most extensively researched drug transport nano-vehicle platforms imparts therapeutic chemicals with the requisite longevity for enhanced distribution and bioavailability. It is being used along with hydrophilic and hydrophobic bioactive drugs. The biocompatibility and biodegradability are added features of these classes of nano-vehicles. The membrane architecture of the carriers, which is remarkably comparable to cell membranes and aids in the integration of pharmaceuticals into them, is responsible for their versatility [11, 12].

8.2.4 *Alginate*

Simultaneous lowering of serum glucose levels along with enhanced serum insulin levels has been observed in diabetic rats due to the administration of insulin-appended alginate nanoparticles in which nicotinamide is used as a permeation agent [13, 14]. Apart from this, alginate-based nanocarriers are being used for the release of a drug called venlafaxine (VLF) through the intranasal route for treating depression [14], loading cisplatin (carcinogen drug) aiming the nonsmall lung cancerous cells [15]. Synthesis of alginate nanoparticles having chitosan coating to increase the daptomycin permeation in the ocular epithelium intended to achieve an antibacterial effect [16].

The nanomaterials of synthetic origin are also reported as nano-carriers for site-specific drug delivery. These include

8.2.5 *Carbon-Based Nanomaterials (CBNs)*

CBNs are widely used as nanocarriers owing to their distinct physicochemical properties and structural dimensions [17, 18]. A family of graphene, carbon nanotubes, fullerenes, mesoporous carbon, and nano-diamonds were employed for the targeted delivery of drugs for malignant cancer [19, 20]. These scaffolds are also utilized for cell culture growth as well as diagnostic devices [21] for in vivo as well as in vitro tumor imaging along with cellular dynamics [22]. A dual-ligand-functionalized nano-diamond, such as cetuximab-NDs-cisplatin bio-conjugated material, has been found to suppress the growth of HepG2 cells [23].

8.2.6 Metal–Organic Framework (MOFs)

MOFs are a class of molecules successfully employed as a nanocarrier for steady drug release of anticancer drugs [24], metabolic labeling of molecules [25], antimicrobial agents [26], hormones [27], and antiglaucoma medications [28]. ZIFs (Zeolite Imidazolate Frameworks) – a subcategory of MOFs which have been successfully employed for drug delivery [29]. ZIF-8 is one of the members of this family used for a system wherein pH-dependent drug release is warranted. This unique mechanism is owed to its property of being unstable under acidic conditions [30]. Similarly, ZIF-8 has been practiced for the delivery of 5-fluorouracil (5-FU) [31], camptothecin [32], and ceftazidime [33] as a pH-responsive nano-vehicle for the targeted, controlled, nontoxic drug delivery. The other members of the MOF family are also effective and were utilized for transporting the drug at the required site, such as MIL-101 [29, 34], NU-1000 [35], UiO-68 [36], and PCN-333 [37].

8.2.7 Nanocapsules

Polymeric nanocapsules have attracted the interest of the scientific fraternity for drug delivery, owing to the unique core–shell microstructure. Nanocapsules consist of vesicular systems with an inner liquid core composed of an oily or an aqueous core containing a thin wall of polymer. This offers an increased drug-loading capacity simultaneously reducing the polymeric matrix of the nanoparticle [38]. The encapsulated drug is protected from the hostile tissue environment along with untimely burst release of the drug due to pH, enzymes, temperature, and various other reasons due to the thin polymeric shell. The versatility of the nanocapsules lies in their ability to tailor the size, shape, and quantity as well. They can be made to suit the desired complex application with specific biochemical, optical, magnetic, and electrical properties, overcoming physiological barriers and the hostile environment of the diseased tissues [39]. Nanocapsules are employed not only to encapsulate proteins, peptides, hormones, metabolites, etc. but also for a variety of biomedical applications such as, anticancer therapy, immunotherapy, and anti-inflammation therapy [40–42]. Nanocapsules are among the most sought-after candidates for drug delivery due to the enhanced permeability and solubility of the drug. The enhanced solubility is attributed to the very high surface-to-volume ratio [43]. Nanocapsules were reported to utilize the targeted delivery of protein for treating tumor cells [44].

8.2.8 Dendrimers

Dendrimers were first introduced in 1985 [45, 46], since then it has attracted the interest of the scientific community. A well-defined size and globular shape with a hyperbranch offer a unique possibility of tethering a drug molecule and guided delivery of the same at the desired site. A polymerization approach has been adopted in order to construct dendrimers. Monomers are used as building blocks. The tribody structure of dendrimer consists of a core, branches resembling trees, protruding out the core, and the terminal functional group. This class of molecules is either been synthesized by a divergent or convergent approach [47]. PAMAM-type dendrimers having carboxyl or hydroxyl functional groups are used for ocular delivery of bioactive drugs, as a result of its enhanced retention of pilocarpine within the eyes [48]. A multifold cytotoxicity enhancement has been observed for the conjugates paclitaxel with PAMAM G4 dendrimer having hydroxyl groups at the terminus and bis (PEG) polymer. Paclitaxel is known for poor solubility, but in this case, not only cytotoxicity but also solubility has been found to be increased [49, 50]. An investigation has been carried out on doxycycline-conjugated PAMAM dendrimer for the potential cellular binding, migration of T47D breast cancer cells and BT-549-Luc, and cytotoxicity [51]. Berberine, a cyclic natural alkaloid containing nitrogen, is poorly studied due to its wretched pharmacokinetic behavior. The drastic boost in anticancer activity against MCF-7 and MDA-MB-468 breast cancer cells has been observed when berberine was conjugated with G4 PAMAM [52]. These conjugates are safe and biocompatible as well. The other applications of dendrimer-based nanocarrier include enhanced permeation of dendrimers for passive targeting of drugs to tumor tissues [53].

8.2.9 Nanogels

Nanogels have great potential as a promising system for carrying an active ingredient for site-specific control release. These nanogels are three-dimensional cross-linked polymeric networks with high drug encapsulation capacity owing to their hydrophilic or amphiphilic macromolecular chains which are able to hold a great amount of water and swell, keeping their original structure intact [54]. The water incorporating capacity by the nanogels can be attributed to the presence of various hydrophilic moieties along the chain, such as $-CONH_2-$, $-OH$, $-SO_3H$, and $-CONH-$. This ability of nanogels helps in diffusing and exchanging the metabolic and biomolecules in the tissue fluids and organs to maintain the biochemical balance [55]. Nanogels have been reported to tether to the desired active biomolecule, which upon complexation with nucleic acid/genetic materials, such as lipoplexes or polyplexes, enhances the transfection of nucleic acid in the cells and ensures its stability [56]. Further nanogel consisting of ethylenediamine (ED) functionalized PFMA was investigated as potent siRNA and pDNA carriers [57]. The pluronic nanogels bearing precoated PEI and conjugated with heparin, containing vascular endothelial growth factor (bFGF) and pDNA encoding VEGF165 genes encapsulated in the heparin-conjugate, have been studied for differentiation and proliferation [58, 59]. Arjunglucoside-I capsulated PNIPAAm/VP nanogels were observed to increase the therapeutic efficacy against parasites when compared to PLA. It is also found that nanogels are effective in lowering hepatotoxicity as well as nephrotoxicity of the drug [60]. In general, nanogels among the other nanocarriers stand out as promising nonviral carriers due to high transfection efficiency, high extracellular stability, low toxicity, and immunogenicity.

8.2.10 Polymeric Micelles

Self-assembly of amphiphilic block copolymers (ABCs) in the spherical, colloidal, supramolecular nanostructure is termed polymeric micelles. A unique structure of this category of carriers with a high loading capacity of the inner core offers a means to site-specific drug delivery of the drug molecules which are otherwise poorly water soluble. The structure of polymeric micelles consists of ha hydrophilic outer core and a hydrophobic inner core in an aqueous environment [61, 62]. In the presence of an aqueous medium, spontaneous rearrangement of the amphiphilic molecules in a supramolecular core takes place, wherein the water-insoluble drug can be introduced in a hydrophobic core. The hydrophilic shell enables prolonged blood circulation by inhibiting opsonins from adsorption on the micelle surface. The pharmacokinetic properties of micelles are size, shape, and surface property dependent, whereas on, the other hand, they are independent of the type and property of the loaded drug [63]. PEG is commonly used in hydrophilic segments in micelle for drug delivery owing to its high water solubility, nontoxicity, and neutral nature [64]. Embelin is an alkyl-substituted hydroxyl benzoquinone bioactive drug molecule known for its hepato-protective effects [65], antidiabetic and antitumor activity [66], and anti-inflammatory potential [67]. In spite of its high bio-activity, poor solubility in water is one of the major drawbacks of its restricted use in drug formulations. Hence, PEG-conjugated embelin not only increases its solubility but also forms micelles while retaining its antitumor activity [68]. It was also observed that paclitaxel was delivered using PEG5K-embelin2 micelles yielding excellent antitumor activity as compared to Taxol in murine models of breast and prostate cancers [69, 70]. Adriamycin (ADR)-conjugated PEG–poly (aspartic acid) block copolymers (PEG-P[Asp(ADR)]) show excellent in vivo antitumor activity by forming the micelles upon exposure to an aqueous medium [71]. Interestingly, vitamin B_{12} tethered at hydrophilic chains of amphiphilic graft copolymer dextran-g-polyethylene oxide cetyl ether, which is also known as DEX-g-PEO-C16, increases drug permeability 2–3 times when the said drug incorporated in the micelles. It is observed that enhancement in the drug permeability is due to receptor-mediated intestinal absorption through a complex formation [72]. It has been noticed that polymeric micelles are being accumulated at the many diseased tissues such as myocardial infarction tissues [73], in solid tumors which suggests that micelles are having a great potential therapeutic activity and efficiency on many other diseases.

8.2.11 Ceramic Nanoparticles

Ceramic nanoparticles comprise inorganic materials such as zirconia, alumina, silica, and titania or a combination of these materials with metals, metal oxides, and metal sulfides. These nanoparticles possess properties like low electrical and thermal conductivity, high stiffness, high elastic modulus, and corrosion resistivity. Ceramics nanoparticles are reported to be more stable, easy to manufacture, customizable, possessing long degradation time, and bioavailability [74, 75]. Apart from these properties, variable size, shape, and porosity make them potential candidates as a nanocarrier [76]. Due to the structural uniqueness, these particles protect the encapsulated molecules such as drug molecules, enzymes, and proteins from denaturalization as a result of external pH and temperature [77]. In addition, ceramic nanoparticles can be decorated with various organic functional groups to enhance their therapeutic usage and controlled, prolonged drug release [78, 79]. Based on the architectural differences, the other members of this nanoscopic ceramic family are ceramic nano-scaffold and nano-clay [80]. Silica nanoparticles have been covalently linked to the cationic surface to enhance its effective binding, condensation, as well as protection of plasmid DNA in DNA transfection. These are also reported to be used in delivering proteins and genes [81, 82]. Silica functionalized iron oxide nanoparticles were successfully employed to deliver haloalkane dehalogenase [83]. 5-Fluorouracil along with amoxicillin-conjugated iron-impregnated hydroxyapatite nanoparticles were found to be ameliorating bioactivity along with extended drug release [84]. Superparamagnetic iron hydroxyapatite nanoparticles are found to be biocompatible and enhance osteoblastic cell proliferation upon introducing a static magnetic field [85]. Based on the utility of hydroxyapatite vehicles for the transportation of malarial merozoite surface protein-119 (MSP-119), one can conclude that hydroxyapatite nanocarriers can be an excellent immuno-adjuvant that can be exploited as an antigen carrier for the immuno-potentiation [86]. Silica-doped nanoparticles were employed for photodynamic therapy for treating cancer [87]. The double-shelled hollow mesoporous silica spheres utilized as a carrier for hydrophilic and hydrophobic anticancer drug transportation vehicles highlighted their biocompatibility of steady release of hydrophilic anticancer drugs, e.g., irinotecan and high loading capacity of hydrophobic anticancer drug, e.g., docetaxel along with enhanced anticancer activity [88]. The other systems wherein ceramic nanocarriers have been used are widely reported in the literature [89].

8.2.12 Nanocrystals and Nanosuspensions

Nanocrystals often known as nanosuspensions are nanoscopic biphasic, colloidal dispersion of aggregates in an aqueous medium of a very large number of molecules stabilized either with a thin coating of surfactants or polymers or both. The minute surfactants quantities are beaded in nanocrystals for electrostatic surface stabilization. The average particle size of the nano-suspensions may range between 200 and 600 nm [90]. The problems associated with the site-specific drug delivery and timely releases of the desired drugs were overcome by the use of nanocrystals in a cost-effective manner without compromising the therapeutic utility of the drug. Typically, the problems associated with nanocarriers are their bioavailability, stability, solubility, and in turn high doses, etc. Nanocrystals offer solutions to all these above-mentioned problems due to the higher ratio of volume to the surface area of the nano-suspensions leading to satisfactory therapeutic concentrations along with low doses by altering pharmacokinetics, and bioavailability subsequently increasing drug safety and efficacy [91]. These nanocrystals can accommodate a large amount of drug with minimum dose volume, as is required in parenteral and ophthalmic drug delivery systems. A bottom-up approach (controlled precipitation/crystallization) or top-down approach (nanosizing) has been adapted to produce nano-suspensions with desired particle size and distribution. The ease of scaling-up of nanocrystals has been widely reported [92]. It has been anticipated that nanocrystals serving as a carrier for drugs like amphotericin B and tacrolimus after oral administration are found to be a potent candidate for targeting the mucosa of the gastrointestinal tract while aiming the mononuclear phagocytic system (MPS) cells for the treatment of leishmaniasis and fungal mycobacterial infections [93]. It has been observed that paclitaxel nano-suspensions increase the safety profile many folds than the commercial Taxol injections [94]. With curcumin nano-suspension, excellent cytotoxicity is reported in Hela and MCF-7 cells, in comparison to curcumin solution. The smaller size of the suspension not only enhances the solubility rate but also keeps the crystalline nature intact thereby improving the physical stability of curcumin [95]. Bioavailability has been improved from 5.2% to 82.3% for oral administration of gonadotropin inhibitor Danzaon in nano-suspension form, whereas oral absorption was found to have increased in case of a nano-suspension of amphotericin in comparison with its regular commercial formulation [96]. A successful formulation of cross-linked polymer nano-suspensions of dexamethasone has been adopted for increased anti-inflammatory activity in the model of rabbit eye irritation [97]. In order to check the possibility of curing HIV, indinavir is loaded as a nano-suspension in bone marrow-derived macrophages. The virus-infected cells in lymph nodes plasma, spleen, etc. were drastically reduced when the formulation was injected into HIV-1-challenged humanized mice thereby protecting CD4 (+) T-cell [98]. The drug diffusion in the skin has been found to be increased when drug nanoparticles are used in the creams [90, 98].

8.2.13 Nanowires

Assemblies of atoms in a wire fashion known as nanowires have gained the attention of the scientific fraternity owing to their versatility, bioavailability, ease of handling for drug delivery [99], and structural resemblance to an extracellular matrix such as collagen fibers [100]. Increased surface area, wide prospects of functionalization of outer surfaces, and very good mechanical performance make them particularly an ideal candidate for biomedical devices and tissue engineering motifs apart from controlled release of drugs to ailing tissues in a fiber diameter-dependent manner [101, 102]. Nanofibers have displayed a high capability of encapsulation and drug loading. Various techniques for synthesizing nanowires such as phase separation [103], electrospinning [104], template synthesis [105], melt-blown [106], and self-assembly [107] can be employed, while electrospinning technique is widely used due to its ability of large-scale production and its effectiveness of fabrication of nanowires of natural polymers and synthetic polymers [108]. Nanowires have the potential in the biomedical field for local chemotherapy [109, 110] and for drug delivery in cancer cells [111]. A hydrophobic, anti-fungal agent, itraconazole has been carried as a topical delivery system by nonbiodegradable polyurethane-based nanofibers [112]. Similarly, silver nanowires (AgNPs) were incorporated into zein nanofibers resulting in excellent bactericidal activity for *S. aureus* and *E. coli* [113]. Silver nanowires are potent for internalization into A549 and MRC-5 cells devoid of any cytotoxicity, hence furnishing an application of silver nanowires being a biocompatible carrier for lung cancer therapies [114]. The iron oxide surface decorated with doxorubicin by a linker which is pH-sensitive resulted in complete cell death (approx. 90%) when it is applied to cancer cells [115]. Further study showed that nanofibers are safe to a cell, which is concluded by in vitro cell adhesion and proliferation for the duration of 3 days. Additionally, within the 8 to 72 h, the mPEG-PLA nanowire showed excellent zero-order drug release profiles [116]. Owing to the anisotropy of the nanowires, the superparamagnetic iron oxide nanoparticles allow for deeper tumor targeting along with higher drug loading [117].

8.2.14 Quantum Dots

Quantum dots comprise clumps of atoms having an outer shell made up of a variety of materials [118]. Quantum dots, popularly known as semiconductor nanocrystals, as well as having special optical properties (e.g., tunable emission, brightness, and photostability), make them a potent candidate for applications in cellular imaging and diagnosis and drug delivery by dissolving, dispersing, adsorption, and coupling, etc. [119]. Owing to their outstanding optical properties, these inorganic moieties, i.e., QDs glow or fluoresce brightly when shined with laser, which is otherwise not visible. The physical and chemical characteristics like the rate of dissolution, saturation solubility, surface hydrophilicity, and hydrophobicity along with biological responses of the drugs will be altered due to the carrier [120]. The potentiality of the quantum dots lies in their ability to fictionalize with biomolecules such as nucleic acids, cells, and proteins. Light emitted by these quantum dots at a variable wavelength ranging from UV to IR makes cellular or subcellular structures visible. Their reduced toxicity to the organic dyes can be attributed to the inert coating present on the inner surface, thereby reducing the toxicity and resulting in increased efficacy and improved therapeutic properties, hence a potential candidate as a diagnostic tool and nanocarrier for controlled drug release [43, 121].

8.3 Therapeutic Nanoparticles Targeted Delivery Applications

The two fundamental categories of nanomaterials are nanostructured and nanocrystalline nanomaterials. The three categories of nanostructured materials are lipid-based, polymer-based, and nonpolymeric, nanoparticles. Polymer-based nanoparticles include nanogels, micelles, dendrimers, protein nanoparticles, and drug conjugates. Nonpolymeric nanoparticles include quantum dots, metallic nanoparticles, carbon nanotubes, nanodiamonds, and silica-based nanoparticles. Liposomes and solid lipid nanoparticles (Fig. 8.1) are two forms of lipid-based nanoparticles. All of this has previously been covered. Polymer- or lipid-based nanoparticles make up the bulk of clinically approved therapeutic nanoparticles thus far [45, 122]. Nanocrystalline particles, which are generated by the combination of medicinal substances in crystalline form, are utilized in some clinical applications in addition to polymer-based, nonpolymeric, or lipid-based nanostructured particles.

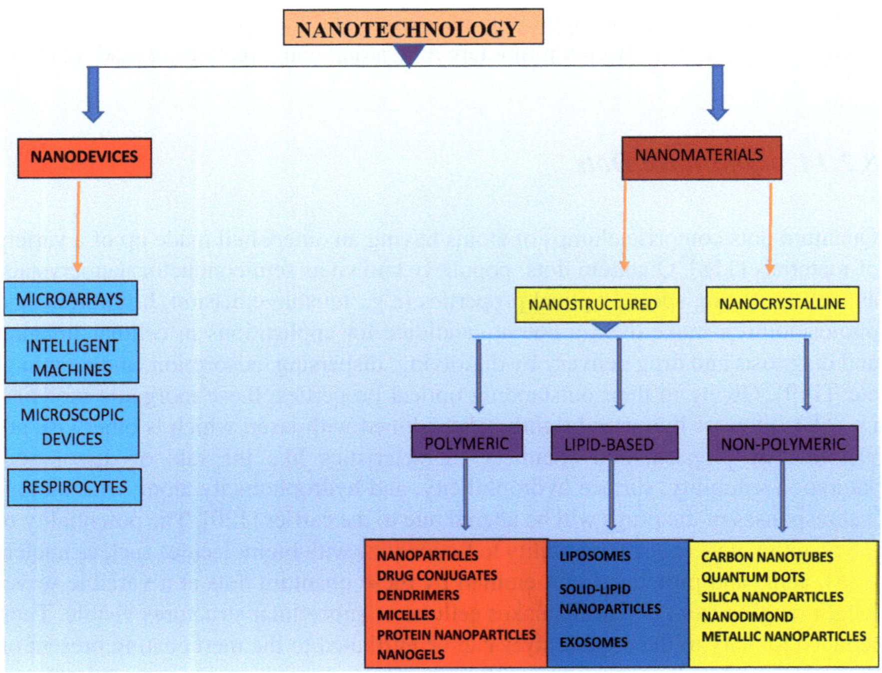

Fig. 8.1 Nanotechnology components that are used in medicinal applications

We described the many types of therapeutically employed nanoparticles, their therapeutic specificity, and current administration techniques in complex pathophysiological diseases in this section (Table 8.1).

8.4 Challenges in the Manufacturing of Nanomedicines

The treatments involving nanocarriers have a high added value than the standard treatments. It is crucial to consider cost-management approaches for the manufacturing process. The high cost of scaled conversion and the manufacturing method itself is the key hurdle to large-scale nanomedicine production. Mass production, the analytical procedures for in-process quality control, and exceptionally precise finished products need time and so are expensive [8, 147]. Nanomedicines must be produced using a process that is both reliable and uniform in quality for every batch. The lack of uniformity in production techniques and quality control testing is another concern. Quality control testing at in-process and end-product stages, regardless of the technology utilized to create most nanomedicines, demands the employment of sophisticated equipment and analytical techniques. By substantiating the selection of analytical methods and accurately interpreting the results, the establishment of guidelines and standard reference materials are critical in supporting the registration of nanomedicines. Most analytical methods recommended for in-process quality control and final products do not have standard protocols or descriptions in pharmacopeias, making innovative nanomedicines difficult to characterize. The particle size of nanomedicines is a critical quality attribute (CQA) of nanomedicine. At the nanoscale, relatively small variations in large size distribution or a particle size might affect a substance's bioavailability. Furthermore, due to the small particle size, particle shape, and form, special analytical processes are required [14, 122, 148]. Other CQAs for nanomedicines include surface coating, morphology, and charge since shape has been shown to alter cell uptake rates, uptake processes, level of uptake and intracellular distribution, and hence toxicity profile. According to a study, the cellular absorption of triangle-shaped nanoparticles was the highest, whereas star-shaped nanoparticles had the lowest [145, 148]. To confirm nanomedicine functionality, manufacturing process repeatability, the functional performance of the systems and its stability, specific analytical techniques such as SEM or TEM (morphological evaluation and size estimation), AFM (morphological evaluation, size, and shape), zeta potential analyzer, laser granulometry, and light scattering (particle size and size deviation) are required. Some of the procedures listed are extremely time-consuming, labor-intensive, and require expensive equipment, making them impractical to use in traditional manufacturing lines [149, 150].

Another problem with nanomedicines is the requirement to register them. The Food and Drug Administration (FDA), on the other hand, sees four major challenges in registering nanomedicines: the first is the regulatory status's inadequacy; the second is the potentiality of these drugs for unidentified risks in the post-market; the third is ensuring the development of new risk–benefit measures followed by

Table 8.1 Some approved and under clinical trials therapeutic nanoparticles

Nanostructure	Nanoparticle Formulation	Active Drug	Indications	Whether Approved or Not	References
Liposome	Nanoliposomes	Irinotecan	Colorectal & Pancreatic Cancer	Yes| (FDA 2015) (Europe 2016)	[123]
Lipid-based	Lipid nanoparticles	Transthyretin targeted siRNA	Amyloidosis (transthyretin-mediated)	Yes| (FDA 2018)	[124]
Liposome	Sphingomyelin & cholesterol	Vincristine sulfate	Lymphoid leukemia (acute)	Yes| (FDA 2012)	[125]
Polymer-based	Glatiramer (l-glutamic acid polymer with l-alanine, l-lysine, and l-tyrosine)	MAB	Multiple sclerosis	Yes| (FDA 2015)	[126]
Liposome	1-Palmitoyl-2-oleoyl-sn-glycero-3-phosphocholine and 1,2-Dioleoyl-sn-glycero-3-phospho-l-serine liposomes	Mifamurtide	Nonmetastasizing osteosarcoma	Yes| (Europe 20O9)	[127]
Liposome	Distearoylphosphatidylcholine, Distearoylphosphatidylglycerol, cholesterol	Daunorubicin Cytarabine	Acute myeloid leukemia	Yes| (FDA 2017)	[128]
Nanocrystal	Hydroxyapatite	–	Osteoinductive bone graft substitute	Yes| (FDA 2009)	[129]
Metallic nanoparticle	Superparamagnetic iron oxide nanoparticle (SPION) covered with dextran	–	Anemia in chronic kidney disease	Yes| (FDA 2009) (Europe 2012)	[130]
Nanocrystal	Paliperidone palmitate	Paliperidone	Schizophrenia	Yes| (FDA 2009) (Europe 2011)	[131]
Metallic nanoparticle	Nanoparticles of ferric oxide core-carboxymaltose shell	–	Iron deficiency anemia in chronic kidney disease	Yes| (FDA 2013)	[132]
Protein-drug conjugate	Maytansine derivative, DM1	Trastuzumab	HER2+ breast cancer	Yes| (FDA 2013)	[133]
Protein-drug conjugate	Glycopegylated coagulation factor IX	Factor IX	Hemophilia	Yes| (FDA 2017)	[134]
Metallic nanoparticle	Nanoparticles of superparamagnetic iron oxide coated with amino silane	–	Glioblastoma, prostate, pancreatic cancer	Yes| (Europe 20O9)	[135]

Nanocrystal	Dantrolene sodium	Dantrolene	Malignant hyperthermia	Yes| (FDA 2014)	[136]
Protein-drug conjugate	PEGylated factor VIII	Factor VIII	Hemophilia	Yes| (FDA 2015)	[137]
Protein-drug conjugate	Albumin	Paclitaxel	Lung cancer, pancreatic cancer	Yes| (FDA 2012) (Europe 2008)	[138]
Protein-drug conjugate	PEGylated uricase	Pegloticase	Gout	Yes| (FDA 2010) (Europe 2013)	[139]
	PEGylated interferon β-1a	Interferon β-1a	Multiple sclerosis	Yes| (FDA 2014) (Europe 2014)	[140]
Dendrimer	Polyethylene glycol (PEG)-platinum	α-Cyclodextrin	Cancer	Clinical/phase I	[141]
Micelle	PEG-polyaspartate polymeric micelle	Paclitaxel	Cancer	Clinical/phase III	[142]
Carbon nanotube	PEGylated single-walled CNT	Cisplatin	Cancer	Preclinical phase	[143]
Metallic nanoparticle	Iron oxide magnetic silica-gold nanoparticles	Doxorubicin	Cancer	Clinical/phase I	[144]
Silica based nanoparticle	Transferrin mesoporous silica	Lactobionic acid, doxorubicin	Cancer	Clinical/phase 0	[145]
Nanodiamonds	PEGylated nanodiamonds	Irinotecan, curcumin	Cancer, autoimmune diseases	Preclinical	[146]

assessments of clinical trials and further research protocols to determine the clinical efficacy and the fourth is consumer labeling of nanomedicine products. The FDA has released nanomedicine recommendations that stress the need for precise nanocarrier characterization and how these features affect system safety, efficacy, and quality. Few studies have been done to show how these features affect patient safety. As a result, the FDA's remark is only somewhat useful to a potential nanomedicine developer. Predicting the link between nanocarrier behavior in vitro and in vivo is one of the most difficult tasks. Tissue accumulation cellular interactions, transit, and biocompatibility are the pivotal elements that need to be investigated with the help of in vivo models, and the expenses of these trials are not small. The accumulation capacity of nanocarriers and their release profile over time must be idealized due to the likelihood of nanocarrier persistence in the circulation or deposition in tissues and organs, as well as a general lack of knowledge of long-term consequences. Finally, nanotechnology is a relatively young science, whereas the process of selling a revolutionary drug takes many years. Nanomedicine's long-term effects on animals are unknown, thus demanding proper pharmacovigilance studies before and after commercialization [150, 151].

8.5 Future Prospective for Controlled Site-Specific Target Delivery of Drugs

Future research trends in the administration and delivery of bioactive compounds will be closely connected to revolutionary manufacturing procedures that combine different structural designs incorporating novel aspects of nanotechnology. To attain this aim, the advantages of two or more types of biomaterials will be merged, which will improve nanoparticle characteristics and summatively improve the advantages of two or more types of biomaterials. The loading of bioactive compounds on nanoparticles and their efficacy are largely benefited and improved by the effective delayed release and in situ delivery control. Target delivery is one of the expected qualities that have yet to be completely developed, and as a result, its future implementation possibilities remain an issue. Resources and industrial techniques that would promote the synthesis of edible nanoparticles with increased qualities are continually being investigated in this regard. Improved specificity and flexibility of the nanoparticles would pave a way for orally administering bioactive chemicals via edible nanoparticles, and it will be a beneficial strategy. Due to precision tissue-targeted administration, this will allow us to make use of phytochemicals' true ability to prevent and treatment of specific pathophysiological conditions [72, 151]. The bioactivity of the drug is being influenced by gut metabolism after oral administration needs a modification of current definitions of "bioaccessibility" and "bioavailability," with a focus on upper gastrointestinal absorption. Indeed, understanding the properties of this physiological process, as well as the derivatives that are being produced during gastrointestinal digestion, is crucial for the design and development of nanoparticle systems. With this goal in mind, selecting nanoparticles that improve absorption in the gastrointestinal tract may no longer be the gold standard, due to the superb bio efficacy of the additional compounds synthesized, for example, in the large intestine as a result of the local microbiota's metabolism, making their absorption in the upper gastrointestinal undesirable. Furthermore, any changes in nanoparticle structure caused by digestive circumstances should be considered when assessing the impact of nanoparticles and/or nanovesicles on the efficacy of bioactive substances. Indeed, altering the structure of nanoparticles might affect their efficiency being as a vehicle of bioactive nutrients and non-nutrients, thus leading to incorrect findings. As a result, assuming that dilution has a substantial influence on the stability of nanoemulsions, liposomes, and micellesm, when combined with digesting fluids, is important. Furthermore, gastrointestinal fluid pH and ions conditions, as well as enzymatic actions in these compartments, may jeopardize nanoparticle stability and development as bioactive drug carriers [94, 152]. As a result of these constraints, it is critical to evaluate nanoparticle behavior in complex matrices and biological systems in order to offer reliable information on their practical utility.

In spite of the rational value of nanotechnology in enhancing the biocompatibility, bioavailability, and reduced toxicity of bioactive molecules, nanoparticles are produced utilizing physicochemical methods, which involve the use of costly and dangerous chemicals, particularly in the case of metallic biomaterials, limiting their precise application in vivo. These limits also include the environmental effect of residues formed during nanoparticle synthesis, which necessitates the use of green methods that do not involve the use of harmful chemicals to manufacture metal nanoparticles. In this sense, employing environmentally friendly and biocompatible chemicals to synthesize nanoparticles should help to reduce the negative effects of these procedures. The interactions between carriers and loaded molecules with gut commensals, enzymes, as well as the ramifications of these interactions on stability and bioefficacy are yet to be explored in the final intestinal stage, despite the fact that they appear to pose additional obstacles [72, 153]. Indeed, applying what has been learned about the utilization of nanoparticles as vehicles for bioactive compounds will allow researchers to decipher and exploit the reciprocal interactions between these molecules, enzymes, and gut microbiota. Another issue that has to be answered is the extent to which nanoparticles alter the pharmacokinetics of bioactive phytochemicals, which might lead to novel uses, such as internalization speed accelerators. This issue must be clarified in order to properly construct the sampling time for determining the pharmacokinetics, bioavailability, and bioactivity of the bioactive substances under research. In this frame, it becomes obvious that the interaction between encapsulated compounds, matrix material, and body enzymes\proteins in complex biological systems, such as mammals, warrants additional investigation in terms of metabolism and bioefficacy [154].

8.6 Conclusion

Since its inception, nanotechnology has been hailed as a promising method for preventing and treating a variety of human health problems. In fact, the industry has used this technique to look for novel options for drug administration based on the alteration of bioactive substances' solubility. Much advancement has been achieved over the past few years in evaluating the input that nanomaterials and nanoparticles give in a variety of pathophysiological circumstances. Encapsulating disease-modifying drugs, phytochemicals, and novel bioactive peptides within the right nanoparticles improves their bioavailability significantly by rendering them safe from decomposition during storage and gastrointestinal path. This fosters cellular absorption through improved retention time with the wall of the intestine, resulting in increased mucus immersion as well as intestinal permeation. It also enhances dispersion in aqueous environments, extends residence duration inside the bodily circulation, modulates its rate of discharge, and delivers them to a specific place. In this framework, this chapter considers the extent to which nanoparticles might boost the therapeutic index of the drug while avoiding toxicity and delivering these bioactive substances to target tissues impacted by certain pathophysiological and chemical conditions.

References

1. Patra JK, Das G, Fraceto LF, Campos EVR, Rodriguez-Torres MP, Acosta-Torres LS, Diaz-Torres LA, Grillo R, Swamy MK, Sharma S, Habtemariam S, Shin H-S (2018) Nano-based drug delivery systems: recent developments and future prospects. J Nanobiotechnol 16:1–33
2. Portero A, Remunan-Lopez C, Criado M, Alonso M (2002) Reacetylated chitosan microspheres for controlled delivery of anti-microbial agents to the gastric mucosa. J Microencapsul 19:797–809
3. Fernández-Urrusuno R, Calvo P, Remuñán-López C, Vila-Jato JL, Alonso MJ (1999) Enhancement of nasal absorption of insulin using chitosan nanoparticles. Pharm Res 16: 1576–1581
4. De Campos AM, Sánchez A, Alonso MJ (2001) Chitosan nanoparticles: a new vehicle for the improvement of the delivery of drugs to the ocular surface. Application to cyclosporin A. Int J Pharm 224:159–168
5. Al-Qadi S, Grenha A, Carrión-Recio D, Seijo B, Remuñán-López C (2012) Microencapsulated chitosan nanoparticles for pulmonary protein delivery: in vivo evaluation of insulin-loaded formulations. J Control Release 157:383–390
6. Silva MM, Calado R, Marto J, Bettencourt A, Almeida AJ, Gonçalves L (2017) Chitosan nanoparticles as a mucoadhesive drug delivery system for ocular administration. Mar Drugs 15:370–386
7. Pistone S, Goycoolea FM, Young A, Smistad G, Hiorth M (2017) Formulation of polysaccharide-based nanoparticles for local administration into the oral cavity. Eur J Pharm Sci 96:381–389
8. Liu S, Yang S, Ho PC (2018) Intranasal administration of carbamazepine-loaded carboxymethyl chitosan nanoparticles for drug delivery to the brain. Asian J Pharm Sci 13: 72–81

9. Elseoud WSA, Hassan ML, Sabaa MW, Basha M, Hassan EA, Fadel SM (2018) Chitosan nanoparticles/cellulose nanocrystals nanocomposites as a carrier system for the controlled release of repaglinide. Int J Biol Macromol 111:604–613
10. Shah AA, Gupta A (2021) Antioxidants in health and disease with their capability to defend pathogens that attack apple species of Kashmir. In: Ekiert HM, Ramawat KG, Arora J (eds) Plant antioxidants and health. Reference series in phytochemistry. Springer, Cham
11. Hansen K, Kim G, Desai KG, Patel H, Olsen KF, Curtis-Fisk J, Tocce E, Jordan S, Schwendeman SP (2015) Feasibility investigation of cellulose polymers for mucoadhesive nasal drug delivery applications. Mol Pharm 12:2732–2741
12. Bozzuto G, Molinari A (2015) Liposomes as nanomedical devices. Int J Nanomedicine 10: 975–999
13. Akbarzadeh A, Rezaei-Sadabady R, Davaran S, Joo SW, Zarghami N, Hanifehpour Y, Samiei M, Kouhi M, Nejati-Koshki K (2013) Liposome: classification, preparation, and applications. Nanoscale Res Lett 8:102–111
14. Mohan A, Narayanan S, Sethuraman S, Krishnan UM (2014) Novel resveratrol and 5-fluorouracil coencapsulated in PEGylated nanoliposomes improve chemotherapeutic efficacy of combination against head and neck squamous cell carcinoma. Bio Med Res Int 2014: 1–14
15. Dimov N, Kastner E, Hussain M, Perrie Y, Szita N (2017) Formation and purification of tailored liposomes for drug delivery using a module-based micro continuous-flow system. Sci Rep 7:120–145
16. Patil NH, Devarajan PV (2016) Insulin-loaded alginic acid nanoparticles for sublingual delivery. Drug Deliv 23:429–436
17. Haque S, Md S, Sahni JK, Ali J, Baboota S (2014) Development and evaluation of brain targeted intranasal alginate nanoparticles for treatment of depression. J Psychiatr Res 48:1–12
18. Menzel C, Jelkmann M, Laffleur F, Bernkop-Schnürch A (2017) Nasal drug delivery: design of a novel mucoadhesive and in situ gelling polymer. Int J Pharm 517:196–202
19. Román JV, Galán MA, del Valle EMM (2016) Preparation and preliminary evaluation of alginate crosslinked microcapsules as potential drug delivery system (DDS) for human lung cancer therapy. Biomed Phys Eng Expr 2:035015
20. Rauti R, Musto M, Bosi S, Prato M, Ballerini L (2019) Properties and behavior of carbon nanomaterials when interfacing neuronal cells: how far have we come? Carbon 143:430–446
21. Beg S, Rahman M, Jai A, Saini S, Hasnain MS, Swain S (2018) Emergence in the functionalized carbon nanotubes as smart nanocarriers for drug delivery applications. Fullerenes, Graphenes and Nanotubes: a Pharmaceutical Approach 2018:105–133. Editor A M Grumezescu
22. Liang F, Chen B (2009) A review on biomedical applications of single-walled carbon nanotubes. Curr Med Chem 17:10–24
23. Pan D, Zhang J, Li Z, Wu M (2010) Hydrothermal route for cutting graphene sheets into blue-luminescent graphene quantum dots. Adv Mater Weinheim 22:734–738
24. Li D, Chen X, Wang H, Liu J, Zheng M, Fu Y (2017) Cetuximab-conjugated nanodiamonds drug delivery system for enhanced targeting therapy and 3D Raman imaging. J Biophotonics 10:1636–1646
25. Cai W, Gao H, Chu C, Wang X, Wang J (2017) Engineering phototheranostic nanoscale metal–organic frameworks for multimodal imaging-guided cancer therapy. ACS Appl Mater Interfaces 9:2040–2051
26. Mao D, Hu F, Kenry JS, Wu W, Ding D, Kong D, Liu B (2018) Metal–organic-framework-assisted in vivo bacterial metabolic labeling and precise antibacterial therapy. Adv Mater 30: 1706831
27. Simon-Yarza T, Mielcarek A, Couvreur P, Serre C (2018) Nanoparticles of metal–organic frameworks: on the road to in vivo efficacy in biomedicine. Adv Mater 30:1707365

28. Gandara-Loe J, Ortuño-Lizarán I, Fernández-Sanchez L, Alió JL, Cuenca N, Vega-Estrada A, Silvestre-Albero J (2019) Metal–organic frameworks as drug delivery platforms for ocular therapeutics. ACS Appl Mater Interfaces 11:1924–1931
29. Sava Gallis DF, Butler KS, Agola JO, Pearce CJ, McBride AA (2019) Antibacterial countermeasures via metal–organic framework-supported sustained therapeutic release. ACS Appl Mater Interfaces 11:7782–7791
30. Chen B, Yang Z, Zhu Y, Xia Y (2014) Zeoliticimidazolate framework materials: recent progress in synthesis and applications. J Mater Chem A 2:16811–16831
31. Vasconcelos IB, da Silva TG, Militão GCG, Soares TA, Rodrigues NM (2012) Cytotoxicity and slow release of the anti-cancer drug doxorubicin from ZIF-8. RSC Adv 2:9437–9442
32. Sun C-Y, Qin C, Wang X-L, Yang G-S, Shao K-Z (2012) Zeoliticimidazolate framework-8 as efficient pH-sensitive drug delivery vehicle. Dalton Trans 41:6906–6909
33. Zhuang J, Kuo C-H, Chou L-Y, Liu D-Y, Weerapana E, Tsung C-K (2014) Optimized metal–organic-framework nanospheres for drug delivery: evaluation of small-molecule encapsulation. ACS Nano 8:2812–2819
34. Taylor-Pashow KML, Della Rocca J, Xie Z, Tran S, Lin W (2009) Post synthetic modifications of iron-carboxylate nanoscale metal−organic frameworks for imaging and drug delivery. J Am Chem Soc 131:14261–14263
35. Wang X-G, Dong Z-Y, Cheng H, Wan S-S, Chen W-H (2015) A multifunctional metal–organic framework based tumor targeting drug delivery system for cancer therapy. Nanoscale 7:16061–16070
36. Chen Y, Li P, Modica JA, Drout RJ, Farha OK (2018) Acid resistant mesoporous metal–organic framework toward oral insulin delivery: protein encapsulation, protection, and release. J Am Chem Soc 140:5678–5681
37. He C, Lu K, Liu D, Lin W (2014) Nanoscale metal–organic frameworks for the co-delivery of cisplatin and pooled siRNAs to enhance therapeutic efficacy in drug-resistant ovarian cancer cells. J Am Chem Soc 136:5181–5184
38. Lian X, Huang Y, Zhu Y, Fang Y, Zhao R (2018) Enzyme-MOF nanoreactor activates nontoxic paracetamol for cancer therapy. Angew Chem Int Ed 57:5725–5730
39. Pohlmann AR, Weiss V, Mertins O, Da Silveira NP, Guterres SS (2002) Spray-dried indomethacin-loaded polyester nanocapsules and nanospheres: development, stability evaluation and nanostructure models. Eu J Pharma Sci 16:305–312
40. Radhika PR, Sasikanth, Sivakumar T (2011) Nanocapsules: A new approach in drug delivery. Int J Pharm Sci Res 2:1426–1429
41. Chiang CS, Hu SH, Liao BJ, Chang YC, Chen SY (2014) Enhancement of cancer therapy efficacy by trastuzumab-conjugated and pH-sensitive nanocapsules with the simultaneous encapsulation of hydrophilic and hydrophobic compounds. Nanomed 10:99–107
42. Baran ET, Özer N, Hasirci V (2002) Poly (hydroxybutyrate-co-hydroxyvalerate) nanocapsules as enzyme carriers for cancer therapy: an in vitro study. J Microencapsul 19:363–376
43. Kim CS, Mout R, Zhao Y, Yeh Y-C, Tang R, Jeong Y, Duncan B, Hardy JA, Rotello VM (2015) Co-delivery of protein and small molecule therapeutics using nanoparticle-stabilized nanocapsules. Bioconjug Chem 26:950–954
44. Rawat M, Singh D, Saraf S, Saraf S (2006) Nanocarriers: promising vehicle for bioactive drugs. Biol Pharm Bull 29:1790–1798
45. Sukhorukov GB, Rogach AL, Garstka M, Springer S, Parak WJ, Munõz-Javier A, Kreft O, Skirtach AG, Susha AS, Ramaye Y (2007) Multifunctionalized polymer microcapsules: novel tools for biological and pharmacological applications. Small 3:944–955
46. Tomalia D, Baker H, Dewald J, Hall M, Kallos G, Martin S, Roeck J, Ryder J (1985) Smith PA new class of polymers: starburst-dendritic. Polym J 17:117–132
47. Newkome GR, Yao Z, Baker GR, Gupta VK (1985) Micelles. Part 1. Cascade molecules: a new approach to micelles. A [27]-arborol. J Org Chem 50:2003–2004

48. Elham A, Sedigheh FA, Abolfaz A, Morteza M, Hamid TN, Sang WJ, Younes H, Kazem N, Roghiyeh P (2014) Dendrimers: synthesis, applications, and properties. Nanoscale Res Lett 9: 247
49. Vandamme TF, Brobeck L (2005) Poly (amidoamine) dendrimers as ophthalmic vehicles for ocular delivery of pilocarpine nitrate and tropicamide. J Control Release 20:23–38
50. Jones CF, Campbell RA, Franks Z, Gibson CC, Thiagarajan G, Vieira-de-Abreu A, Sukavaneshvar S, Mohammad SF, Li DY, Ghandehar H, Weyrich AS, Brooks BD, Grainger DW (2012) Cationic PAMAM dendrimers disrupt key platelet functions. Mol Pharm 9:1599–1611
51. Khandare JJ, Jayan S, Singh A, Chandna P, Yang W, Vorsa N, Minko T (2006) Dendrimer versus linear conjugate: influence of polymeric architecture on the delivery and anticancer effect of paclitaxel. Bioconjug Chem 17:1464–1472
52. Chittasupho C, Anuchapreeda S, Sarisuta N (2017) CXCR4 targeted dendrimer foranti-cancer drug delivery and breast cancer cell migration inhibition. Eur J Pharm Biopharm 119:310–321
53. Gupta L, Sharma AK, Gothwal A, Khan MS, Khinchi MP, Qayum A, Singh SK, Gupta U (2017) Dendrimer encapsulated and conjugated delivery of berberine: a novel approach mitigating toxicity and improving in vivo pharmacokinetics. Int J Pharm 528:88–99
54. Saovapakhiran A, D'Emanuele A, Attwood D, Penny J (2009) Surface modification of PAMAM dendrimers modulates the mechanism of cellular internalization. Bioconjug Chem 20:693–701
55. Patri AK, Kukowska-Latallo JF, Baker Jr JR (2005) Targeted drug delivery with dendrimers: comparison of the release kinetics of covalently conjugated drug and non-covalent drug inclusion complex. Adv Drug Deliv Rev 57:2203–2214
56. Zhong Q, Krishna Rao KSV (2016) Recent advances in stimuli-responsive poly (amidoamine) dendrimer nanocarriers for drug delivery. Indian J Adv Chem Sci 4:195–207
57. Yamashita S, Katsumi H, Hibino N, Isobe Y, Yagi Y, Kusamori K, Sakane T, Yamamoto A (2017) Development of PEGylated carboxylic acid-modified polyamidoamine dendrimers as bone-targeting carriers for the treatment of bone diseases. J Control Release 262:10–17
58. Iordana NA, Gabriela R, Alina D, Loredana EN, Chiriac AP (2017) Basic concepts and recent advances in nanogels as carriers for medical applications. Drug Deliv 24:539–557
59. Sunasee R, Wattanaarsakit P, Ahmed M (2012) Biodegradable and nontoxic nanogels as nonviral gene delivery systems. Bioconjug Chem 23:1925–1933
60. Li R, Wu W, Song H (2016) Well-defined reducible cationic nanogels based on functionalized low-molecular-weight PGMA for effective pDNA and siRNA delivery. Acta Biomater 41: 282–292
61. Yang H, Choi J, Park J (2014) Differentiation of endothelialprogenitor cells into endothelial cells by heparin-modified supramolecular pluronic nanogels encapsulating bFGF and complexed withVEGF165 genes. Biomaterials 35:4716–4728
62. Tyagi R, Lala S, Verma AK, Nandy AK, Mahato SB, Maitra A, Basu MK (2005) Targeted delivery of arjunglucoside I using surface hydrophilic and hydrophobic nanocarriers to combat experimental leishmaniasis. J Drug Target 13:161–171
63. Qiu L, Zheng C, Jin Y, Zhu K (2007) Polymeric micelles as nanocarriers for drug delivery. Expert Opin Ther Pat 17:819–830
64. Kataoka K (2004) Smart polymeric micelles as nanocarriers for gene and drug delivery. In: International Conference on MEMS, NANO and Smart Systems (ICMENS'04), pp. 4–5
65. Li Y, Zhang T, Liu Q, He J (2019) PEG-Derivatized dual-functional Nanomicelles for improved cancer therapy. Front Pharmacol 10(808):1–14
66. Singh D, Singh R, Singh P, Gupta RS (2009) Effects of embelin on lipid peroxidation and free radical scavenging activity against liver damage in rats. Basic Clin Pharmacol Toxicol 105: 243–248
67. Bhandari U, Jain N, Pillai KK (2007) Further studies on antioxidant potential and protection of pancreatic beta-cells by Embeliaribes in experimental diabetes. Exp Diabetes Res 2007:15803

68. Li Y, Li D, Yuan S, Wang Z, Tang F, Nie R (2013) Embelin induced MCF-7 breast cancer cell apoptosis and blockade of MCF-7 cells in the G2/M phase via the mitochondrial pathway. Oncol Lett 5:1005–1009
69. Chitra M, Sukumar E, Suja V, Devi CS (1994) Antitumor, anti-inflammatory and analgesic property of embelin, a plant product. Chemotherapy 40:109–113
70. Huang Y, Lu J, Gao X, Li J, Zhao W, Sun M et al (2012) PEG derivatized embelin as a dual functional carrier for the delivery of paclitaxel. Bioconjug Chem 23:1443–1451
71. Lu J, Huang Y, Zhao W, Marquez RT, Meng X, Li J (2013) PEG derivatized embelin as a nanomicellar carrier for delivery of paclitaxel to breast and prostate cancers. Biomaterials 34: 1591–1600
72. Yokoyama M, Okano T, Sakurai Y, Ekimoto H, Shibazaki C, Kataoka K (1991) Toxicity and antitumor activity against solid tumors of micelle-forming polymeric anticancer drug and its extremely long circulation in blood. Cancer Res 51:3229–3236
73. Francis MF, Cristea M, Winnik F (2005) Exploiting the vitamin B12 pathway to enhance oral drug delivery via polymeric micelles. Biomacromolecules 6:2462–2467
74. Lukyanov AN, Hartner WC, Torchilin VP (2004) Increased accumulation of PEG–PE micelles in the area of experimental myocardial infarction in rabbits. J Control Release 4: 187–193
75. Kwon G, Suwa S, Yokoyama M, Okano T, Sakurai Y, Kataoka K (1994) Enhanced tumor accumulation and prolonged circulation times of micelle forming poly(ethylene oxide-aspartate)block copolymer-adriamycin conjugates. J Control Release 29:17–23
76. Lei Y, Sheldon BW, Webster TJ (2010) Nanophase ceramics for improved drug delivery: current opportunities and challenges. Am Ceram Soc Bull 89:24–33
77. Armatas GS, Kanatzidis MG (2006) Mesostuctured germanium oxide with cubic pore symmetry. Nature 441:1122–1125
78. Luo D, Han E, Belcheva N, Saltzman WM (2004) A self-assembled, modular DNA delivery system mediated by silica nanoparticles. J Control Rel 95:333–341
79. Jain TK, Roy I, De TK, Maitra A (1998) Nanometer silica particles encapsulating active compounds: A novel ceramic drug carrier. J Am Chem Soc 120:11092–11095
80. Fadeel B, Garc'ıa-Bennett AE. (2010) Better safe than sorry: understanding the toxicological properties of inorganic nanoparticles manufactured for biomedical applications. Adv Drug Deliv Rev 62:362–374
81. Singh D, Singh S, Sahu J, Srivastava S, Sing RM (2016) Ceramic nanoparticles: recompense, cellular uptake and toxicity concerns. Artif Cells Nanomed Biotechnol 44:401–409
82. Johnson AK, Zawadzka AM, Deobald LA (2008) Novel method for immobilization of enzymes to magnetic nanoparticles. J Nanopart Res 10:1009–1025
83. Sarath Chandra V, Baskar G, Suganthi R (2012) Blood compatibility of iron-doped nanosize hydroxyapatite and its drug release. ACS Appl Mater Interfaces 4:1200–1210
84. Panseri S, Cunha C, D'Alessandro T (2012) Intrinsically superparamagnetic Fe-hydroxyapatite nanoparticles positively influence osteoblast-like cell behaviour. J Nanobiotechnol 10:32
85. Goyal AK, Khatri K, Mishra N (2009) Development of self-assembled nanoceramic carrier construct(s) for vaccine delivery. J Biomater Appl 24:65–84
86. Roy I, Ohulchanskyy T, Pudavar H (2003) Ceramic-based nanoparticles entrapping water-insoluble photosensitizing anticancer drugs: a novel drug-carrier system for photodynamic therapy. J Am Chem Soc 125:860–7865
87. Chen Y, Chen H, Ma M (2011) Double mesoporous silica shelled spherical/ellipsoidal nanostructures: synthesis and hydrophilic/hydrophobic anticancer drug delivery. J Mater Chem 21:5290–5298
88. Saraf S (2009) Process optimization for the production of nanoparticles for drug delivery applications. Expert Opin Drug Deliv 6:187–196
89. Chingunpituk J (2007) Nanosuspension technology for drug delivery. Walailak J Scie Tech 4: 139–153

90. Nagarwal RC, Kumar R, Dhanawat M, Das N, Pandit JK (2011) Nanocrystal technology in the delivery of poorly soluble drugs : an overview. Curr Drug Deliv 8:398–406
91. Merisko-Liversidge E, Liversidge GG, Cooper ER (2003) Nanosizing: a formulation approach for poorly-water-soluble compounds. Eur J Pharm Sci 18:113–120
92. Gao Y, Li Z, Sun M, Guo C, Yu A, Xi Y (2011) Preparation and characterization of intravenously injectable curcumin nanosuspension. Drug Deliv 18:131–142
93. Prabhakar C, Krishna KB (2011) A review on nanosuspensions in drug delivery. Int J Pharma Bio Sci 2:549–558
94. McMillan J, Batrakova E, Gendelman HE. Cell delivery of therapeutic nanoparticle. Prog Mole Bio Trans Sci (2011) 104: 571–572. Elsevier, New York, NY, USA
95. Constantinides PP, Chaubal MV, Shorr R (2008) Advances in lipid nanodispersions for parenteral drug delivery and targeting. Adv Drug Del Rev 60:757–767
96. Martin CR, Kohli P (2003) The emerging field of nanotube biotechnology. Nat Rev Drug Discov 2:29–37
97. Shan XQ, Liu CS, Li FQ, Ouyang C, Gao Q, Zheng KS (2015) Nanoparticles vs. nanofibers: a comparison of two drug delivery systems on assessing drug release performance in vitro. Designed Monomers and Poly 18:678–689
98. Thuy TTN, Chiranjit G, Seong-Gu H, Noppavan C, Jun SP (2012) Porous core/sheath composite nanofibers fabricated bycoaxial electrospinning as a potential mat for drug release system. Int J Pharm 439:296–306
99. Tatsuya O, Kengo T, Satoru K (2010) Time-programmed dual release formulation by multi-layered drug-loaded nanofiber meshes. J Control Release 143:258–264
100. Huang Z-M, Zhang Y-Z, Kotaki M, Ramakrishna S (2003) A review on polymer nanofibers by electrospinning and their applications in nanocomposites. Compos Sci Technol 63:2223–2253
101. Che G, Lakshmi B, Martin C, Fisher E, Ruoff RS (1998) Chemical vapor deposition based synthesis of carbon nanotubes and nanofibers using a template method. Chem Mater 10:260–267
102. Hassan MA, Yeom BY, Wilkie A, Pourdeyhimi B, Khan SA (2013) Fabrication of nanofiber melt blown membranes and their filtration properties. J Membr Sci 427:336–344
103. Hartgerink JD, Beniash E, Stupp SI (2001) Self-assembly and mineralization of peptideamphiphile nanofibers. Science 294:1684–1688
104. Bhardwaj N, Kundu SC (2010) Electrospinning: a fascinating fiber fabrication technique. Biotechnol Adv 28:325–347
105. Wen C-Y, Wu L-L, Zhang Z-L, Liu Y-L, Wei S-Z, Hu J, Tang M, Sun E-Z, Gong Y-P, Yu J (2013) Quick-response magnetic nanospheres for rapid, efficient capture and sensitive detection of circulating tumor cells. ACS Nano 8:941–949
106. Zamani M, Prabhakaran MP, Ramakrishna S (2013) Advances in drug delivery via electrospun and electrosprayed nanomaterials. Int J Nanomedicine 8:2997–3017
107. Liu S, Zhou G, Huang Y, Xie Z, Jing X (2003) Biodegradable electrospun fibers as drug delivery system for local cancer treatment: a rugged path to the bright future: original research article: biodegradable electrospun fibers for drug delivery. J Control Release 2014:52–53
108. Verreck G, Chun I, Rosenblatt J, Peeters J, Dijck AV, Mensch J, Noppe M, Brewster ME (2003) Incorporation of drugs in an amorphous state into electrospun nanofibers composed of a water-insoluble, nonbiodegradable polymer. J controlled release: official journal of the Controlled Release Society 92:349–360
109. Dashdorj U, Reyes MK, Unnithan AR, Tiwari AP, Tumurbaatar B, Park CH, Kim CS (2015) Fabrication, and characterization of electrospunzein/ag nanocomposite mats for wound dressing applications. Int J Bio Macromol 80:1–7
110. Singh M, Movia D, Mahfoud OK, Volkov Y, Prina-Mello A (2013) Silver nanowires as prospective carriers for drug delivery in cancer treatment: an in vitro biocompatibility study on lung adenocarcinoma cells and fibroblasts. Eur J Nanomed 5:195–204
111. Martínez-Banderas AI, Aires A, Quintanilla M, Holguín-Lerma JA, Lozano-Pedraza C, Teran FJ, Moreno JA (2019)´ Perez JE, Ooi BS, Ravasi T, Merzaban JS, Cortajarena AL, Kosel J,

iron-based Core-Shell nanowires for combinatorial drug delivery and Photothermal and magnetic therapy. ACS Appl Mater Interfaces 11:43976–43988
112. Shana X, Liub C, Lic F, Ouyanga C, Gaoa Q, Zheng K (2015) Nanoparticles vs nanofibers: a comparison of two drug delivery systems on assessing drug release performance in vitro. Designed Monomers and Polymers 18:678–689
113. Shen S, Wu Y, Liu Y, Wu D (2017) High drug-loading nanomedicines: Progress, current status, and prospects. Int J Nanomedicine 12:4085–4109
114. Dwaine FE, Christopher GT (2003) Nanotechnology and medicine. Expert Opin Biol Ther 3: 655–663
115. Abbaspourrad A, Datta SS, Weitz DA (2013) Controlling release from pH responsive microcapsules. Langmuir 29:12697–12702
116. Oluwole DO, Nyokong T (2015) Physicochemical behavior of nanohybrids of mono and tetra substituted carboxyphenoxyphthalocyanine covalently linked to GSH-CdTe/CdS/ZnS quantum dots. Polyhedron 87:8–16
117. Schneider R, Schmitt F, Frochot C, Fort Y, Lourette N, Guillemin F (2005) Design, synthesis, and biological evaluation of folic acid targeted tetraphenylporphyrin as novel photosensitizers for selective photodynamic therapy. Bioorg Med Chem 13:2799–2808
118. Yoo HS, Park TG (2004) Folate-receptor-targeted delivery of doxorubicin nano-aggregates stabilized by doxorubicin-PEG-folate conjugate. J Control Release 100:247–256
119. Ju J, Chen W (2015) Graphene quantum dots as fluorescence probes for sensing metal ions: synthesis and applications. Curr Org Chem 19:1150–1162
120. Yao X, Niu X, Ma K, Huang P, Grothe J, Kaskel S, Zhu Y (2017) Graphene quantum dots-capped magnetic mesoporous silica nanoparticles as a multifunctional platform for controlled drug delivery, magnetic hyperthermia, and photothermal therapy. Small 13:1602225
121. Justin R, Tao K, Román S, Chen D, Xu Y, Geng X, Ross IM, Grant RT, Pearson A, Zhou G (2016) Photoluminescent and superparamagnetic reduced graphene oxide–iron oxide quantum dots for dual modality imaging, drug delivery and photothermal therapy. Carbon 97:54–70
122. Biswas A, Khandelwal P, Das R, Salunke G, Alam A, Ghorai S, Chattopadhyay S, Poddar P (2017) Oxidant mediated one-step complete conversion of multi-walled carbon nanotubes to graphene quantum dots and their bioactivity against mammalian and bacterial cells. J Mater Chem B 5:785–796
123. Zhang H (2016) Onivyde for the therapy of multiple solid tumors. Onco Targets 9:3001–3007
124. Akinc A, Maier MA, Manoharan M, Fitzgerald K, Jayaraman M, Barros S, Ansell S, Du X, Hope MJ, Madden TD et al (2019) The Onpattro story and the clinical translation of nanomedicines containing nucleic acid-based drugs. Nat Nanotechnol 14:1084–1087
125. Silverman JA, Deitcher SR (2013) Marqibo(R) (vincristine sulfate liposome injection) improves the pharmacokinetics and pharmacodynamics of vincristine. Cancer Chemother Pharm 71:555–564
126. Komlosh A, Weinstein V, Loupe P, Hasson T, Timan B, Konya A, Alexander J, Melamed-Gal S, Nock S (2019) Physicochemical and biological examination of two Glatiramer acetate products. Biomedicine 7:49
127. Venkatakrishnan K, Liu Y, Noe D, Mertz J, Bargfrede M, Marbury T, Farbakhsh K, Oliva C, Milton A (2014) Pharmacokinetics and pharmacodynamics of liposomal mifamurtide in adult volunteers with mild or moderate renal impairment. Br J Clin Pharm 77:986–997
128. Anselmo AC, Mitragotri S (2019) Nanoparticles in the clinic: an update. Bioeng Transl Med 4: e10143
129. Choi YH, Han HK (2018) Nanomedicines: current status and future perspectives in aspect of drug delivery and pharmacokinetics. J Pharm Investig 48:43–60
130. Bullivant JP, Zhao S, Willenberg BJ, Kozissnik B, Batich CD, Dobson J (2013) Materials characterization of Feraheme/ferumoxytol and preliminary evaluation of its potential for magnetic fluid hyperthermia. Int J Mol Sci 14:17501–17510

131. Kaduk JA, Dmitrienko AO, Gindhart AM, Blanton TN (2017) Crystal structure of paliperidone palmitate (INVEGA SUSTENNA®), C39H57FN4O4. Powder Diffract 32: 222–227
132. Lyseng-Williamson KA, Keating GM (2009) Ferric Carboxymaltose. Drugs 69:739–756
133. Kim MT, Chen Y, Marhoul J, Jacobson F (2014) Statistical modeling of the drug load distribution on trastuzumab emtansine (Kadcyla), a lysine-linked antibody drug conjugate. Bioconjug Chem 25:1223–1232
134. Farjadian F, Ghasemi A, Gohari O, Roointan A, Karimi M, Hamblin MR (2019) Nanopharmaceuticals and nanomedicines currently on the market: challenges and opportunities. Nanomedicine (Lond) 14:93–126
135. Weissig V, Pettinger TK, Murdock N (2014) Nanopharmaceuticals (part 1): products on the market. Int J Nanomedicine 9:4357–4373
136. Patra JK, Das G, Fraceto LF, Campos EVR, Rodriguez-Torres MDP, Acosta-Torres LS, Diaz-Torres LA, Grillo R, Swamy MK, Sharma S et al (2018) Nano based drug delivery systems: recent developments and future prospects. J Nanobiotechnol 16:71
137. Chelle P, Yeung CHT, Croteau SE, Lissick J, Balasa V, Ashburner C, Park YS, Bonanad S, Megías-Vericat JE, Nagao A et al (2020) Development and validation of a population-pharmacokinetic model for Rurioctacog alfa Pegol (Adynovate®): A report on behalf of the WAPPS-Hemo investigators ad hoc subgroup. Clin Pharmacokinet 59:245–256
138. Miele E, Spinelli GP, Miele E, Tomao F, Tomao S (2009) Albumin-bound formulation of paclitaxel (Abraxane ABI-007) in the treatment of breast cancer. Int J Nanomedicine 4:99–105
139. Pardridge WM (2012) Drug transport across the blood-brain barrier. J Cereb Blood Flow Metab 32:1959–1972
140. Patel HN, Patel PM (2013) Dendrimer applications - A review. Int J Pharma Bio Sci 4:454–463
141. Wang X, Wang C, Zhang Q, Cheng Y (2016) Near infrared light-responsive and injectable supramolecular hydrogels for on-demand drug delivery. Chem Commun (Camb) 52:978–981
142. Sendi P, Proctor RA (2009) Staphylococcus aureus as an intracellular pathogen: the role of small colony variants. Trends Microbiol 17:54–58
143. Bhirde AA, Patel S, Sousa AA, Patel V, Molinolo AA, Ji Y, Leapman RD, Gutkind JS, Rusling JF (2010) Distribution and clearance of PEG-single-walled carbon nanotube cancer drug delivery vehicles in mice. Nanomed (Lond Engl) 5:1535–1546
144. Kharlamov AN, Feinstein JA, Cramer JA, Boothroyd JA, Shishkina EV, Shur V (2017) Plasmonic photothermal therapy of atherosclerosis with nanoparticles: long-term outcomes and safety in NANOM-FIM trial. Futur Cardiol 13:345–363
145. Zhou J, Li M, Lim WQ, Luo Z, Phua SZF, Huo R, Li L, Li K, Dai L, Liu J et al (2018) A transferrin-conjugated hollow Nanoplatform for redox-controlled and targeted chemotherapy of tumor with reduced inflammatory reactions. Theranostics 8:518–532
146. Madamsetty VS, Pal K, Keshavan S, Caulfield TR, Dutta SK, Wang E, Fadeel B, Mukhopadhyay D (2019) Development of multi-drug loaded PEGylated nanodiamonds to inhibit tumor growth and metastasis in genetically engineered mouse models of pancreatic cancer. Nanoscale 11:22006–22018
147. Costa J, Silva N, Sarmento B, Pintado M (2015) Potential chitosan-coated alginate nanoparticles for ocular delivery of daptomycin. Eur J Clin Microbiol Infect Dis 34:1255–1262
148. Luo Y, Li M, Hu G, Tang T, Wen J, Li X, Wang L (2018) Enhanced photocatalytic activity of sulfur-doped graphene quantum dots decorated with TiO_2 nanocomposites. Mater Res Bull 97: 428–435
149. Huang J, Deng Y, Ren J, Chen G, Wang G, Wang F, Wu X (2018) Novel in situ forming hydrogel based on xanthan and chitosan re-gelifying in liquids for local drug delivery. Carbohydr Polym 186:54–63

150. Wang C, Jin J, Sun Y, Yao J, Zhao G, Liu Y (2017) In-situ synthesis and ultrasound enhanced adsorption properties of MoS_2/graphene quantum dot nanocomposite. Chem Eng J 327:774–782
151. Ruzycka-Ayoush M, Kowalik P, Kowalczyk A, Bujak P, Nowicka AM, Wojewodzka M, Kruszewski M, Grudzinski IP (2021) Quantum dots as targeted doxorubicin drug delivery nanosystems in human lung cancer cells. Cancer Nano 12:1–27
152. Peynshaert K, Soenen SJ, Manshian BB, Doak SH, Braeckmans K, De Smedt SC (2017) Coating of quantum dots strongly defines their effect on lysosomal health and autophagy. Acta Biomater 48:195–205
153. Wang Y, Yang C, Hu R, Toh HT, Liu X, Lin G (2015) Assembling Mn: ZnSe quantum dots-siRNA nanoplexes for gene silencing in tumor cells. Biomater Sci 3:192–202
154. Demir GM, Ilhan M, Akkol EK (2017) Effect of paclitaxel loaded chitosan nanoparticles and quantum dots on breast cancer. PRO 1:1074

Chapter 9
Natural Plant-Derived Bioactive Compounds as Health Promoters

Sunidhi Shreya, Priya Arya, and Amit Gupta

Abstract Bioactive compounds in the form of primary and secondary metabolites were isolated and purified from plant products, including fungi, microbes, etc. In the literature, these metabolites are described as having immunobiological properties on organisms that may be used in the food and pharmaceutical industries. Identification, isolation, and finding novel properties of such molecules may be helpful in finding various applications in the fields of cosmetics, materials science, bioremediation, etc. So, these bioactive molecules were extracted and obtained from plant products, and this is the only option for preventing the burden of human diseases. These molecules are one of the ideal sources for drug development, and their concentration is very low, as reported in the literature. Nowadays, various methods are applied to develop some effective method or protocol for extraction along with isolation and characterization of bioactive molecules. Today, conventional methods are being replaced because they are time-consuming, and these are mainly through green solvents (ionic liquids, eutectic solvents, etc.) and nonconventional techniques (electric fields, microwaves, etc.). All these methods were characterized and optimized for their strategies, which mainly boost the commercial values of agrowastes along with organic residues, promoting a sustainable circular economy. Inspite of the development of microfluidics, nanoencapsulation, and metabolic engineering, which may be able to help improve the screening process along with the extraction, stability, and functionality of compounds. In this chapter, we focused on extraction and characterization methods of bioactive compounds and discuss about its promising health beneficial attributes of bioactive components.

Keywords Bioactive · Medicinal plant · Extraction · Isolation · Characterization

S. Shreya · P. Arya · A. Gupta (✉)
Department of Microbiology, Graphic Era (Deemed to be) University, Dehradun, India
e-mail: dr.amitgupta.bt@geu.ac.in

R. Soni et al. (eds.), *Microbial Bioactive Compounds*,
https://doi.org/10.1007/978-3-031-40082-7_9

9.1 Introduction

Extraction is one of the major methods for the extraction of molecules from medicinal plant products. It is crucial to isolate and extract the desired chemical component, which is available in crude form in plant material. Further standardization and characterization are required for getting the pure molecule and evaluating its biological properties. In this method, various steps were followed, i.e. prewashing, preparation of a homogenous sample from dried plant material, solvent system, etc. Proper precautions must be taken so as not to lose active constituents during plant extract preparation. The most important criteria for the selection of plant material are based on traditional uses. In addition, the selection of the solvent system (polar or nonpolar solvents) is also important and required for targeting the bioactive compounds (hydrophilic or lipophilic) [1, 2]. In hydrophilic extraction, it is preferable to use polar solvents like methanol, ethanol, ethyl acetate, etc., whereas in the case of lipophilic compounds, it is preferable to use dichloromethane or a mixture of dichloromethane and methanol in a ratio of 1:1. In contrast, hexane is also used to remove chlorophyll content. Depending on the suitability of the solvent system, we can easily target that compound in polar or nonpolar form using a suitable extraction method. In simple words, extraction efficiency is totally dependent on particle size because of the enhancement in solvent penetration and properly diffusing the solutes. If someone is working on fine particle size, in this case, we consume maximum absorption of solute in solids and face difficulty in subsequent filtration. If samples were exposed to high temperatures, which may ultimately increase the solubility rate, but somehow lost the extracts and obtained undesirable impurities, extraction efficiency is totally dependent on time duration, and it may increase within a certain time range. In contrast, if we enhance the time duration during extraction, it will not affect the sample at all until solute equilibrium is reached inside and outside the solid material. The extraction methods [3, 4] are totally dependent on two factors:

- Conventional methods (maceration, percolation, and decoction) preferably use organic solvents and require a large volume of solvent and a long extraction time. The simplest method is maceration, which is mainly applied to thermolabile components. This is a very simple extraction method, and its major disadvantage is its long extraction time and extremely low efficiency (Table 9.1).
- Some modern methods, such as supercritical fluid extraction, pressurized liquid extraction, and microwave-assisted extraction, have also been applied in natural product extraction, and they offer some advantages such as lower or minimized consumption of organic solvent, shorter or reduced extraction times, and higher selectivity. These modern methods have some advantages over conventional methods where they meet the concept of green processes. In short, these technologies are energy-efficient alternatives and produce maximum yield with less extraction time.

Table 9.1 Extraction methods for natural products

Types of method	Required solvent	Temperature	Time	Pressure	Consumption of organic solvent	Extraction of polar natural products
Maceration	Aqueous (water) and nonaqueous solvent	Room temperature	Extensive	Atmospheric	Huge	Dependent on extracting solvent
Percolation	Aqueous (water) and nonaqueous solvent	Occasionally under heat and normally, kept at room temperature	Extensive	Atmospheric	Huge	Dependent on extracting solvent
Decoction	Water	Under heat	Average	Atmospheric	None	Polar compounds

9.2 Plants' Major Types of Bioactive Compounds

9.2.1 Flavonoids

Flavonoids are the most abundant and diversified class of bioactive molecules, frequently referred to as phytonutrients or phytochemicals, and are the primary elements of polyphenols. They have a chemical structure that includes a double-bonded diphenylpropane structure characterized by two benzene rings (rings A and B) that are linked by a three carbon chain, which forms a closed pyran ring (heterocyclic ring with oxygen, the C ring) with the benzene A ring. Flavonoids are naturally found as glycosylated or esterified conjugates, although they can also exist as aglycones, particularly as the outcome of food. Flavonoids are found in all fruits and vegetables and, combined with carotenoids, are responsible for their distinct hues. Apples, oranges, carrots, onions, tomatoes, the latter, parsley, lemon, beans, and other fruits and vegetables are high in flavonoids. Flavonoids are compounds found in various plants that have been demonstrated to have beneficial effects on human health. These effects are mainly due to their antioxidant and anti-inflammatory properties, which help combat the damaging effects of free radicals on cells and tissues. Flavonoids have also been shown to possess anti-ageing and anti-carcinogenic properties. They can positively impact the nervous system and regulate the activities of certain enzymes and receptors. Proanthocyanidins, which are oligomers of flavonoids, have a similar structure and effects. Both flavonoids and proanthocyanidins are pigments that occur naturally in various plants [5, 6].

9.2.2 Terpenoids

Terpenoids represent a significant cluster of natural products, mainly coming from plants but also produced by other organisms like bacteria and yeast in primary or secondary metabolism. They are made up of two five-carbon units called isoprenoids. Terpenoids offer therapeutic benefits in the management of various types of diseases, including cancer, as well as displaying antimicrobial, antihyperglycaemic, and anti-inflammatory properties. Moreover, they have insecticidal properties and are anti-allergenic, antispasmodic, antiviral, and immune modulators. Monoterpenoids have two isoprene units, while sesquiterpenoids have three. They are considered lipophilic in nature and have a high volatilization rate. They have intense smells and tastes. Their activities vary significantly, and a variety of them have been used in herbal medicines. Alzheimer's patients have high acetylcholinesterase activity, which is decreased by some bicyclic monoterpenoids. Diterpenoids are made up of 4 isoprene units (each with 20 carbons). They are extremely lipophilic and have intense tastes, but they are not volatile and hence odourless. Diterpenoids are also often found in resins. The resins are complicated, lipid-soluble combinations that often contain both nonvolatile and volatile chemicals. Resins released by wood are the most common; however, resins are also found in herbaceous plants. All of them are sticky, and how fluid they are depending on how many volatile compounds they contain. They get harder when exposed to the air. The majority of resins have antibacterial and wound-healing properties. α-terpineol contains insecticidal and skin penetration enhancing effects. Sesquiterpenes have antiallergic and anti-inflammatory qualities [7, 8].

9.2.3 *Alkaloids*

Alkaloids contain nitrogen, which causes the alkalinity of the compound. They are bitter in taste and heterocyclic in nature. These substances are often generated by a wide variety of plant species, mostly blooming plants and certain animals. Like inorganic alkalis, alkaloids react with acids to produce salts. In acid-base methods, these nitrogen atoms have the ability to serve as bases. In their purest form, alkaloids are typically odourless, pigmentless crystalline solids; however, occasionally they can also be yellowish liquids. It demonstrated a wide range of therapeutic qualities. Although several of them have local anaesthetic characteristics, their usefulness for therapeutic purposes is constrained. Although morphine is a potent opioid used to treat pain, its utility is constrained by its propensity for addiction. Quinine is an effective antimalarial drug. The human diet includes several alkaloids in both food and beverage forms. Alkaloids are compounds found in plants that are consumed by humans, including tomatoes, potatoes, cacao seeds, coffee seeds, and tea leaves. Alkaloids can directly affect the human brain or stimulate human organs like the central nervous system. Nicotine is a powerful stimulant and an extremely addictive chemical that is derived from the tobacco plant (*Nicotiana tabacum*). The antiparasitic, antiplasmodial, anticorrosive, antioxidative, antibacterial, anti-HIV, and insecticidal actions of alkaloids are only a few of their numerous beneficial uses [9, 10].

9.2.4 *Betalains*

These are red and yellow in colour and contain nitrogen. The reddish-violet betacyanins and the yellow betaxanthins are two different types of indole-derived pigments known as betalains. The hues of these pigments are determined by the resonance of double bonds in the betalain structure. They can be included in an aqueous food system since they are water-soluble. Several cacti, pear, red and yellow beetroot, and amaranth are food sources of betalains. The health benefits of betalains are cumulative due to their antioxidant, anticancer, anti-lipidemic, and antibacterial properties. They are harmless when consumed, making them potentially useful as functional foods and a prospective replacement for supplemental medicines in diseases like cancer and hypertension that are examples of oxidative stress, inflammation, and dyslipidemia-related diseases [11].

9.2.5 *Glucosinolates*

The glucosinolates include sulphur-rich aglycones generated from amino acids. Certain pungent substances and their secondary metabolites, which include sulphur and nitrogen, naturally contain substances called glucosinolates. Sulphur and nitrogen are found in glucosinolates. They are produced from glucose and an amino acid. Cruciferous plants, such as wasabi, broccoli, cabbage, and kale, are the major sources of glucosinolate. The most physiologically active breakdown products of glucosinolates are isothiocyanates. Numerous glucosinolates and their physiologically active by-products, in particular the isothiocyanates, have been demonstrated in studies to have defensive properties against cancer and dementia and are well recognized for killing a variety of cancer cells without harming healthy cells. They lessen the chance of developing dementia and delay the elderly's rate of cognitive deterioration [12].

9.3 Bioactive Components and Epigenetic Modifications

In the literature, bioactive components used as nutrients in food may directly impact human health and also show effectiveness against intracellular pathogens. These bioactive compounds directly target the DNA expression in a cell or tissue and modify the genome, which indicates enhancing or declining the fabrication of specific proteins in a cell. So, these reorganizations were reported in DNA or histones and considered epigenetic marks, and the phenomenon is called epigenomics. In short, epigenetic modifications occur in chromatin without any effect on the nucleotide sequence. These changes modified the pattern of gene expression, but this expression may be progressive and somehow reversible. These epigenetic marks are transmitted from cell to cell or from one generation to another as cell division occurs. According to the literature, these epigenetic marks are totally dependent on two factors, i.e., DNA methylation and histone modification [13, 14]. In DNA methylation, proteins were chemically tagged with methyl groups of DNA bases, which make DNA more or less accessible to transcriptional apparatus and change the expression pattern of specific genes. Similarly, modifications in histones are also reported because DNA is always intact and shows its modifications chemically in histones. So, DNA is wrapped around histones, which directly affect the structure of DNA and also have an effect on proteins with reference to transcriptional activity.

- **Folates** (a water-soluble vitamin, foliate, and folic acid) are directly obtained from food and its supplements. The most familiar examples are lemons, oranges, and tomato juice, mushrooms, yeast, bananas, spinach, etc. In general, folate is available on the market in the form of capsules that are prevalent and enhance folate levels in the blood; more importantly, the use of these capsules in pregnant women has been linked to neural tube effects. This folate is mainly involved in the synthesis of DNA, including its repair, along with its methylation. During dietary digestion, folate undergoes a series of reactions, first being converted into tetrahydrofolate (remethylation of homocysteine to methionine, precursor of S-adenosyl-L-methionine). Thereafter, methyl group transfer occurred and converted S-adenosyl-L-methionine into S-adenosyl-L-homocysteine (SAH), an inhibitor of the methylation reactions [15]. This chemical reaction is having some significant impacts, especially in those patients with folate deficiency, and changes are occurring in specific proto-oncogenes because of DNA methylation patterns. In short, the recommendation of the uptake of folate capsules in the diet must be scrutinized and considered as one of the most active bioactive compounds pertaining to reducing the abnormal proto-oncogene expression in cellular events.
- **Vitamin A** (a fat-soluble vitamin; retinoids) is reported in green vegetables, fish, meat, etc. This vitamin plays an important role in cell growth, differentiation, reproduction, vision, and immune function. In addition, this vitamin is metabolized intracellularly in the form of retinal and retinoic acid (the active metabolite of vitamin A) and supports several physiological functions. In general, when this bioactive compound in the form of a vitamin is absorbed and migrated to the nucleus, it binds to receptors (nuclear retinoic acid receptors, RARs) and is then characterized in the form of RAR (α, β, and γ) which heterodimerize with retinoid X receptors (RXRs) [16]. Due to this complex, which ultimately binds with peculiar elements and declines at the gene level, our interest at the transcriptional level with reference to biological and pharmacological responses has declined in the case of disease conditions.
- **Vitamin D3** (cholecalciferol) is reported in eggs, fish, milk, etc. and, most importantly, is directly obtained from sun exposure. Contrasting appearances of vitamin D are available, e.g., in humans, the existence of two vitamins is reported, i.e., vitamin D2 (reported in plants) and vitamin D3 (reported in human skin when exposed to sunlight). In the literature, this vitamin D3 played an important role in calcium homeostasis and then converted vitamin D3 into calcitriol (the active form), which showed its importance as a bioactive component in human nutrition [17].

9.4 Promising Health Beneficial Attributes of Bioactive Components

Bioactive components were isolated, characterized, and purified from plant products using organic solvents, and they are called secondary metabolites. These metabolites have promising therapeutic applications, i.e., antioxidant properties. In the literature, phenolics from plant products may be considered one of the phytochemical or bioactive compounds that may help to maintain better human health. So, these phenolics, which are more commonly reported in fruits and vegetables (orange and yellow-coloured), may contain enough lipophilic molecules called carotenoids. These carotenoids are widely applied and used for various industrial purposes, especially food in the form of pigments, and promoted as health dietary agents. One of the most familiar examples is related to protection against cardiovascular diseases, sunburn, cataracts, etc. due to the usage of these bioactive compounds like zeaxanthin, β-cryptoxanthin, and lutein. In addition, carotenoids are gaining more interest because of their strong antioxidant properties, which may be helpful in reducing the burden of disease. Similarly, polyphenols are reported as natural antioxidants that are mainly derived from fruits, cereals, vegetables, spices, etc. Several classes of polyphenols (phenolic acids, flavonoids, and anthocyanins) are also reported, which show several immunobiological properties [18, 19].

In the literature, these phenolic components from plant products are called free radical inhibitory or antioxidant agents because of their ability to release electrons or hydrogen atoms. A large number of antioxidative compounds were reported from fruit and vegetables, which may be helpful in reducing the cases of disorders, especially in the heart, cancer, arthritis, etc. Similarly, bioactive compounds in the form of peptides were reported in wheat, cereals, and rice, which may show antihypertensive activity [18–20]. According to the literature, several plant-derived bioactive compounds have some biological properties, i.e., antidiabetic and anticancer activity. Some of the most common health-promoting attributes of bioactive components are:

Anticancer Medicinal plant-derived drugs play a paramount role in pharmaceutical or human health care. In nature, several prophylactic or therapeutic compounds with chemical diversity were reported in several species of plants, animals, and microorganisms. In the literature, plant-derived candidates were reported and claimed to have anticancer activity, but most of them are still in clinical trials. To date, researchers working on structural modifications of compounds derived from plants have shown chemical diversity, which may further improve bioactive molecules, which is helpful for the process of drug development. Numerous studies were conducted related to anticancer in order to identify some novel phytochemical or bioactive compounds from plant products. So, these bioactive molecules, alone or in combination with other drugs, showed additive or synergistic effects, giving some options or an ideal candidate against cancer therapies. The most familiar examples of phytochemicals, especially anthraquinone, reported in plants, e.g., *aloe vera*, have anticancer potential. *Aloe vera* leaves contain bioactive molecules, i.e., aloe-emodin and emodin, derivatives of anthraquinone, which have antiproliferative effects in cancer model studies. In addition, emodin showed effective results with reference to declining androgen receptors and prostate cancer growth. Both derivatives of anthraquinone that arrest cell cycle analysis (enhancement in p53 expression and upregulation of p21) played an important role in the induction of apoptosis and disrupted the membrane potential of mitochondria, cytochrome c release, and caspase activation [20–22].

Today, cancer is one of the serious health concerns that are reported in both developing and developed countries. This disease may be due to unwanted cell growth and irregular cell division. In this regard, medicinal plants containing bioactive molecules have been used by villagers since ancient times to treat cancer. The most familiar examples of phytochemicals from plants reported as having anticancer activity are curcumin, vinblastine, vincristine, camptothecin, etc. In contrast, terpenoids and flavonoids also showed promising results against cancer.

Antidiabetes One of the chronic diseases, diabetes mellitus, is one of the metabolic disorders seen in the pancreas, which is mainly seen through β-cells that have a condition called hyperglycaemia. This condition may arise due to a deficiency or decline in the level of insulin production by the pancreas, and several medications are required to control and lower the blood glucose level to a normal level. However, several drugs were available, but they showed several side effects and caused several serious consequences. In this regard, traditional methods were adopted using bioactive elements from plants and played an important role as alternative medicine. In the literature, bioactive compounds from plants were reported as antidiabetic agents and showed more promising activity in terms of their safety and efficacy. The most familiar example is seen in the case of *Momordica charantia*, i.e., bitter melon, which may have contained a number of bioactive compounds like momordin, momordicosides, polypeptide-p, saponins, etc., which are totally similar to insulin-like proteins and are responsible for declining blood glucose levels. These compounds were reported in the callus, seeds, and fruits of *Momordica charantia*, which are totally similar to human insulin and showed an antidiabetic effect in animal model studies. In addition, bitter melon may have enhanced the tolerance rate of glucose levels in diabetic mice and has also been reported in humans. Similarly, ginseng (family *Araliaceae*), a traditional plant, is reported as an antidiabetic plant because of a specific type of saponin called ginsenosides. In the literature, ginsenoside Rb2 was more effective in decreasing blood glucose levels, and these studies were conducted in streptozotocin-induced diabetic rats. The roots of this plant also showed antidiabetic properties because of bioactive compounds like ginsenosides (triterpene glycosides or saponins), panaxans, vanillic acid, salicylates, etc. All parts of the ginseng plant contained alkaloids, polypeptides, vitamins, proteins, phenols, etc. Moreover, ginseng is also available in red form, i.e., as red ginseng extracts, where ginsenoside (Rg2 and Rg3) concentrations are still higher than normal ginseng. This red ginseng significantly reduced blood glucose levels and enhanced the level of plasma insulin in streptozotocin-induced diabetic rats [22–

24]. Another antidiabetic plant like *Tinospora cordifolia* (family *Menispermaceae*) commonly known as Guduchi, reported polysaccharide content which is directly correlated with β-cell regenerative properties and claimed its antidiabetic medicine having some side effects. This activity is mainly due to its existence of glycosides, terpenoids, flavonoids, alkaloids, phenolic constituents, polysaccharides, etc.

Gut Health Food containing diverse varieties of bioactive components and bioactivators is reported in fruits, including phytochemicals. Diverse varieties of bioactive compounds were reported in different types of fruits, i.e., seed, peel, and fruit. Their concentrations of bioactive molecules may have varied within the same fruit due to geographical locations. All these variations are mainly due to some factors that are directly influenced by nutritional food components, i.e., environmental conditions, ripening stage, season, soil texture, etc. In the literature, fruit is rich in polyphenols, one of the largest groups of bioactive compounds, and their structure may have totally varied from one fruit to another. Most of the fruits claimed their antioxidant and anti-inflammatory potential and showcased their importance with reference to human health care. Several studies related to preclinical and clinical research were conducted related to the isolation and purification of bioactive compounds from fruits and showed their activity against human health disorders of digestive, reproductive, and cardiovascular disease. In short, these bioactive compounds from fruits may upregulate their antioxidant and anti-inflammatory potential to mitigate health ailments [25, 26].

Antithrombotic Bioactive substances have an impact on human health on a biological level. An example of this is an antithrombotic, which helps alleviate excessive bleeding. Platelet aggregation, which occurs when blood clots begin to develop as a result of platelets beginning to accumulate together in the circulation, may be prevented by polyphenols by slowing down the platelet aggregation process [24–27].

Manage Blood Pressure Bioactive chemicals found in a variety of fruits, vegetables, tea, and mineral water, according to epidemiological research, may help prevent high blood pressure. The inclusion of a variety of foods high in flavonoids in the diet may be a useful strategy for lowering blood pressure [26–28].

Anti-inflammatory Inflammation is a biological reaction to an infection, harm, or discomfort. Chronic inflammatory conditions, including arthritis, allergies, atherosclerosis, and possibly cancer, appear to be linked to one another. Nitric oxide (NO) is one of the main mediators of inflammation. The overproduction of NO is inhibited by bioactive elements such as polyphenols, bioactive peptides, etc. Consequently, bioactive substances can reduce inflammation [22–29].

Antioxidative The majority of the antioxidant properties of foods and plants are contributed by bioactive substances like polyphenols and carotenoids. Carotenoids' pigment has the capacity to function as antioxidants and thereby shielding cells from photooxidation. Additionally, research has been done on how carotenoids react with radical species, and they are well-known for their ability to quench singlet oxygen [22–30].

Eye Health Eye illnesses may be prevented or delayed in their course by bioactive substances found in plants and diets. According to recent research, consuming at least 6 milligrams of lutein (carotenoids) daily can reduce the prevalence of macular degeneration by 43%. Increasing lutein and zeaxanthin in the diet can also help decrease or stop present eye damage and prevent current issues from getting worse [30, 31].

Cardiovascular Disease Anthocyanins, polyphenols, flavonoids, and other plant-based bioactive chemicals may play a significant role in declining the rate of cardiovascular disease, according to epidemiological research. The likelihood of cardiovascular disease is reduced when people include foods with high anthocyanin and flavonoid content in their diet. The development of atherosclerosis and CVD is significantly influenced by oxidative stress and inflammation. The preventive effects of many bioactive substances against atherosclerosis and CVD may be related to their anti-inflammatory, antioxidative, and metabolic capabilities. It has been demonstrated in several studies that anthocyanin-rich berries are linked to improved heart health, substantial reductions in glucose metabolism, lipid peroxidation, LDL oxidation, dyslipidaemia, and total plasma antioxidant capacity. According to a number of cohort studies and randomized trials, flavonoids may reduce the incidence of CVD in part through improving LDL cholesterol levels, endothelial function, and sensitivity to insulin. Intake of six types of flavonoids, including anthocyanidins, proanthocyanidins, flavonols, flavanones, flavones, and flavan-3-ols, has also been linked to significant reductions in the risk of CVD [29–33].

Anticarcinogenic Numerous epidemiological studies that initially discovered a consistent link between a high intake of fruits and vegetables and a decreased risk of acquiring a variety of illnesses, including various types of cancer, have since revealed the health-promoting effects of bioactive compounds. These studies also discovered that bioactive compounds may inhibit the growth and progression of various cancer cells. Rich in phenolic compounds and with strong antioxidant qualities, eggplant might be useful in detoxifying free radicals. As a result, eggplants could potentially be used as a preventative food to lower the risk of cancer. Plant derivatives like polyphenols can restore adverse or unhealthy epigenetic alterations in cancer cells, obstruct carcinogenesis, stop the spread of metastatic cancer, or make tumour cells more sensitive to radiation and chemotherapy. The NF-κB pathway, which is known to control cell transformation, angiogenesis, cell proliferation, inflammation, invasion/metastasis, and survivability of cancer cells, is said to be inhibited by flavonoids found in fruits and vegetables. In addition, it has been demonstrated that the carotenoids and terpenoids found in plants have anti-inflammatory and anticancer properties. They hinder NF-κB signalling pathways, which play a crucial role in tumours and inflammatory diseases [31–34].

9.5 Characterization of Bioactive Molecules

Bioactive chemicals are compounds derived from natural sources that are not often found in food and are typically present in modest amounts. They have been recognized as essential elements associated with healthy living and illness prevention. These chemicals are the subject of significant research to determine their influence on health. They possess the power to influence one or more metabolic pathways, leading to improved health outcomes. Plants are the main source of bioactive substances. Natural bioactive compounds, commonly referred to as plant secondary metabolites, are vital for species competition, protection, attraction, and signalling but are not absolutely necessary for plant life. These substances are essential to animal life and hold considerable promise for improving human health, especially when it comes to their pharmacological or toxicological impacts on microbial infections and illnesses. Additionally, they have medicinal antioxidant qualities. Noncommunicable illnesses are spreading more widely as the population ages and becomes less active. As a result, consumers now have a greater need for natural goods as they look for long-term solutions to enhance their quality of life through customized nutrition. The necessity of a balanced intake is stressed since the quantity of bioactive compounds from plants consumed is frequently crucial in deciding whether the effect is beneficial or detrimental. These substances have a variety of effects, ranging from being very damaging to being beneficial or healing. In this regard, we explore characterization procedures for extracting bioactive compounds from plant products [4–8] (Fig. 9.1).

Nuclear Magnetic Resonance Spectroscopy (NMR Spectroscopy) It is a type of spectroscopy that relies on how atom nuclei absorb radiation from the electromagnetic spectrum in the range of 4 to 900 MHz. For figuring out the structure of bioactive (organic) molecules, NMR has emerged as the method of choice over the

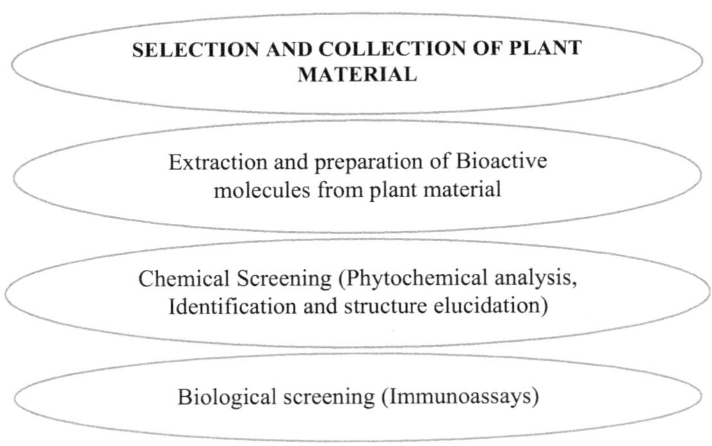

Fig. 9.1 Process of bioactive compounds from plant products

past 50 years. It is the only spectroscopic technique for which thorough investigation of the sample and deciphering of the full spectrum are often anticipated. Numerous nuclei have spin, and all nuclei have electrical charges, according to the basic principle of NMR. The base energy can move to a higher energy level (often a single energy gap) when an external magnetic field is supplied. Once the spin rebounds to its base level, energy is emitted at the exact same frequency that was used for the energy transfer, which occurs at a wavelength that is equivalent to radiofrequencies. To produce an NMR spectrum for the target nucleus, the signal that corresponds to this transfer is processed and measured in a variety of ways. The bioactive metabolite is put in a magnetic field, and the NMR signal is generated by radio waves excitation of the sample's nuclei, which is detected by sensitive radio receivers. A molecule's atom's intramolecular magnetic field can alter the resonance frequency, providing information about the molecule's electronic structure and its many functional groups. Proton and carbon-13 NMR spectroscopy are the two most commonly used forms of NMR; however, they may be used with any material that has nuclei with spin. The study of the manner in which electromagnetic radiation interacts with matter is known as spectroscopy. NMR spectroscopy uses NMR phenomena to investigate the material's physicochemical and biological characteristics [35, 36].

Mass Spectrometry One of the methods related to analytical chemistry for determining the bioactive compounds present is the mass-to-charge ratio and abundance of gas-phase ions. This technique involves bombarding the bioactive molecule (sample) with electrons; the bioactive molecules are mainly transformed into rapidly moving positive ions, including minor fragments, when they pass through a detector after moving through the electric and magnetic fields; and the charged particles are sorted according to their masses. Hence, they are detected, and the signals are recorded. A relative abundance in the form of a graph plotted against the mass/charge ratio (m/e) is called a mass spectrum. The masses of particles and molecules in a sample, their elemental or isotopic signatures, and the chemical structures of molecules and other chemical compounds may all be inferred from these spectra. A sample, which might be solid, liquid, or gas, is often ionized during a technique, such as by being bombarded with electrons. Some molecules in the sample might break into charged pieces as a result. Then, by accelerating them and exposing them to an electric or magnetic field, these ions are sorted based on their mass-to-charge ratio. Ions with the same mass-to-charge ratio will deflect to the same degree. A system that can detect charged particles, such as an electron multiplier, is used to find the ions [37, 38].

9.6 Chromatography Techniques

9.6.1 Thin-Layer Chromatography (TLC)

This technique conveniently delivers qualitative information, and quantitative data may be obtained by paying close attention to the little details. To separate and distinguish between different chemicals of interest, scientists use thin-layer chromatography [39]. A thin coating of silica is fixed to glass or aluminium to provide support in the construction of a TLC plate. The mobile phase is the solvent mixture, while the stationary phase is the silica gel. The compound of interest is, to varying degrees, soluble in the ideal solvent solution. The partition equilibrium of the mixture's constituent parts leads to separation. The simplest use of the method involves placing a small portion of the sample combination to be separated in a narrow zone or area towards one end of the TLC plate and then letting it dry. Making sure that the sample area is not submerged in the solvent, the strip or plate is next inserted into the solvent mixture, with this end sinking in. The test mixture divides into distinct components as the solvent goes towards the opposite end of the strip. This is called the development of TLC plates. Several things affect the separation.

- Solubility: A substance will travel along the plate more quickly; the more soluble it is in a solvent.
- Attractions between the elements and the silica; the compound travels less when it interacts with the silica.
- Compound size; bigger compounds travel along the plate more slowly.

9.6.2 HPLC (High-Performance Liquid Chromatography) and HPTLC (High-Performance Liquid Chromatography)

This is an analytical method that is applied for the separation and estimation of bioactive molecules from plant material. Utilizing the fundamentals or basic techniques of column chromatography, where bioactive compounds are easily separated, and using the spectroscopy method in order to identify and quantify them. The development of column chromatography from low-pressure compatible glass columns to high-pressure compatible metal columns occurred in the 1960s. In simple terms, HPLC is a greatly enhanced variation of column liquid chromatography. A solvent is pushed through a column at high pressures of up to 400 atmospheres rather than being permitted to slide through it under gravity. In a separation column with a stationary phase (granular substance with very tiny porous particles) and a mobile phase (solvent or solvent mix), it may proceed with the purification processes. The mobile phase is pushed along the separation column under high pressure. The sample is in the form of a bioactive metabolite, which may be delivered into the mobile phase flow from the pump to the separation column through a valve with a linked sample loop, such as a tiny tube or a stainless steel capillary [40]. So, due to interactions with the stationary phase, components of several bioactive molecules are maintained at variable degrees and then migrate at various speeds across the column. Following its exit from the column, each chemical or bioactive molecule is identified by an appropriate detector, which sends a signal to the computer's HPLC programme. A chromatogram is produced in the computer's HPLC software at the conclusion of this procedure or run. The various compounds may be recognized and measured using the chromatogram. An excellent substitute for high-performance liquid chromatography (HPLC) and gas chromatography (GC), the HPTLC technique, is an automated, high-tech variation of thin-layer chromatography with improved and enhanced separation efficiency and detection limits. High-performance thin-layer chromatography is often referred to as planar chromatography or flat-bed chromatography. High-performance thin-layer chromatography (HPTLC), an advancement of thin-layer chromatography (TLC), is a reliable, easy-to-use, quick, and effective technique for quantitative analysis of substances [41]. An analytical method termed HPTLC is based on traditional liquid chromatography (TLC), but it has been modified to allow for quantitative evaluation of the compounds and to improve the separation of the compounds' resolution. Some of the modifications allow for elevated resolution, such as the use of higher-quality TLC plates with smaller particles in the stationary phase. Repeated plate development utilizing a multiple development device can further enhance the separation. As a result, HPTLC provides a higher limit of detection (LOD) and improved resolution. Similar to the manner in which TLC separates matter by adsorption, HPTLC operates on the same principles. Through capillary action, the solvent or mobile phase flows. The adsorbent (stationary phase) is approached by the analytes with affinities that vary. Moving closer to the stationary phase is the component with stronger affinity. Towards the stationary phase, a low-affinity component moves quickly. So, the components are separated on a chromatographic plate.

9.6.3 Non-chromatographic Techniques

Phytochemical Screening Assay The word "phytochemicals", which refers to substances obtained from plants, is frequently used to refer to the numerous secondary metabolic products that are present in plants. A crucial tool in bioactive component analysis, the phytochemical screening assay, is a rapid, easy, and affordable process that provides the researcher with a quick answer regarding the different kinds of phytochemicals present in a combination. Following the extraction of the crude extract or active fraction from the plant material, phytochemical screening may be carried out using the proper assays to determine the types of phytochemicals present in the extract combination or fraction [42].

Immunoassay Analyses of bioactive substances are increasingly using immunoassays, which employ monoclonal antibodies against drugs as well as low molecular-weight natural bioactive molecules. For receptor-binding inquiries, enzyme tests, and analytical procedures, they exhibit great specificity and sensitivity. MAb-based enzyme-linked immunosorbent assays (ELISA) are frequently more sensitive than traditional HPLC techniques. Hybridoma technology is a process for producing monoclonal antibodies in specialized cells [43].

Fourier-Transform Infrared Spectroscopy (FTIR) For the characterization and identification of chemicals or functional groups (chemical bonds) found in an unknown mix of plants, FTIR has proven to be an invaluable instrument. A chemical "fingerprint" may be made from the FTIR spectra of pure substances since they are often so distinctive. By comparing an unknown compound's spectrum to a collection of known compounds, it is possible to determine the spectrum of the majority of common plant chemicals [37, 38]. There are several techniques to prepare samples for FTIR. The simplest method for liquid samples is to put one drop of the sample between two sodium chloride plates. A thin layer is created between the plates by the drop. An alternative is to dissolve solid samples in a solvent, such as methylene chloride, and then pour the resulting solution onto a single salt plate. A thin coating of the original substance is then left on the plate once the solvent has evaporated.

9.7 Future Prospects

Bioactive substances function as health promoters and are present in food, animal products, and nature. In vitro and in vivo are common laboratory techniques that are often used in order to demonstrate a bioactive substance. Bioactive compounds are the secondary metabolites produced by plants. They produce food that improves health by changing metabolic processes. Depending on the compound type, quantity, or bioavailability of the compound, the effects could be beneficial or detrimental. These substances have a variety of impacts, ranging from beneficial health maintenance to therapeutic benefits to being harmful or even lethal. The quantity of bioactive substances consumed frequently determines whether the impact is beneficial or harmful. They exhibit advantageous properties like antioxidative behaviour, antimicrobial impact, and organic pigmentation, among others. Recently, some indications have emerged regarding the involvement of specific bioactive substances in the postponement of illnesses like cancer and cardiovascular diseases. Although dietary nutrients are indispensable for survival, the necessity of bioactive compounds has not been established since the body can operate without them or their functions are concealed by nutrients that perform the same task. Numerous plant components that have the potential to have positive impacts on health are included in bioactive compounds. They are present in trace amounts and typically have pharmacological effects. These bioactive substances may have negative health impacts or positive health effects, depending on the dose. These effects have already been researched and studied using animal models and cell and tissue cultures. But in the present situation, an extensive amount of epidemiological evidence indicates that bioactive substances provide a wide range of health benefits for people, including the ability to prevent cancer, improve eye health, reduce the risk of cardiovascular disease, control blood pressure, and more.

References

1. Kalogeropoulos N, Chiou A, Pyriochou V, Peristeraki A, Karathanos VT (2012) Bioactive phytochemicals in industrial tomatoes and their processing byproducts. LWT - Food Sci Techn 49(2):213–216. https://doi.org/10.1016/j.lwt.2011.12.036
2. Karimi R, Azizi MH, Xu Q (2019) Effect of different enzymatic extractions on molecular weight distribution, rheological and microstructural properties of barley bran β-glucan. Int J Biol Macromolecules 126:298–309. https://doi.org/10.1016/j.ijbiomac.2018.12.165
3. Habeebullah K, Sattari Z, Al-Haddad S, Fakhraldeen S, Al-Ghunaim A, Al-Yamani F et al (2020) Enzyme-assisted extraction of bioactive compounds from Brown seaweeds and characterization. J Appl Phycol 32(1):615–629. https://doi.org/10.1007/s10811-019-01906-6
4. Khairil Anuar M, Mohd Zin Z, Juhari NH, Hasmadi M, Smedley KL, Zainol MK (2020) Influence of pectinase–assisted extraction time on the antioxidant capacity of Spent Coffee Ground (SCG). Food Res 4:2054–2061
5. Mulinacci N, Prucher D, Peruzzi M, Romani A, Pinelli P, Giaccherini C, Vincieri FF (2004) Commercial and laboratory extracts from artichoke leaves: estimation of caffeoyl esters and flavonoidic compounds content. J Pharm Biomed Anal 34:349–357
6. Altemimi AW, Watson DG, Kinsel M, Lightfoot DA (2015) Simultaneous extraction, optimization, and analysis of flavonoids and polyphenols from peach and pumpkin extracts using a tlc-densitometric method. Chem Cent J 9:1–15
7. Thoppil RJ, Bishayee A (2011) Terpenoids as potential chemopreventive and therapeutic agents in liver cancer. World J Hepatol 3:228–249
8. Wang Z, Yeast T, Han H, Jetter R (2010) Cloning and characterization of oxidosqualene cyclases from Kalanchoe daigremontiana: enzymes catalyzing up to ten rearrangement steps yielding friedelin and other triterpenoids. J Biol Chem 285:29703–29712
9. Hedhili S, Courdavault V, Giglioli-Guivarch N, Gantet P (2007) Regulation of terpene moiety biosynthesis of *Catharanthus roseus* terpene indole alkaloids. Phytochem Rev 6:341–351
10. Decendit A, Liu D, Ouelhazi L, Doireau P, Mérillon J, Rideau M (1992) Cytokinin-enhanced accumulation of indole alkaloids in *Catharanthus roseus* cell cultures: the factor affecting the cytokinin response. Plant Cell Rep 11:400–403
11. Osorio-Esquivel O, Alicia-Ortiz-Moreno Alvarez VB, Dorantes-Alvarez L, Giusti MM (2011) Phenolics, betacyanins and antioxidant activity in Opuntia joconostle fruits. Food Res Int 44: 2160–2168
12. Halkier BA, Gershenzon J (2006) Biology and biochemistry of glucosinolates. Annu Rev Plant Biol 57:303–333. https://doi.org/10.1146/annurev.arplant.57.032905.105228
13. Ajila CM, Naidu KA, Bhat SG, Prasada Rao UJS (2007) Bioactive compounds and antioxidant potential of mango peel extract. Food Chem 105:982–988
14. Alupului A, Calinescu I, Lavric V (2012) Microwave extractions of active principles from medicinal plants. UPB Science Bulletin 74:129–142
15. Stefanska B, Salame P, Bednarek A, Fabianowska-Majewska K (2012) Comparative effects of retinoic acid, vitamin D and resveratrol alone and in combination with adenosine analogues on methylation and expression of phosphatase and Tensin homologue tumour suppressor gene in breast cancer cells. Br J Nutr 107:781–790. https://doi.org/10.1017/S0007114511003631

16. Vu HT, Scarlett CJ, Vuong QV (2018) Phenolic compounds within banana peel and their potential uses: a review. J Funct Foods 40:238–248
17. Pereira A, Maraschin M (2015) Banana (musa spp) from peel to pulp: ethnopharmacology, source of bioactive compounds and its relevance for human health. J Ethnopharmacol 160:149–163
18. Lu JJ, Bao JL, Chen XP, Huang M, Wang YT (2012) Alkaloids isolated from natural herbs as the anticancer agents Evid based complement. Alternat Med 2012:485042. https://doi.org/10.1155/2012/485042
19. Finkeldey L, Schmitz E, Ellinger S (2021) Effect of the intake of Isoflavones on risk factors of breast cancer—a systematic review of randomized controlled intervention studies. Nutrients 13: 2309. https://doi.org/10.3390/nu13072309
20. Kunej T, Godnic I, Ferdin J, Horvat S, Dovc P, Calin GA (2011) Epigenetic regulation of MicroRNAs in cancer: an integrated review of literature. Mutat Res Mol Mech Mutagen 717: 77–84. https://doi.org/10.1016/j.mrfmmm.2011.03.008
21. El Omari N, Bakha M, Imtara H, Guaouguaoua FE, Balahbib A, Zengin G, Bouyahya A (2021) Anticancer mechanisms of phytochemical compounds: focusing on epigenetic targets. Environ Sci Pollut Res 28:47869–47903. https://doi.org/10.1007/s11356-021-15594-8
22. Mondal A, Gandhi A, Fimognari C, Atanasov AG, Bishayee A (2019) Alkaloids for cancer prevention and therapy: current Progress and Future perspectives. Eur J Pharmacol 858:172472. https://doi.org/10.1016/j.ejphar.2019.172472
23. Goldberg RB, Kendall DM, Deeg MA, Buse JB, Zagar AJ, Pinaire JA, Tan MH, Khan MA, Perez AT, Jacober SJ et al (2005) A comparison of lipid and glycemic effects of pioglitazone and rosiglitazone in patients with type 2 diabetes and dyslipidaemia. Diabetes Care 28:1547–1554. https://doi.org/10.2337/diacare.28.7.1547
24. Yakubu OE, Imo C, Shaibu C, Akighir J, Ameh DS (2020) Effects of methanolic leaf and stem-bark extracts of *Adansonia digitata* in Alloxan-induced diabetic Wistar rats. J Pharmacol Toxicol 15:1–7. https://doi.org/10.3923/jpt.2020.1.7
25. Sturm C, Wagner AE (2017) Brassica-derived plant bioactives as modulators of Chemopreventive and inflammatory Signalling pathways. Int J Mol Sci 18:1890. https://doi.org/10.3390/ijms18091890
26. Broom LJ, Kogut MH (2018) Gut immunity: its development and reasons and opportunities for modulation in monogastric production animals. Anim Health Res Rev 19:46–52. https://doi.org/10.1017/S1466252318000026
27. Memarzia A, Khazdair MR, Behrouz S, Gholamnezhad Z, Jafarnezhad M, Saadat S, Boskabady MH (2021) Experimental and clinical reports on anti-inflammatory, antioxidant, and immunomodulatory effects of Curcuma longa and curcumin, an updated and comprehensive review. Biofactors 47:311–350. https://doi.org/10.1002/biof.1716
28. Potoka DA, Nadler EP, Upperman JS, Ford HR (2002) Role of nitric oxide and peroxynitrite in gut barrier failure. World J Surg 26:806–811. https://doi.org/10.1007/s00268-002-4056-2
29. Maldonado Galdeano C, Cazorla SI, Lemme Dumit JM, Vélez E, Perdigón G (2019) Beneficial effects of probiotic consumption on the immune system. Ann Nutr Metab 74:115–124. https://doi.org/10.1159/000496426
30. Pickard JM, Zeng MY, Caruso R, Nunez G (2017) Gut microbiota: role in pathogen colonization, immune responses, and inflammatory disease. Immunol Rev 279:70–89. https://doi.org/10.1111/imr.12567
31. Collado MC, Meriluoto J, Salminen S (2007) Role of commercial probiotic strains against human pathogen adhesion to intestinal mucus. Lett Appl Microbiol 45:454–460. https://doi.org/10.1111/j.1472-765X.2007.02212.x
32. Bonnefont-Rousselot D (2016) Resveratrol and cardiovascular diseases. Nutrients 8:250. https://doi.org/10.3390/nu8050250
33. Hibino S, Kawazoe T, Kasahara H, Itoh S, Ishimoto T, Sakata-Yanagimoto M, Taniguchi K (2021) Inflammation-induced tumorigenesis and metastasis. Int J Mol Sci 22:5421. https://doi.org/10.3390/ijms22115421

34. Voorrips LE, Goldbohm RA, van Poppel G, Sturmans F, Hermus RJ, van den Brandt PA (2000) Vegetable and fruit consumption and risks of colon and rectal cancer in a prospective cohort study: The Netherlands cohort study on diet and cancer. Am J Epidemiol 152:1081–1092
35. Dhaiwal K, Chahal KK, Kataria D, Kumar A (2017) Gas chromatography-mass spectrometry analysis and *in vitro* antioxidant potential of ajwain seed (*Trachyspermum ammi* L.) essential oil and its extracts. J Food Biochem 3:e12364
36. Chauhan NS, Sharma V, Thakur M, Dixit VK (2010) Curculigo orchioides: the black gold with numerous health benefits. Zhong Xi Yi Jie He Xue Bao 7:613–623
37. Hameed IH, Ibraheam IA, Kadhim HJ (2015) Gas chromatography mass spectrum and fourier-transform infrared spectroscopy analysis of methanolic extract of *Rosmarinus officinalis* leaves. J Pharmacogn Phytother 7(6):90–106
38. Hameed IH, Jasim H, Kareem MA, Hussein AO (2015) Alkaloid constitution of *Nerium oleander* using gas chromatography-mass spectroscopy (GC-MS). J Med Plants Res 9(9): 326–334
39. Kenyon AS, Flinn PE, Layloff TP (1995) Rapid screening of pharmaceuticals by thin-layer chromatography: analysis of essential drugs by visual methods. J AOAC Int 78:41–49. https://doi.org/10.1093/jaoac/78.1.41
40. Staszek D, Orłowska M, Waksmundzka-Hajnos M, Sajewicz M, Kowalska T (2013) Marker fingerprints originating from TLC and HPLC for selected plants from the *Lamiaceae* family. J Liq Chromatogr Relat Technol 36:2463–2475. https://doi.org/10.1080/10826076.2013.790770
41. Morlock G, Schwack W (2006) Determination of isopropyl-thioxanthone (ITX) in milk, yoghurt and fat by HPTLC-FLD, HPTLC-ESI/MS and HPTLC-DART/MS. Anal Bioanal Chem 385:586–595. https://doi.org/10.1007/s00216-006-0430-5
42. Savithramma N, Rao ML, Suhrulatha D (2011) Screening of medicinal plants for secondary metabolites. Middle East J Sci Res 8:579–584
43. Reis SRIN, Valente LMM, Sampaio AL, Siani AC, Gandini M, Azeredo EL, D'Avila LA, Mazzei JL, Henriques MGM, Kubelka CF (2008) Immunomodulating and antiviral activities of *Uncaria Tomentosa* on human monocytes infected with dengue Virus-2. Int Immunopharmacol 8(3):468–476. https://doi.org/10.1016/j.intimp.2007.11.010

Chapter 10
Prolific Microbial Agents as Key Products for Sustainable Agriculture

Viabhav Kumar Upadhayay, Yogesh Dashrath Naik, Nishant Ranjan, Chandranshu Kastury, Shivam Shekhar, Shailesh Kumar, and Vandna Jaggi

Abstract The current agricultural system is confronted with the challenge of excessive reliance on chemical-based fertilizers and pesticides. While these inputs have revolutionized agriculture, they also pose significant environmental risks. As a result, the utilization of agriculturally important microorganisms has become imperative to ensure sustainable agriculture in an environmentally friendly manner. These microorganisms can serve as biofertilizers, offering a wide range of plant growth-stimulating traits such as nitrogen fixation, nutrient solubilization, synthesis of siderophores and phytohormones, etc. By establishing symbiotic relationships, they enhance soil fertility, improve nutrient availability, and promote plant growth, thereby reducing the rely on synthetic fertilizers. Moreover, beneficial microorganisms act as natural adversaries to pests, providing an alternative to chemical pesticides. Microbes also enhance crop resilience to abiotic stresses such as drought and salinity through the production of stress-tolerant compounds, modulation of plant hormones, and improved nutrient uptake efficiency. Furthermore, they contribute to climate-smart agriculture by sequestering carbon in the soil, thereby mitigating greenhouse gas emissions. The use of microbial consortia further enhances plant

V. K. Upadhayay (✉)
Department of Microbiology, College of Basic Sciences & Humanities, Dr. Rajendra Prasad Central Agricultural University, Samastipur, Bihar, India

Y. D. Naik
Department of Agricultural Biotechnology and Molecular Biology, College of Basic Sciences & Humanities, Dr. Rajendra Prasad Central Agricultural University, Samastipur, Bihar, India

N. Ranjan · S. Shekhar
College of Basic Sciences & Humanities, Dr. Rajendra Prasad Central Agricultural University, Samastipur, Bihar, India

C. Kastury · S. Kumar
Department of Botany, Plant Physiology and Biochemistry, College of Basic Sciences & Humanities, Dr. Rajendra Prasad Central Agricultural University, Samastipur, Bihar, India

V. Jaggi
Department of Microbiology, College of Basic Sciences & Humanities, G. B. Pant University of Agriculture and Technology, Pantnagar, India

R. Soni et al. (eds.), *Microbial Bioactive Compounds*,
https://doi.org/10.1007/978-3-031-40082-7_10

growth, disease suppression, and stress tolerance. Additionally, microorganisms play an imperative role in biofortifying food crops, improving nutrient absorption, and addressing malnutrition. In summary, microorganisms offer diverse applications in sustainable agriculture, providing transformative solutions for crop-based food production.

Keywords Abiotic stress · Agriculturally important microorganisms · Beneficial microorganisms · Biofertilizers · Crops · Microbial consortia

10.1 Introduction

Sustainable agriculture has attracted a lot of attention recently as a way to meet the rising global food demand while reducing the negative environmental effects of traditional farming practices, which rely on chemical fertilizers and pesticides. In this context, the incorporation of microorganisms into agricultural systems has emerged as a promising and comprehensive strategy, in which agriculturally important microbes can be used as biofertilizers and biopesticide agents and ensure eco-friendly agricultural practices [1, 2]. However, the intricate interplay among microorganisms, plants, and soil forms the basis of the microbial-plant-soil nexus, which is fundamental to sustainable agriculture. In order to fully explore the potential of agriculturally important microorganisms for sustainable crop production, it is important to understand their mechanisms of action. These microorganisms possess diverse traits and functions that contribute to improved plant growth, improved nutrient availability, and improved soil health [3, 4]. By elucidating these mechanisms, targeted strategies can be developed to optimize their effectiveness and application in agricultural systems. The utilization of microorganisms as biofertilizers is a noteworthy application in the realm of sustainable agriculture. Microorganisms possessing diverse plant growth-promoting characteristics, including nitrogen fixation and phosphate solubilization, play a vital role in enhancing soil fertility and nutrient accessibility [5]. Biofertilizers facilitate nutrient uptake and augment crop productivity by creating symbiotic relationships with plants, thereby mitigating the need for synthetic fertilizers and their associated environmental hazards [6]. In addition, microorganisms have enormous potential as biocontrol agents in pest control. Beneficial microorganisms can act as natural enemies of pests, inhibiting their growth, reproduction, and pathogenicity [7]. This biological control approach offers an environmentally friendly alternative to chemical pesticides and minimizes negative impacts on ecosystems and human health. The ability of crops to withstand abiotic stresses such as drought, salinity, and extreme temperatures is essential for ensuring sustainable agricultural production [8]. Various mechanisms have been identified through which microorganisms can enhance plant tolerance and adaptation to stress [3, 8, 9]. Some of the methods used to enhance plant growth and resilience include producing stress-tolerant compounds, modifying plant hormone levels, and enhancing nutrient uptake efficiency [10]. Integrating these stress-tolerant microorganisms into agricultural practices offers prospective options for maintaining crop yields in challenging environments. In addition, the application of microbial-based products shows potential for implementing climate-smart agricultural practices. Microbial organisms have the ability to sequester carbon within the soil, thereby decreasing the amount of greenhouse gas especially carbon dioxide and serving as a means of climate change mitigation [11]. Microbial consortia, consisting of compatible microorganisms, have demonstrated increased synergistic impacts on plant growth and productivity. Microbial consortia are now an efficient way for improving crop growth, disease suppression, and stress tolerance in plants compared to single microbial inoculants, by utilizing the various functional properties of different microorganisms [12]. Microbial inoculants are also important for biofortifying food crops, which can help alleviate global malnutrition and enhance human health [13–15]. They have the potential to improve the absorption and availability of vital nutrients, including iron, zinc, and selenium, in crops [16]. This improvement enhances the nutritional quality of crops, addressing nutrient deficiencies in vulnerable populations.

This chapter sheds light on the diverse roles of microorganisms as key components in sustainable agriculture and highlights their immense potential for various applications. Comprehensive understanding and effective use of microorganisms are critical to unlocking transformative and long-lasting solutions for sustainable plant-based food production.

10.2 The Microbial-Plant-Soil Nexus: A Holistic Approach to Sustainable Agriculture

The microbial-plant-soil nexus embodies a comprehensive and scientifically sound tactic to achieving the goals of sustainable agriculture. It recognizes the complicated interactions between microorganisms, plants, and soil within agricultural ecosystems [17]. This approach highlights the use of beneficial microbes in augmenting plant growth, soil health, and promoting sustainable agricultural practices. Microorganisms play a key and indispensable role in nutrient cycling, a fundamental process in agricultural systems [18]. They actively contribute in the decomposition of complex organic matter, breaking it down into simpler forms that can be readily absorbed by plants [19]. Moreover, microorganisms contribute to nitrogen fixation, converting atmospheric N into a biologically usable form for plants. Through these processes, microorganisms improve nutrient accessibility, thereby promoting plant growth and reducing the need for synthetic fertilizers [20]. Furthermore, microorganisms exert a profound influence on soil structure and health. They actively contribute to the formation of soil aggregates, which enhance soil structure, porosity, and water infiltration [21]. These improvements in soil structure, in turn, enhance soil fertility and nutrient retention capacity. Additionally, certain microbes possess plant growth-promoting traits such as phytohormone production, nutrient solubilization, and facilitation of nutrient uptake by plants [2]. The microbial-plant-soil nexus also plays a crucial role in disease suppression within agricultural systems. Beneficial microorganisms act as natural antagonists against plant pathogens by actively competing for resources, producing antimicrobial compounds, and inducing systemic resistance in plants [22]. These activities effectively reduce the occurrence and severity of plant diseases. By harnessing biocontrol agents derived from microorganisms, farmers can minimize the use of chemical pesticides, which can have detrimental effects on the environment and human health [7]. Various strategies can be implemented to fully harness the potential of the microbial-plant-soil nexus. One such strategy involves the application of microbial biofertilizers, which contain beneficial microorganisms that enhance nutrient availability and promote plant growth [2]. These biofertilizers can be applied to seeds, roots, or soil to establish thriving and beneficial microbial communities, ultimately enhancing overall crop productivity. Furthermore, biopesticides derived from microorganisms offer an environmentally friendly alternative for pest control. Microbial-based biopesticides, such as *Bacillus thuringiensis* (Bt) and entomopathogenic fungi, specifically target pests while minimizing harm to nontarget organisms and reducing the risk of pest resistance development [23]. Integrating the microbial-plant-soil nexus into sustainable agricultural systems necessitates the adoption of soil management practices that support microbial activity [17]. This includes minimizing soil disturbance, maintaining soil organic matter through cover cropping and crop rotation, and reducing the use of chemical inputs that can disrupt microbial communities. By implementing these practices, farmers can foster a healthy and productive microbial ecosystem within the soil.

10.3 Understanding the Mechanisms of Action of Agriculturally Important Microorganisms

Agriculturally important microorganisms show imperative contribution in augmenting plant growth through diverse arrays of mechanisms. These microorganisms exert both direct and indirect influences on plant growth and development. Directly, they facilitate plant growth via phytostimulation and bio-fertilization. On the contrary, in the indirect way, they function as "bio-pesticides" or "biocontrol" agents [24]. The direct mechanisms employed by agriculturally important microorganisms encompass the facilitation of nutrient uptake and enhancement of nutrient availability. They possess the capacity to fix nitrogen [25], solubilize phosphorus and other essential mineral nutrients [26], and mineralize organic compounds [24]. Moreover, these microorganisms produce phytohormones such as "IAA," "ethylene," "cytokinins," and "gibberellins," which elicit plant growth responses [27]. The production of siderophores, which facilitate iron uptake, can be regarded as both a direct and indirect mechanism [28]. In addition to their direct effects, agriculturally important microorganisms exhibit indirect mechanisms that contribute to the promotion of plant growth. These encompass the production of antibiotics and hydrolytic enzymes, which help in combating plant pathogens and supporting plant health [29]. They also have the capability to induce systemic resistance in plants, thereby enhancing defense mechanisms against pathogens [8]. Furthermore, these microorganisms secrete exopolysaccharides (EPS), which foster soil aggregation and improve soil structure, ultimately benefiting plant growth [30].

The multifaceted activities of agriculturally important microorganisms establish them as precious contributors to plant growth and development. By harnessing these beneficial interactions, these microorganisms present promising prospects for promoting sustainable agriculture and ensuring ecological balance.

10.4 Microbial Agents as Biofertilizers for Improving Crop Productivity

Biofertilizers, classified as organic fertilizers, consist of microbial strains possessing plant growth-promoting characteristics. The excessive use of chemical-based fertilizers in recent years has raised concerns about their detrimental effects on the environment. Consequently, there is a growing public interest in adopting eco-friendly strategies. Utilizing biofertilizers is revolutionizing agricultural practices by providing an environmentally sustainable approach and reducing dependence on agrochemicals. This transformation hinges upon the careful selection of microbial strains to ensure optimal results. Microbial strains, including bacteria, fungi, and mycorrhizae such as "*Bacillus*," "*Rhizobium*," "*Lactobacillus*," '*Azotobacter*," "*Pseudomonas*," "photosynthetic bacteria," "*Trichoderma* sp., "*Glomus* sp.," "*Gigaspora* sp.," "*Pezizella* sp.," and "yeasts" exhibit a wide range of capabilities, such as nitrogen fixation, solubilization of phosphate, zinc, iron, and potassium, as well as the production of phytohormones and cellulolytic enzymes [2, 31–33]. These strains are primarily utilized as biofertilizers [2, 33]. Through the processes of nitrogen fixation, phosphate, potassium, and zinc solubilization, secretion of plant growth-regulating substances like hormones and vitamins and facilitation of organic matter biodegradation, biofertilizers play a crucial role in augmenting the plant growth and also contribute in maintaining soil health [2, 34]. Biofertilizers are widely acknowledged as microbial inoculants that effectively enhance nutrient availability in the soil, addressing the multifaceted challenges stemming from intensive chemical fertilizer usage [35]. In addition to their role in facilitating nutrient uptake by plants, biofertilizers exert a significant influence on various vital plant physiological processes, including the augmentation of water absorption and the promotion of photosynthetic rates [36]. Extensive research has documented the capacity of biofertilizers to enhance both abiotic and biotic stress tolerance in plants [8, 37]). Moreover, they play a pivotal role in the bioremediation of pesticides, contributing to the mitigation of their environmental impact [38–40]. Functioning as effective biocontrollers and biofertilizers exhibit noteworthy antagonistic properties against a diverse range of soil-borne plant pathogens, encompassing *Rhizoctonia* root rot, chill wilt, *Pythium* root rot, mung bean root rot, and parasitic nematodes [2, 41]. Advancements in bioformulation technologies are imperative for the successful commercialization of proficient microbial strains that possess biocontrol and plant growth-promoting capabilities. Several essential characteristics define an exemplary biofertilizer: (1) it must demonstrate environmental friendliness; (2) the microbial strains employed in its formulation must be nonpathogenic; (3) it should provide crops with high-quality nutrients; and (4) it should exhibit an extended shelf life [2, 42]. The meticulous selection of microorganisms possessing desirable traits stands as a pivotal factor in biofertilizer production. A comprehensive understanding of the interactions between microorganisms, crops, and the environment is vital to enhance crop growth [2, 43]. Microbes utilized in bioformulations undergo rigorous testing under in situ and in vivo conditions to ascertain the preservation of desired properties and the attainment of desired outcomes [44]. Furthermore, the chosen microbes for biofertilizer formulation should exhibit genetic stability, target specific crops, maintain synchrony with the native microbial population, and demonstrate survivability even in the absence of a host [5, 45]. The development of biofertilizers has traditionally focused on single microbial strains, but recent research emphasizes the advantages of employing multiple strains in the form of microbial consortia. These consortia act synergistically through diverse mechanisms, resulting in heightened effectiveness for crop enhancement [46, 47]. The process of biofertilizer development is intricate and requires rigorous assessments to meet stringent quality standards. Ensuring the viability of microorganisms is of paramount importance, enabling them to sustain soil fertility even after extended periods of storage [2, 48]. Biofertilizers can be formulated as dried powder, granules, or liquid, utilizing different carrier materials to support microbial growth and facilitate efficient delivery [49]. Liquid biofertilizers, in particular, can incorporate specialized cell protectants to extend their shelf life and require lower application dosages compared to other formulations [50]. The selection of an appropriate carrier material is a critical consideration, as it plays a significant role in preserving cell viability during storage and transportation [51]. The ideal carrier material should be nontoxic, possess high moisture absorption and water retention capacities, have a prolonged shelf life, and be easily processable [52]. Encapsulation of the inoculants with the carrier material ensures convenient handling, efficacy, and long-term storage capabilities [53].

The evaluation of biofertilizer efficacy can vary based on crop specificity. Various strategies can be employed, including seed treatment and soil application, for the utilization of specific biofertilizers [54]. In the context of paddy cultivation, seedling treatment with biofertilizers emerges as the preferred approach [55]. Following the application of biofertilizers to the soil, seeds, or roots, microorganisms establish colonization in the vicinity of the roots, thereby promoting growth in the targeted crop [56]. The root exudates excreted by plants facilitate the proficient colonization of microorganisms within the rhizosphere, optimizing their establishment and function [45].

10.5 Microbes as Biocontrol Agents and Their Potential for Pest Management

Pest management plays a pivotal role in modern agriculture and ecosystem preservation. However, traditional methods of pest control relying on chemical pesticides have proven to be environmentally harmful, posing risks to ecosystems, human health, and nontarget organisms [57–59]. Consequently, there is a growing interest in developing sustainable and eco-friendly alternatives for pest management. Microbes, including bacteria, fungi, viruses, and nematodes, have emerged as promising biocontrol agents due to their effective pest control capabilities while minimizing the negative impacts associated with conventional approaches [60]. For over a century, the study of microbes and their role in the health of living beings has been widely recognized. In modern agriculture, microbes have gained prominence as natural pesticides when combined with hybrid seeds, high-yield varieties, and regular irrigation, making them a leading trend in the agricultural sector [61]. Researchers are exploring sustainable methods to safeguard crops from insects and pathogens while enhancing soil health by harnessing the power of beneficial microorganisms. These microorganisms serve as natural biocontrol agents, inhibiting the growth of harmful pests and diseases while promoting plant growth and development [62, 63]. Among the bacteria used in agriculture, *Bacillus thuringiensis* (Bt) has long been employed due to its insecticidal proteins, making it a valuable and environmentally friendly biopesticide. Recent studies have suggested its potential use as a biofertilizer to enhance plant growth and its application in the development of transgenic plants (Liliana [64, 65]). *Pseudomonas chlororaphis* isolates are also utilized as biopesticides, providing protection to plants against a wide range of microbial pathogens, insects, and nematodes [66]. Entomopathogenic fungi (EPF) offer an environmentally sustainable approach to biocontrol against insect pests [67]. With over 700 species identified from approximately 90 different genera, these fungi have the ability to infect and induce disease in insects under favorable conditions [67]. Notable strains include *Beauveria bassiana*, *Metarhizium anisopliae*, *Hirsutella*, *Isaria*, *Lecanicillium*, and *Beauveria* [68, 69]. These fungi produce spores that attach to the pest's cuticle, penetrate it, and ultimately lead to the pest's demise. Fungal biocontrol agents are particularly effective against pests such as aphids, whiteflies, and thrips, and they offer a lower risk of developing resistance compared to chemical pesticides [67]. Insect-specific viruses, such as nucleopolyhedroviruses and granuloviruses, have shown great potential as biocontrol agents. These viruses selectively infect and eliminate their host insects [70, 71]. Microbial biocontrol agents provide numerous benefits for pest management. They demonstrate precise targeting, effectively controlling pests while minimizing harm to beneficial organisms. Moreover, these agents are environmentally friendly and pose no toxicity risks to humans. They also have the advantage of rapid degradation, reducing the potential for persistent residues in soil, water, and food [60]. Furthermore, their utilization supports sustainability and organic farming practices by decreasing reliance on synthetic pesticides. They can be seamlessly integrated into pest management programs, complementing other control methods.

10.6 Microbial Role in Enhancing Crop Resilience to Abiotic Stresses

Abiotic stress has emerged as a significant global concern, causing substantial agricultural losses on a widespread scale [72]. It encompasses the detrimental effects of nonliving environmental factors that impose stress on various species. These factors comprise extreme light conditions (both high and low), radiation (UV-B and UV-A), temperature fluctuations (both high and low), water-related challenges (drought, flooding, and submergence), chemical influences (heavy metals and pH), salinity resulting from excessive Na + levels, deficiency or excess of essential nutrients, gaseous pollutants (such as ozone and sulfur dioxide), mechanical factors, and other less common stressors [73]. These stressors can manifest individually or in combination. In agricultural settings, crops and plants regularly encounter stress due to a complex interplay of these factors, resulting in distinct effects [72, 74]. The accumulation of heavy metals in plants has detrimental consequences for their growth, photosynthetic activity, and crop yield [75]. Salinity stress disrupts various physiological processes, including seed germination, seedling establishment, vegetative growth, ionic toxicity, osmotic pressure, and oxidative damage [76–78]. Drought stress adversely affects key components of photosynthesis, such as photosystem-I and photosystem-II, and impairs the functionality of enzymes like ascorbate peroxidase, glutathione reductase, and superoxide dismutase [79]. Cold stress induces cell and tissue dehydration, crystallization of cellular water, reduced membrane conductivity, increased leakage of reactive electrolytes, decreased weight, and lower relative water content, ultimately leading to poor crop yield [80]. Plants require a unique response tailored to their environmental conditions to adapt to specific abiotic stress conditions. Recent research indicates that each abiotic stress situation necessitates a precise, personalized plant response, and the interaction of two or more stress factors may require a distinct response [72]. When two or more stresses occur simultaneously, an opposing response may be required. For instance, a common field scenario involves the combination of heat and drought stress. Under heat stress conditions, plants open stomata to cool the leaves through transpiration. However, when heat stress is combined with drought stress, plants are unable to open stomata, resulting in higher leaf temperatures [81]. Microbes employ various biochemical and molecular mechanisms to mitigate the adverse impacts of different abiotic stresses on plant growth and development [82].

Plants receive protection against abiotic stressors through various mechanisms employed by microorganisms. These include the synthesis of phytohormones, osmolytes, and exopolysaccharides (EPS), as well as the activity of 1--aminocyclopropane-1-carboxylate (ACC) deaminase and the induction of stress-responsive genes (Upadhayay et al., 2023). Plant-associated microorganisms, such as endophytes, arbuscular mycorrhizal fungi, and plant growth-promoting rhizobacteria [82], have been recognized for their ability to enhance crop yield and improve stress tolerance. Plant growth-promoting rhizobacteria (PGPR) play a vital role in this regard by producing phytohormones like indole-3-acetic acid (IAA), cytokinins, and abscisic acid. They also produce antioxidants such as superoxide dismutase (SOD), peroxidase (POD), ascorbate peroxidase (APX), catalase (CAT), and glutathione reductase (GR). Additionally, PGPR possess the enzyme ACC deaminase, which aids in the degradation of the ethylene precursor ACC, thereby helping to alleviate the adverse effects of abiotic stress and induce systemic tolerance [83, 84]. The presence of PGPR with ACC deaminase enzyme enables the regulation of ethylene production by converting ACC into alpha-ketobutyrate and ammonia, providing relief from stress [85]. AMF colonization has shown increased tolerance to water stress by enabling hyphae to reach water sources inaccessible to non-colonized plants via soil pores that are inaccessible to root hairs. Khalvati et al. [86] demonstrated water transport to the roots under drought conditions. Kavroulakis et al. [87] observed tolerance to water stress in *Solanum lycopersicum* cv ACE 55 by *Fusarium solani*, resulting in increased net CO_2 assimilation rate, enhanced antioxidant activity, and stomatal conductance. Mathur et al. [88] demonstrated drought resistance in *Triticum aestivum* through the colonization of *Rhizophagus intraradices* and *Funneliformis* spp., leading to increased relative water content, chlorophyll content, and restoration of electron transport in PS-II. *Bacillus* sp. and *Enterobacter* sp. provide drought tolerance in *Triticum aestivum* and *Zea mays* through the production of indole-3-acetic acid and salicylic acid [89]. *Funneliformis mosseae* enhances tolerance to low temperatures in *Solanum melongena* L. by improving photochemical reactions, activating the antioxidant defense system, accumulating protective molecules, and reducing membrane damage [90]. *Rhizoglomus intraradices* enhances salinity tolerance in *Pisum sativum* by improving nutrient uptake and promoting the accumulation of compatible osmolytes [91]. *Kocuria rhizophila* regulates plant hormones like abscisic acid and indole-3-acetic acid and improves nutrient acquisition, thereby providing salinity tolerance in Zea mays [92]. Table 10.1 illustrates the use of a various microbial strains to successfully mitigate a variety of stressors faced by plants.

10.7 Microbial Products and Soil Carbon Sequestration: A Pathway to Climate-Smart Agriculture

Climate change is a crucial global issue that has garnered significant attention from the scientific community worldwide. The primary driver behind climate change is human activities, which have contributed to a steady increase in global temperatures. Since the late nineteenth century, the Earth's average surface temperature has risen by approximately 0.9 °C, largely attributed to the substantial surge in carbon dioxide

Table 10.1 Microbial-mediated approaches for mitigating abiotic stress in plants

Microorganism	Plant	Type of stress	Stress mitigation	References
Bacillus megaterium PB50	Rice	Drought stress	Improved growth under osmotic stress, protected from physical stress by stomatal closure	Arun et al. [93]
Arthrobacter woluwensis (AK1)	Soybean	Salinity stress	Salt-tolerant gene GmST1 is expressed with 42.85% expression	Khan et al. [13, 14]
Bacillus megaterium and *Pantoea agglomerans*	*Vigna radiata*	Drought and aluminum stress	The consortium decreased Al uptake and increased abiotic stress tolerance	Silambarasan et al. [94]
Compost + PGPR	Tomato	Drought stress	Enhancement in the plant growth, accumulation of osmolytes and minerals, decrease patterns in activity of antioxidant enzymes	Tahiri et al. [95]
Enterobacter cloacae PM23	Maize	Salinity stress	Augmentation in radial scavenging capacity, relative water content, soluble sugar, phenolic content, flavonoid content, and accumulation of osmolytes (glycine betaine, proline, etc.)	Ali et al. [96]
S. putrefaciens and *C. dubliniensis*	Pearl millet	Drought stress	Increase in relative water content, improvement in the level of proline accumulation, enhancement in the expression level of genes related to phytohormone biosynthesis, and drought-responsive transcription factors	Manjunatha et al. [97]
L. fusiformis and *L. sphaericus*	Maize	Cold stress	Increase in level of osmolytes, phytohormones, and phenolics, improvement in the activity of antioxidant enzymes	Jha and Mohamed [98]

(CO_2) emissions resulting from human-induced activities [99]. The period of industrialization, which commenced in the 1750s, witnessed a rapid and substantial rise in atmospheric CO_2 concentration from 277 to 400 parts per million (ppm) [99, 100]. Since around 1920, fossil fuel combustion has become the dominant contributor to CO_2 emissions, disrupting the natural carbon cycle and necessitating the implementation of carbon sequestration measures [99, 101]. Carbon sequestration refers to the process of capturing and storing atmospheric CO2 in the soil over an extended period. This method is predominantly achieved by incorporating crop residues and organic matter into the soil [102]. Additionally, indirect sequestration can occur through chemical reactions that transform CO_2 into inorganic compounds like "calcium carbonate ($CaCO_3$)" or "magnesium carbonates ($MgCO_3$)" [99, 103]. On the other hand, direct sequestration involves the fixation of CO2 into plant biomass through photosynthesis [104]. Various natural elements function as either carbon sources or sinks, depending on their capacity to absorb or release carbon. "Organic matter decomposition," "respiration and digestion activities," "volcanoes," and "water bodies" serve as natural carbon sources [105], while forests, photosynthesis, Earth's crust, soil, oceans, and freshwater bodies act as carbon sinks [99, 106]. Maintaining a balanced carbon cycle necessitates a proportional release of carbon from sources and sinks [107]. Carbon sequestration is influenced by a multitude of factors. These factors include the rate of production and decomposition of soil organic matter, the composition of the parent material, the position of the landscape, temperature and precipitation patterns, the presence of living organisms, and various management practices [99, 108]. Among these factors, SOM plays a significant role in modifying soil carbon stocks, thereby affecting the potential for soil sequestration [109]. Numerous processes contribute to the release and transport of SOM within the soil, influencing its physical, chemical, and biological characteristics, and ultimately impacting the potential for carbon sequestration [108]. The microbes found in rhizospheric soil, specifically known as "plant growth-promoting rhizobacteria" or "PGPR," have the capability to enhance soil microbial functioning, creating a positive contribution to global climate change. PGPR constitutes a significant portion of the overall microbial community and plays a crucial role in carbon sequestration [110]. The mechanisms by which PGPR mitigate climate change and sequester carbon involve multiple pathways [99]. PGPR plays a vital role in nutrient cycles, including those of "C" and "N" [111]. They exert improvement in the production of "glomalin" in the rhizospheric milieu by promoting mycorrhizal colonization [112]. Glomalin acts as an important reservoir of C and N in the soil [113]. Moreover, PGPR has the ability to directly enhance plant growth and allocate more C to plant biomass, thereby facilitating effective carbon recycling [99]. Additionally, research has shown that PGPR affects soil quality by regulating the amount of C in micro- and macroaggregates [112]. Soil microbial activities are directly or indirectly influenced by elevated temperature and carbon dioxide levels. High-temperature conditions enhance microbial activities, creating a positive feedback loop for climate change. Similarly, low moisture conditions can have comparable effects [114]. The maintenance of ecosystem C aggregation relies on achieving a balance between plant productivity and heterotrophic respiration, which is accomplished through the decomposition of SOM [99, 115].

Several studies have extensively documented the beneficial impacts of elevated CO_2 levels on plant growth, as well as the increased input of photosynthetic C into soils [116]. These increased carbon inputs can promote microbial growth, leading to an increase in soil microbial communities under elevated CO_2 [117]. Consequently, this may accelerate soil organic matter decomposition, potentially resulting in net carbon losses in the soil. Elevated carbon dioxide levels also stimulate rhizosphere priming effects, enhancing the decomposition of soil organic matter through microbial activity [118, 119]. The enzymatic activity of PGPR facilitates the decomposition of soil organic matter [120]. Moisture levels play a vital role in shaping the activities of microbial communities involved in climate change processes. In different soil environments, microbial activity tends to increase under conditions of drought and water stress. This response is primarily attributed to the decrease in water levels and the introduction of O_2 into previously oxygen-depleted soils [99]. Peatlands and wetlands are recognized as crucial reservoirs that store substantial amounts of C in terrestrial ecosystems [99, 121]. Consequently, the heightened degradation of resilient and stable organic matter under dry conditions can have significant implications for the global C cycle dynamics [122].

10.8 Microbial Consortia: An Effective Way for Plant Growth

The rhizosphere, a thriving area of soil, is teeming with a variety of different microorganisms. These subterranean microbes interact in complex ways both with each other and with plant roots, mutually benefiting plant growth. Plant signalling molecules such as root exudates play a crucial role and shape the rich spectrum of microbial diversity in this zone. The plant growth-promoting rhizomicrobes are a selected group of rhizosphere inhabitants that contribute to plant development through an impressive repertoire of mechanisms [12, 123]. From phosphate solubilization to nitrogen fixation to the production of plant growth hormones and antimicrobial compounds, these PGPRs serve as excellent substitutes for chemical inputs that often upset the delicate balance of soil biological and chemical properties. While biofertilizers typically feature a single microbial strain, pioneering research shows that the application of co-inoculation or consortium biofertilizers containing two or more microbial strains consistently produce more profound benefits for plant growth [12, 124]. Numerous scientific studies have advocated the utilization of microbial consortia as a promising approach to enhance plant growth, health, and survival, both under challenging environmental conditions and in natural settings [46, 47, 125, 126]. These consortia have been observed to stimulate plant roots, inducing the secretion of increased amounts of amino acids, growth regulators, and sugars. Moreover, they enhance the plant roots' ability to efficiently utilize minerals and other constituents present in the rhizosphere [46, 47, 124, 126]. This symbiotic interaction contributes to improved nitrogen fixation, thereby enabling plants to adapt to changes in environmental conditions more effectively [126]. Within the framework of field experiments, the inoculation of rice crops with a consortium comprising three distinct bacterial strains, namely, "*Burkholderia ubonensis* (la3c3)," "*Burkholderia vietnamiensis* (la1a4)," and "*Citrobacter bitternis* (p9a3m)," showed noteworthy improvement in both grain yield and quality and also reduced the use of nitrogen fertilizer by up to 25% [127]. In the study conducted by Kumar et al. [38–40], the tetra combination of *A. chlorophenolicus*, *B. megaterium*, *Enterobacter* sp., and *P. aeruginosa* exhibited significant improvements in plant height, grain yield, and straw yield for wheat under both greenhouse and field conditions. The application of a talc-based formulation including a consortium consisting of *K. pneumoniae*, *Erwinia* sp., and *P. nitritireducens* showed an extraordinary per-plant cumin seed yield (0.42 g). This notable formulation not only elevated essential agronomic parameters such as plant height, dry weight, and

100 seed weight but also resulted in a substantial enhancement in the overall yield, indicating a promising role of bacterial consortium in cumin cultivation [128]. A consortium comprising *Erwinia* sp. (nitrogen fixer), *C. arthrosphaerae* (phosphorus solubilizer), and *P. gessardii* (potassium solubilizer) enhanced growth and physiological parameters, including root/shoot length and biomass, chlorophyll, carotenoids, phenolics, flavonoids, and soluble sugar content in barley crops compared to the untreated control [129]. Tyagi et al. [130] showed that a tri-inoculant formulation ("*Serendipita indica*," '*Rhizophagus intraradices*," and "*Azotobacter chroococcum*") increased root and shoot length, fresh and dry weight, membrane electrolyte leakage, chlorophyll content, relative water content, and antioxidant enzyme activities (POX) significantly increased, CAT, PPO, SOD) in maize plants under drought conditions compared to the uninoculated control. The study by Kapadia et al. [131] showed that a microbial consortium consisting of *Bacillus* sp., *Delftia* sp., *Enterobacter* sp., and *Achromobacter* sp. significantly increased growth and mineral uptake of salt stressed tomatoes. The consortium treatment resulted in increased leaf, shoot and root dry weight, leaf count, shoot length, root length, secondary roots, and improved chlorophyll content compared to the control group, ultimately helping the plants to thrive in a saline environment.

10.9 Contribution of Microorganisms in Biofortification of Food Crops

To address the needs of an expanding world population, it is crucial to implement strategies that optimize biomass productivity. "Green revolution" has contributed a lot in terms of giving to higher crop yields. But a specific type of micronutrient deficiency, known as "hidden hunger," affects nearly half of the global population, leading to malnutrition [132]. In addition to macronutrients (N, P, K) and calories, essential micronutrients such as zinc, iron, selenium, etc. are vital for human health [15, 16]. The widespread presence of micronutrient deficiencies in low- and middle-income countries has a significant impact on human health [16, 133]. The lack of micronutrients poses a significant health burden, especially in regions with inadequate access to proper nutrition. According to the "United Nations System Standing Committee on Nutrition (UNSSCN, [134])," more than 50% of child mortality cases are directly or indirectly attributed to micronutrient deficiencies, which also contribute to major risk factors for maternal mortality [16]. Micronutrients such as iron (Fe), zinc (Zn), and selenium (Se) are essential for vital biological processes and must be obtained through the diet [13–15]. Inadequate intake of these micronutrients can lead to various health problems and increase the risk of developing several diseases [15, 16, 135].

Crop biofortification offers a promising solution to these challenges by enhancing the nutrient content of staple foods, particularly targeting low-income households that struggle to afford a diverse diet. Traditional approaches such as plant breeding, agronomic strategies, and genetic engineering have been employed for biofortification, but their effectiveness has been inconsistent, and they are laborious and costlier approaches [136]. In recent years, the utilization of naturally occurring soil microorganisms, specifically plant growth-promoting microbes like bacteria and mycorrhizal fungi, has emerged as a viable approach for crop biofortification ([3, 4, 45, 55] and 2022d). The interactions between plants and these microbes play a pivotal role in improving soil nutrition and facilitating the movement of micronutrients to different plant parts through processes such as solubilization, mobilization, and translocation of micronutrients [15, 16]. Microorganisms employ diverse strategies to enhance micronutrient uptake in plants, including the production of siderophores and other chelating substances, secretion of organic acids, proton extrusion, modification of root morphology and anatomy, reduction of anti-nutritional factors like phytic acid in food grains, secretion of phenolics and related compounds, and production of phytohormones as signaling molecules [3, 4, 16, 25, 45, 137]. The exploration of these potential plant growth-promoting (PGP) bacteria offers an alternative to chemical crop protection agents while also promoting environmental health and sustainability [13, 14, 138]. This makes them highly suitable for extensive use in organic agriculture. Several microorganisms have been identified and listed in Table 10.2 for their application in crop biofortification.

Table 10.2 Microbial-assisted biofortification of various crops

Microorganism	Micronutrient (s)	Plant	References
Zinc-solubilizing bacterial strains	Zinc	Wheat	Ali et al. [139]
Enterobacter sp. EG16	Selenium	Pak choi (*Brassica rapa* ssp. *chinensis*)	Yuan et al. [140]
Exiguobacterium sp. S17	Selenium	Brassica juncea (Indian mustard)	Marfetán et al. [141]
Consortium of *Bacillus subtilis*, *Bacillus aryabhattai*, and *Paenibacillus polymyxa*	Zinc and iron	Maize	Ahmad et al. [142]
Pseudomonas protegens	Zinc	Wheat	Singh et al. [3]
Three strains of Bacillus subtilis + soil applied iron	Iron	Groundnut	Sarwar et al. [143]
Bacillus altitudinis WR10	Iron	Wheat	Sun et al. [144]
Consortium (*Rhizobium* + plant growth-promoting rhizobacteria)	Iron	Lentil (*Lens esculenta*)	Kumar et al. [38–40]
Bacillus altitudinis	Zinc	Chickpea (*Cicer arietinum* L.)	Kushwaha et al. [145]
Bacillus mojavensis + *Bacillus cereus*	Iron	Sorghum	Mansani et al. [146]
S. marcescens FA-4	Zinc	Rice	Shakeel et al. [147]
Enterobacter sp. MN17 + Zn application	Zinc	*Kabuli* chickpea	Ullah et al. [148]

10.10 Challenges and Opportunities for Commercializing Microbial Products in Sustainable Agriculture

Microbe-based products, including biofertilizers and biopesticides, hold significant potential for sustainable agriculture. However, several challenges hinder their wide acceptance and successful implementation. To address them effectively, it is crucial to understand the issues faced by microbial-based products.

I. *Limited awareness and understanding*: One of the main obstacles faced by microbial-based products is the insufficient awareness and understanding among farmers, agronomists, and policy-makers regarding the advantages and appropriate utilization of these products. It is imperative to develop educational initiatives and outreach programs that effectively disseminate knowledge and enhance confidence in these innovative solutions.

II. *Regulatory hurdles*: Regulatory frameworks for microbial products often lack clarity or exhibit regional variations, which present a challenge in their commercialization. The lengthy and expensive registration processes further hinder manufacturers and limit market access. To overcome these obstacles, it is crucial to streamline regulations and establish clear guidelines that facilitate the smooth commercialization of microbial products.

III. *Product variability and efficacy*: The effectiveness of microbial products can be influenced by factors such as environmental conditions, crop species, and management practices, leading to variability in their performance. Ensuring consistent product efficacy across diverse agricultural systems and improving product stability are crucial challenges that require attention.

IV. *Quality control and standardization*: Maintaining product quality and standardization is critical to the successful commercialization of microbial products. The development of standardized production protocols, quality control measures, and certification programs will increase product reliability and consumer confidence.

V. *Limited scalability and production costs*: Manufacturing microbial products faces the challenge of increasing production and achieving cost efficiencies. Efficient production processes, optimized fermentation techniques, and research into alternative microbial strains are required to overcome these obstacles.

VI. *Market acceptance and competitiveness*: Microbial-based products compete with traditional agrochemicals that already dominate the market. It is crucial to convince farmers of the long-term benefits, cost-effectiveness, and environmental benefits of microbial products in order to achieve market acceptance and encourage their widespread use.

VII. *Research and development gaps*: Sustained research and development efforts are essential to identify and characterize novel microbial strains, refine formulation methods, and gain a comprehensive understanding of the complex interactions between microbes, plants, and the environment. Bridging these knowledge gaps will encourage innovation and accelerate the development of more efficient and impactful microbial products.

It is essential to encourage collaboration between researchers, industry stakeholders, policy-makers, and farmers to resolve these challenges. This collaborative effort will facilitate the exchange of knowledge and spur innovation in the field of microbial products. Increased investments in research, capacity building, and infrastructure are essential for accelerating the development of these products and driving advancements. In addition, public-private partnerships can play a crucial role in facilitating technology transfer, expanding market access, and providing regulatory support. By addressing these obstacles, we can unlock the maximum potential of microbial products and increase their adoption via sustainable agriculture. Utilizing the power of beneficial microorganisms provides numerous opportunities to improve agricultural productivity, reduce reliance on chemical additives, and promote environmentally responsible agricultural practices.

10.11 Conclusion and Future Prospects

In conclusion, the integration of microorganisms into sustainable agriculture presents promising solutions for enhancing crop productivity, soil health, and environmental sustainability. Microbes play pivotal roles as biofertilizers, biocontrol agents, and stress-tolerant enhancers, promoting eco-friendly practices and reducing reliance on synthetic inputs. Their contributions to climate-smart agriculture, microbial consortia, and biofortification of food crops underscore their significance in achieving sustainable and resilient agricultural systems. Unlocking the potential of microorganisms holds tremendous prospects for advancing and transforming sustainable crop-based food production.

Looking ahead, advanced microbial formulations offer exciting prospects for revolutionizing agricultural practices. The incorporation of nanotechnology into microbial formulations, e.g., nanobiofertilizers, can lead to improved nutrient delivery and controlled release systems, thus optimizing nutrient uptake by plants [9]. In addition, the development of encapsulated biofertilizers can protect microorganisms and ensure their viability and stability during storage and application. These formulation advances have the potential to improve the efficiency and potency of microbial products, resulting in higher plant productivity and optimized resource utilization.

References

1. Fenibo EO, Ijoma GN, Nurmahomed W, Matambo T (2022) The potential and green chemistry attributes of biopesticides for sustainable agriculture. Sustainability 14(21):14417
2. Khan A, Panthari D, Sharma RS, Punetha A, Singh AV, Upadhayay VK (2023a) Biofertilizers: a microbial-assisted strategy to improve plant growth and soil health. In: Advanced microbial techniques in agriculture, environment, and health management. Elsevier, pp 97–118

3. Singh J, Singh AV, Upadhayay VK, Khan A, Chandra R (2022) Prolific contribution of pseudomonas protegens in Zn biofortification of wheat by modulating multifaceted physiological response under saline and non-saline conditions. World J Microbiol Biotechnol 38(12): 227. https://doi.org/10.1007/s11274-022-03411-4
4. Upadhayay VK, Khan A, Singh J, Singh AV (2022a) Bacterial assisted improved Zn consignment in root and shoot of rice plant by zinc solubilizing Serratia marcescence bearing plant probiotic traits. Adv Biores 13(1):01–08
5. Kour D, Rana KL, Yadav AN, Yadav N, Kumar M, Kumar V, Vyas P, Dhaliwal HS, Saxena AK (2020) Microbial biofertilizers: bioresources and eco-friendly technologies for agricultural and environmental sustainability. Biocatal Agric Biotechnol 23(101487):101487. https://doi.org/10.1016/j.bcab.2019.101487
6. Ortiz A, Sansinenea E (2022) The role of beneficial microorganisms in soil quality and plant health. Sustainability 14(9):5358
7. Elnahal AS, El-Saadony MT, Saad AM, Desoky ESM, El-Tahan AM, Rady MM et al (2022) The use of microbial inoculants for biological control, plant growth promotion, and sustainable agriculture: a review. Eur J Plant Pathol 162(4):759–792
8. Upadhayay VK, Maithani D, Dasila H, Taj G, Singh AV (2023a) Microbial services for mitigation of biotic and abiotic stresses in plants. In: Advanced microbial techniques in agriculture, environment, and health management. Elsevier, pp 67–81. https://doi.org/10.1016/b978-0-323-91643-1.00003-x
9. Upadhayay VK, Chitara MK, Mishra D, Jha MN, Jaiswal A, Kumari G, Ghosh S, Patel VK, Naitam MG, Singh AK, Pareek N, Taj G, Maithani D, Kumar A, Dasila H, Sharma A (2023b) Synergistic impact of nanomaterials and plant probiotics in agriculture: a tale of two-way strategy for long-term sustainability. Front Microbiol 14:1133968. https://doi.org/10.3389/fmicb.2023.1133968
10. Kumar A, Patel JS, Meena VS, Ramteke PW (2019) Plant growth-promoting rhizobacteria: strategies to improve abiotic stresses under sustainable agriculture. J Plant Nutr 42(11–12): 1402–1415
11. Six J, Frey SD, Thiet RK, Batten KM (2006) Bacterial and fungal contributions to carbon sequestration in agroecosystems. Soil Sci Soc Am J 70(2):555–569
12. Kaushal M, Devi S, Kumawat KC, Kumar A (2023) Microbial consortium: a boon for a sustainable agriculture. In: Parray JA (ed) Climate change and microbiome dynamics. Climate change management. Springer, Cham. https://doi.org/10.1007/978-3-031-21079-2_2
13. Khan A, Singh J, Upadhayay VK, Singh AV, Shah S (2019a) Microbial biofortification: a green technology through plant growth promoting microorganisms. In: Sustainable green technologies for environmental management. Springer, Singapore, pp 255–269
14. Khan MA, Asaf S, Khan AL, Adhikari A, Jan R, Ali S, Imran M, Kim K-M, Lee I-J (2019b) Halotolerant rhizobacterial strains mitigate the adverse effects of NaCl stress in soybean seedlings. Biomed Res Int 2019:9530963. https://doi.org/10.1155/2019/9530963
15. Upadhayay VK, Singh J, Khan A, Lohani S, Singh AV (2019) Mycorrhizal mediated micronutrients transportation in food based plants: a biofortification strategy. In: Mycorrhizosphere and Pedogenesis. Springer, Singapore, pp 1–24
16. Upadhayay VK, Singh AV, Pareek N (2018) An insight in decoding the multifarious and splendid role of microorganisms in crop biofortification. Int J Curr Microbiol Appl Sci 7: 2407–2418. https://doi.org/10.20546/ijcmas.2018.706.286
17. Babalola OO, Fadiji AE, Enagbonma BJ, Alori ET, Ayilara MS, Ayangbenro AS (2020) The nexus between plant and plant microbiome: revelation of the networking strategies. Front Microbiol 11:548037. https://doi.org/10.3389/fmicb.2020.548037
18. Khatoon Z, Huang S, Rafique M, Fakhar A, Kamran MA, Santoyo G (2020) Unlocking the potential of plant growth-promoting rhizobacteria on soil health and the sustainability of agricultural systems. J Environ Manag 273:111118. https://doi.org/10.1016/j.jenvman.2020.111118

19. Etesami H, Hemati A, Alikhani HA (2019) Microbial bioconversion of agricultural wastes for rural sanitation and soil carbon enrichment. Microbial interventions in agriculture and environment. Rhizosphere, Microbiome and Agro-Ecology 2:179–204
20. Rajkumar R, Kurinjimalar C (2021) Microbes and plant mineral nutrition. In: Soni R, Suyal DC, Bhargava P, Goel R (eds) Microbiological activity for soil and plant health management. Springer, Singapore. https://doi.org/10.1007/978-981-16-2922-8_5
21. Kaviya N, Upadhayay VK, Singh J, Khan A, Panwar M, Singh AV (2019) Role of microorganisms in soil genesis and functions. In: Mycorrhizosphere and pedogenesis. Springer, Singapore, pp 25–52
22. Dimkić I, Janakiev T, Petrović M, Degrassi G, Fira D (2022) Plant-associated bacillus and pseudomonas antimicrobial activities in plant disease suppression via biological control mechanisms-a review. Physiol Mol Plant Pathol 117:101754
23. Qadri M, Short S, Gast K, Hernandez J, Wong ACN (2020) Microbiome innovation in agriculture: development of microbial based tools for insect pest management. Front Sustain Food Syst 4:547751
24. Upadhyay SK, Srivastava AK, Rajput VD, Chauhan PK, Bhojiya AA, Jain D, Chaubey G, Dwivedi P, Sharma B, Minkina T (2022) Root exudates: mechanistic insight of plant growth promoting rhizobacteria for sustainable crop production. Front Microbiol 13:916488. https://doi.org/10.3389/fmicb.2022.916488
25. Khan A, Singh AV, Pareek N, Arya P, Upadhayay VK, Kumar Jugran A, Kumar Mishra P, Goel R (2023b) Credibility assessment of cold adaptive pseudomonas jesenni MP1 and P. palleroniana N26 on growth, rhizosphere dynamics, nutrient status, and yield of the kidney bean cultivated in Indian central Himalaya. Frontiers. Plant Sci 14:1042053. https://doi.org/10.3389/fpls.2023.1042053
26. Goswami D, Thakker JN, Dhandhukia PC (2016) Portraying mechanics of plant growth promoting rhizobacteria (PGPR): a review. Cogent Food & Agriculture 2:1127500. https://doi.org/10.1080/23311932.2015.1127500
27. Contreras-Cornejo HA, Ramírez-Ordorica A, Álvarez-Navarrete M, Macías-Rodríguez L (2022) The role of Phytohormones in cross-communication between plants and Rhizomicrobes. In: Microbial cross-talk in the rhizosphere. Singapore, Springer Nature Singapore, pp 59–97
28. Ahmed E, Holmström SJ (2014) Siderophores in environmental research: roles and applications. Microb Biotechnol 7(3):196–208. https://doi.org/10.1111/1751-7915.12117
29. Hakim S, Naqqash T, Nawaz MS, Laraib I, Siddique MJ, Zia R et al (2021) Rhizosphere engineering with plant growth-promoting microorganisms for agriculture and ecological sustainability. Frontiers in Sustainable Food Systems 5:617157
30. Olagoke FK, Bettermann A, Nguyen PTB, Redmile-Gordon M, Babin D, Smalla K et al (2022) Importance of substrate quality and clay content on microbial extracellular polymeric substances production and aggregate stability in soils. Biol Fertil Soils 58(4):435–457
31. Gangwar M, Saini P, Nikhanj P, Kaur S (2017) Plant growth-promoting microbes (PGPM) as potential microbial bio-agents for eco-friendly agriculture. In: Advances in soil microbiology: recent trends and future prospects: Volume 2: soil-microbe-plant interaction. Springer, Singapore, pp 37–55
32. Rehmaan IU, Jan B, Khan NF, Islam T, Rehman S, Sheergojri IA et al (2022) Nitrogen biofertilizers: role in sustainable agriculture. In: Advances in plant nitrogen metabolism. CRC Press, pp 170–184
33. Roshani KA, Singh AV, Upadhayay VU, Prasad B (2020) Development of potential microbial consortia and their assessment on wheat (Triticum aestivum) seed germination. Environ Ecol 38(1):6–16
34. Dhir B (2017) Biofertilizers and biopesticides: eco-friendly biological agents. In: Advances in environmental biotechnology. Springer, Singapore, pp 167–188

35. Shahwar D, Mushtaq Z, Mushtaq H, Alqarawi AA, Park Y, Alshahrani TS, Faizan S (2023) Role of microbial inoculants as bio fertilizers for improving crop productivity: a review. Heliyon 9(6):e16134. https://doi.org/10.1016/j.heliyon.2023.e16134
36. Bhat MA, Mishra AK, Jan S, Bhat MA, Kamal MA, Rahman S et al (2023) Plant growth promoting Rhizobacteria in plant health: a perspective study of the underground interaction. Plan Theory 12(3):629
37. Masmoudi F, Alsafran M, Jabri HA, Hosseini H, Trigui M, Sayadi S et al (2023) Halobacteria-based biofertilizers: a promising alternative for enhancing soil fertility and crop productivity under biotic and abiotic stresses—a review. Microorganisms 11(5):1248
38. Kumar A, Jha MN, Singh D, Pathak D, Rajawat MVS (2021a) Prospecting catabolic diversity of microbial strains for developing microbial consortia and their synergistic effect on lentil (Lens esculenta) growth, yield and iron biofortification. Arch Microbiol 203(8):4913–4928. https://doi.org/10.1007/s00203-021-02446-9
39. Kumar A, Maurya BR, Raghuwanshi R (2021b) The microbial consortium of indigenous rhizobacteria improving plant health, yield and nutrient content in wheat (Triticum aestivum). J Plant Nutr 44(13):1942–1956. https://doi.org/10.1080/01904167.2021.1884706
40. Kumar M, Yadav AN, Saxena R, Paul D, Tomar RS (2021c) Biodiversity of pesticides degrading microbial communities and their environmental impact. Biocatal Agric Biotechnol 31:101883. https://doi.org/10.1016/j.bcab.2020.101883
41. Jiao X, Takishita Y, Zhou G, Smith DL (2021) Plant associated rhizobacteria for biocontrol and plant growth enhancement. Front Plant Sci 12:634796. https://doi.org/10.3389/fpls.2021.634796
42. Mahanty T, Bhattacharjee S, Goswami M, Bhattacharyya P, Das B, Ghosh A, Tribedi P (2017) Biofertilizers: a potential approach for sustainable agriculture development. Environ Sci Pollut Res Int 24(4):3315–3335. https://doi.org/10.1007/s11356-016-8104-0
43. Santoyo G (2022) How plants recruit their microbiome? New insights into beneficial interactions. J Adv Res 40:45–58
44. Mishra J, Arora NK (2016) Bioformulations for plant growth promotion and combating Phytopathogens: a sustainable approach. In: Arora N, Mehnaz S, Balestrini R (eds) Bioformulations: for sustainable agriculture. Springer, New Delhi. https://doi.org/10.1007/978-81-322-2779-3_1
45. Upadhayay VK, Singh AV, Khan A (2022b) Cross talk between zinc-solubilizing bacteria and plants: a short tale of bacterial-assisted zinc biofortification. Front Soil Sci 1:788170. https://doi.org/10.3389/fsoil.2021.788170
46. Maheshwari DK, Das A, Dheeman S, Pandey P (2023a) An overall insight into the attributes, interactions, and future applications of "microbial consortium" for plant growth promotion with contemporary approaches. In: Sustainable agrobiology: design and development of microbial consortia. Springer, Singapore, pp 3–22
47. Maheshwari DK, Kumar S, Kumar P, Kumar S, Dheeman S (2023b) Effect of microbial consortium vs. perfected chemical fertilizers for sustainable crop growth. In: Maheshwari DK, Dheeman S (eds) Sustainable agrobiology. Microorganisms for sustainability, vol 43. Springer, Singapore. https://doi.org/10.1007/978-981-19-9570-5_14
48. Pirttilä AM, Mohammad Parast Tabas H, Baruah N, Koskimäki JJ (2021) Biofertilizers and biocontrol agents for agriculture: how to identify and develop new potent microbial strains and traits. Microorganisms 9(4):817. https://doi.org/10.3390/microorganisms9040817
49. Malusá E, Sas-Paszt L, Ciesielska J (2012) Technologies for beneficial microorganisms inocula used as biofertilizers. TheScientificWorldJOURNAL 2012:491206. https://doi.org/10.1100/2012/491206
50. Allouzi MMA, Allouzi SMA, Keng ZX, Supramaniam CV, Singh A, Chong S (2022) Liquid biofertilizers as a sustainable solution for agriculture. Heliyon e12609:e12609
51. Sahu PK, Brahmaprakash GP (2016) Formulations of biofertilizers – approaches and advances. In: Singh D, Singh H, Prabha R (eds) Microbial inoculants in sustainable agricultural productivity. Springer, New Delhi. https://doi.org/10.1007/978-81-322-2644-4_12

52. Parveen N, Mishra R, Singh DV, Kumar P, Singh RP (2023) Assessment of different carrier materials for the preparation of microbial formulations to enhance the shelf life and its efficacy on the growth of spinach (Spinacia oleracea L.). World J Microbiol Biotechnol 39(7):180
53. Schoebitz M, López MD, Roldán A (2013) Bioencapsulation of microbial inoculants for better soil–plant fertilization. A review. Agron Sustain Dev 33:751–765
54. Kumar S, Sindhu SS, Kumar R (2022) Biofertilizers: an ecofriendly technology for nutrient recycling and environmental sustainability. Curr Res Microb Sci 3:100094
55. Upadhayay VK, Singh AV, Khan A, Singh J, Pareek N, Raghav A (2022c) FE-SEM/EDX based zinc mobilization analysis of Burkholderia cepacia and Pantoea rodasii and their functional annotation in crop productivity, soil quality, and zinc biofortification of paddy. Front Microbiol 13:852192. https://doi.org/10.3389/fmicb.2022.852192
56. Basu A, Prasad P, Das SN, Kalam S, Sayyed RZ, Reddy MS, El Enshasy H (2021) Plant growth promoting rhizobacteria (PGPR) as green bioinoculants: recent developments, constraints, and prospects. Sustainability 13(3):1140
57. Aktar MW, Sengupta D, Chowdhury A (2009) Impact of pesticides use in agriculture: their benefits and hazards. Interdiscip Toxicol 2(1):1–12. https://doi.org/10.2478/v10102-009-0001-7
58. Pathak VM, Verma VK, Rawat BS, Kaur B, Babu N, Sharma A, Dewali S, Yadav M, Kumari R, Singh S, Mohapatra A, Pandey V, Rana N, Cunill JM (2022) Current status of pesticide effects on environment, human health and it's eco-friendly management as bioremediation: a comprehensive review. Front Microbiol 13:962619. https://doi.org/10.3389/fmicb.2022.962619
59. Sharma A, Kumar V, Shahzad B, Tanveer M, Sidhu GPS, Handa N, Kohli SK, Yadav P, Bali AS, Parihar RD, Dar OI, Singh K, Jasrotia S, Bakshi P, Ramakrishnan M, Kumar S, Bhardwaj R, Thukral AK (2019) Worldwide pesticide usage and its impacts on ecosystem. SN Applied Sci 1:1446. (2019). https://doi.org/10.1007/s42452-019-1485-1
60. Ayilara MS, Adeleke BS, Akinola SA, Fayose CA, Adeyemi UT, Gbadegesin LA, Omole RK, Johnson RM, Uthman QO, Babalola OO (2023) Biopesticides as a promising alternative to synthetic pesticides: a case for microbial pesticides, phytopesticides, and nanobiopesticides. Front Microbiol 14:1040901. https://doi.org/10.3389/fmicb.2023.1040901
61. Chattopadhyay P, Banerjee G, Mukherjee S (2017) Recent trends of modern bacterial insecticides for pest control practice in integrated crop management system. 3. Biotech 7:60. https://doi.org/10.1007/s13205-017-0717-6
62. Köhl J, Kolnaar R, Ravensberg WJ (2019) Mode of action of microbial biological control agents against plant diseases: relevance beyond efficacy. Front Plant Sci 10:845. https://doi.org/10.3389/fpls.2019.00845
63. Koza NA, Adedayo AA, Babalola OO, Kappo AP (2022) Microorganisms in plant growth and development: roles in abiotic stress tolerance and secondary metabolites secretion. Microorganisms 10(8):1528. https://doi.org/10.3390/microorganisms10081528
64. Jouzani GS, Valijanian E, Sharafi R (2017) Bacillus thuringiensis: a successful insecticide with new environmental features and tidings. Appl Microbiol Biotechnol 101(7):2691–2711. https://doi.org/10.1007/s00253-017-8175-y
65. Pardo-López L, Soberón M, Bravo A (2013) Bacillus thuringiensis insecticidal three-domain cry toxins: mode of action, insect resistance and consequences for crop protection. FEMS Microbiol Rev 37(1):3–22. https://doi.org/10.1111/j.1574-6976.2012.00341.x
66. Anderson AJ, Kim YC (2018) Biopesticides produced by plant-probiotic pseudomonas chlororaphis isolates. Crop Protection (Guildford, Surrey) 105:62–69. https://doi.org/10.1016/j.cropro.2017.11.009
67. Bamisile BS, Akutse KS, Siddiqui JA, Xu Y (2021) Model application of entomopathogenic fungi as alternatives to chemical pesticides: prospects, challenges, and insights for next-generation sustainable agriculture. Front Plant Sci 12:741804. https://doi.org/10.3389/fpls.2021.741804

68. Vivekanandhan P, Swathy K, Murugan AC, Krutmuang P (2022) Insecticidal efficacy of Metarhizium anisopliae derived chemical constituents against disease-vector mosquitoes. J Fungi (Basel, Switzerland) 8(3):300. https://doi.org/10.3390/jof8030300
69. Wang H, Peng H, Li W, Cheng P, Gong M (2021) The toxins of Beauveria bassiana and the strategies to improve their virulence to insects. Front Microbiol 12:705343. https://doi.org/10.3389/fmicb.2021.705343
70. Ibarra JE, Rincón-Castro D (2008) Insect viruses diversity, biology, and use as bioinsecticides. Tropical Biology and Conservation Management 5:1–10
71. Williams T, López-Ferber M, Caballero P (2021) Nucleopolyhedrovirus coocclusion technology: a new concept in the development of biological insecticides. Front Microbiol 12:810026. https://doi.org/10.3389/fmicb.2021.810026
72. Mittler R (2006) Abiotic stress, the field environment and stress combination. Trends Plant Sci 11(1):15–19. https://doi.org/10.1016/j.tplants.2005.11.002
73. Pereira A (2016) Plant abiotic stress challenges from the changing environment. Front Plant Sci 7:1123. https://doi.org/10.3389/fpls.2016.01123
74. Suzuki N, Rivero RM, Shulaev V, Blumwald E, Mittler R (2014) Abiotic and biotic stress combinations. New Phytol 203(1):32–43. https://doi.org/10.1111/nph.12797
75. Uzu G, Sobanska S, Aliouane Y, Pradere P, Dumat C (2009) Study of lead phytoavailability for atmospheric industrial micronic and sub-micronic particles in relation with lead speciation. Environ Pollut 157(4):1178–1185. https://doi.org/10.1016/j.envpol.2008.09.053
76. Feng ZT, Deng YQ, Fan H, Sun QJ, Sui N, Wang BS (2014) Effects of NaCl stress on the growth and photosynthetic characteristics of Ulmus pumila L. seedlings in sand culture. Photosynthetica 52(2):313–320. https://doi.org/10.1007/s11099-014-0032-y
77. Guo J, Li Y, Han G, Song J, Wang B (2018) NaCl markedly improved the reproductive capacity of the euhalophyte Suaeda salsa. Funct Plant Biol 45(3):350–361. https://doi.org/10.1071/FP17181
78. Zhao KF, Song J, Fan H, Zhou S, Zhao M (2010) Growth response to ionic and osmotic stress of NaCl in salt-tolerant and salt-sensitive maize. J Integr Plant Biol 52(5):468–475. https://doi.org/10.1111/j.1744-7909.2010.00947.x
79. Hoekstra FA, Golovina EA, Buitink J (2001) Review. In TRENDS in plant science (Vol. 6, issue 9). http://plants.trends.com
80. Jia Y, Zou D, Wang J, Sha H, Liu H, Inayat MA, Sun J, Zheng H, Xia N, Zhao H (2017) Effects of γ-aminobutyric acid, glutamic acid, and calcium chloride on Rice (Oryza sativa L.) under cold stress during the early vegetative stage. J Plant Growth Regul 36(1):240–253. https://doi.org/10.1007/s00344-016-9634-x
81. Rizhsky L, Liang H, Mittler R (2002) The combined effect of drought stress and heat shock on gene expression in tobacco. Plant Physiol 130(3):1143–1151. https://doi.org/10.1104/pp.006858
82. Inbaraj MP (2021) Plant-microbe interactions in alleviating abiotic stress—a mini review. In: Frontiers in Agronomy, vol 3, Frontiers Media S.A. https://doi.org/10.3389/fagro.2021.667903
83. Farooq M, Wahid A, Kobayashi N, Fujita D, Basra SMA (2009) Plant drought stress: effects, mechanisms and management. In agronomy for sustainable development (Vol. 29, issue 1, pp. 185–212). https://doi.org/10.1051/agro:2008021
84. Porcel R, María Zamarreño Á, María García-Mina J, Aroca R (2014) Involvement of plant endogenous ABA in bacillus megaterium PGPR activity in tomato plants. http://www.biomedcentral.com/1471-2229/14/36
85. Saleem M, Arshad M, Hussain S, Bhatti AS (2007) Perspective of plant growth promoting rhizobacteria (PGPR) containing ACC deaminase in stress agriculture. In. J Ind Microbiol Biotechnol 34(10):635–648. https://doi.org/10.1007/s10295-007-0240-6
86. Khalvati MA, Hu Y, Mozafar A, Schmidhalter U (2005) Quantification of water uptake by arbuscular mycorrhizal hyphae and its significance for leaf growth, water relations, and gas

exchange of barley subjected to drought stress. Plant Biol 7(6):706–712. https://doi.org/10.1055/s-2005-872893
87. Kavroulakis N, Doupis G, Papadakis IE, Ehaliotis C, Papadopoulou KK (2018) Tolerance of tomato plants to water stress is improved by the root endophyte fusarium solani FsK. Rhizosphere 6:77–85. https://doi.org/10.1016/j.rhisph.2018.04.003
88. Mathur S, Tomar RS, Jajoo A (2019) Arbuscular mycorrhizal fungi (AMF) protects photosynthetic apparatus of wheat under drought stress. Photosynth Res 139(1–3):227–238. https://doi.org/10.1007/s11120-018-0538-4
89. Jochum MD, McWilliams KL, Borrego EJ, Kolomiets MV, Niu G, Pierson EA, Jo YK (2019) Bioprospecting plant growth-promoting Rhizobacteria that mitigate drought stress in grasses. Front Microbiol 10:2106. https://doi.org/10.3389/fmicb.2019.02106
90. Pasbani B, Salimi A, Aliasgharzad N, Hajiboland R (2020) Colonization with arbuscular mycorrhizal fungi mitigates cold stress through improvement of antioxidant defense and accumulation of protecting molecules in eggplants. Sci Hortic 272:109575. https://doi.org/10.1016/j.scienta.2020.109575
91. Parihar M, Rakshit A, Rana K, Tiwari G, Jatav SS (2020) The effect of arbuscular mycorrhizal fungi inoculation in mitigating salt stress of pea (Pisum Sativum L.). Commun Soil Sci Plant Anal 51(11):1545–1559. https://doi.org/10.1080/00103624.2020.1784917
92. Li X, Sun P, Zhang Y, Jin C, Guan C (2020) A novel PGPR strain Kocuria rhizophila Y1 enhances salt stress tolerance in maize by regulating phytohormone levels, nutrient acquisition, redox potential, ion homeostasis, photosynthetic capacity and stress-responsive genes expression. Environ Exp Bot 174:104023. https://doi.org/10.1016/j.envexpbot.2020.104023
93. Arun KD, Sabarinathan KG, Gomathy M, Kannan R, Balachandar D (2020) Mitigation of drought stress in rice crop with plant growth-promoting abiotic stress-tolerant rice phyllosphere bacteria. J Basic Microbiol 60(9):768–786. https://doi.org/10.1002/jobm.202000011
94. Silambarasan S, Logeswari P, Cornejo P, Kannan VR (2019) Role of plant growth–promoting rhizobacterial consortium in improving the Vigna radiata growth and alleviation of aluminum and drought stresses. Environ Sci Pollut Res 26:27647–27659
95. Tahiri A-I, Meddich A, Raklami A, Alahmad A, Bechtaoui N, Anli M, Göttfert M, Heulin T, Achouak W, Oufdou K (2022) Assessing the potential role of compost, PGPR, and AMF in improving tomato plant growth, yield, fruit quality, and water stress tolerance. J Soil Sci Plant Nutr 22(1):743–764. https://doi.org/10.1007/s42729-021-00684-w
96. Ali B, Wang X, Saleem MH, Hafeez A, Afridi MS, Khan S et al (2022) PGPR-mediated salt tolerance in maize by modulating plant physiology, antioxidant defense, compatible solutes accumulation and bio-surfactant producing genes. Plan Theory 11(3):345
97. Manjunatha BS, Nivetha N, Krishna GK, Elangovan A, Pushkar S, Chandrashekar N, Aggarwal C, Asha AD, Chinnusamy V, Raipuria RK, Watts A, Bandeppa S, Dukare AS, Paul S (2022) Plant growth-promoting rhizobacteria Shewanella putrefaciens and Cronobacter dublinensis enhance drought tolerance of pearl millet by modulating hormones and stress-responsive genes. Physiol Plant 174(2):e13676. https://doi.org/10.1111/ppl.13676
98. Jha Y, Mohamed HI (2023) Inoculation with Lysinibacillus fusiformis strain YJ4 and Lysinibacillus sphaericus strain YJ5 alleviates the effects of cold stress in maize plants. Gesunde Pflanzen 75(1):77–95. https://doi.org/10.1007/s10343-022-00666-7
99. Sarfraz R, Hussain A, Sabir A, Ben Fekih I, Ditta A, Xing S (2019) Role of biochar and plant growth promoting rhizobacteria to enhance soil carbon sequestration-a review. Environ Monit Assess 191(4):251. https://doi.org/10.1007/s10661-019-7400-9
100. Misra AK, Bharadwaj V, Rastogi A (2018) Monitoring the carbon dioxide concentration at different land use patterns in Pantnagar (India) and surroundings. Clim Chang Environ Sustain 6(2):119–126
101. Fosbøl PL, Suleman H, Svendsen HF, Hauschild MZ (2022) Introduction to sustainable carbon capture. In: Sustainable Carbon Capture. CRC Press, pp 1–28

102. Hussain S, Hussain S, Guo R, Sarwar M, Ren X, Krstic D et al (2021) Carbon sequestration to avoid soil degradation: a review on the role of conservation tillage. Plan Theory 10(10):2001
103. Yoo Y, Kim I, Lee D, Yong Choi W, Choi J, Jang K, Park J, Kang D (2022) Review of contemporary research on inorganic CO2 utilization via CO2 conversion into metal carbonate-based materials. J Ind Eng Chem 116:60–74. https://doi.org/10.1016/j.jiec.2022.09.007
104. Gayathri R, Mahboob S, Govindarajan M, Al-Ghanim KA, Ahmed Z, Al-Mulhm N et al (2021) A review on biological carbon sequestration: a sustainable solution for a cleaner air environment, less pollution and lower health risks. Journal of King Saud University-Science 33(2):101282
105. Reddy PP (2015) Causes of climate change. In: Climate resilient agriculture for ensuring food security. Springer, New Delhi. https://doi.org/10.1007/978-81-322-2199-9_2
106. Nadeem M, Khalid R, Kanwal S, Mujtaba G, Qadir G, Ahmed M, Hayat R (2023) Soil microbes and climate-smart agriculture. In: Global agricultural production: resilience to climate change. Springer International Publishing, Cham, pp 107–147
107. Tarin MWK, Fan L, Tayyab M, Sarfraz R, Chen L, He T (2018) Effects of bamboo biochar amendment on the growth and physiological characteristics of Fokienia hodginsii. Appl Ecol Environ Res 16(6):8055–8074. https://doi.org/10.15666/aeer/1606_80558074
108. Elbasiouny H, El-Ramady H, Elbehiry F, Rajput VD, Minkina T, Mandzhieva S (2022) Plant nutrition under climate change and soil carbon sequestration. Sustainability 14(2):914
109. Khan N, Jhariya MK, Raj A, Banerjee A, Meena RS (2021) Soil carbon stock and sequestration: implications for climate change adaptation and mitigation. In: Ecological intensification of natural resources for sustainable agriculture. Springer, Singapore, pp 461–489
110. Song W, Tong X, Liu Y, Li W (2020) Microbial community, newly sequestered soil organic carbon, and δ15N variations driven by tree roots. Front Microbiol 11:314. https://doi.org/10.3389/fmicb.2020.00314
111. Rai PK, Singh M, Anand K, Saurabh S, Kaur T, Kour D et al (2020) Role and potential applications of plant growth-promoting rhizobacteria for sustainable agriculture. In: New and future developments in microbial biotechnology and bioengineering. Elsevier, pp 49–60
112. Walley FL, Gillespie AW, Adetona AB, Germida JJ, Farrell RE (2014) Manipulation of rhizosphere organisms to enhance glomalin production and C sequestration: pitfalls and promises. Can J Plant Sci 94(6):1025–1032. https://doi.org/10.4141/cjps2013-146
113. Treseder KK, Turner KM (2007) Glomalin in ecosystems. Soil Sci Soc Am J 71(4):1257–1266
114. Sofi JA, Lone AH, Ganie MA, Dar NA, Bhat SA, Mukhtar M, Dar MA, Ramzan S (2016) Soil microbiological activity and carbon dynamics in the current climate change scenarios: a review. Pedosphere 26(5):577–591. https://doi.org/10.1016/S1002-0160(15)60068-6
115. Bhattacharyya SS, Ros GH, Furtak K, Iqbal HM, Parra-Saldívar R (2022) Soil carbon sequestration–an interplay between soil microbial community and soil organic matter dynamics. Sci Total Environ 152928:152928
116. Bhattacharyya P, Varghese E, Dash PK, Padhy SR, Das A, Santra P, Mohapatra T (2023) Anticipated atmospheric CO2 elevation differentially influenced the soil microbial diversities in crop, grassland, and forest: a meta-analysis. Rhizosphere 25:100630
117. Kuzyakov Y, Horwath WR, Dorodnikov M, Blagodatskaya E (2018) Effects of elevated CO2 in the atmosphere on soil C and N turnover. In: Developments in soil science. Elsevier, Amsterdam, pp 207–219
118. He M, Li Q, Chen L, Qin S, Kuzyakov Y, Liu Y, Zhang D, Feng X, Kou D, Wu T, Yang Y (2023) Priming effect stimulates carbon release from thawed permafrost. Glob Chang Biol. https://doi.org/10.1111/gcb.16750
119. Nie M, Bell C, Wallenstein MD, Pendall E (2015) Increased plant productivity and decreased microbial respiratory C loss by plant growth-promoting rhizobacteria under elevated CO 2. Sci Rep 5:1–6. https://doi.org/10.1038/srep09212
120. Shahbaz M, Kuzyakov Y, Sanaullah M, Heitkamp F, Zelenev V, Kumar A, Blagodatskaya E (2017) Microbial decomposition of soil organic matter is mediated by quality and quantity of crop residues: mechanisms and thresholds. Biol Fertil Soils 53:287–301

121. Orella J, Africa DR, Bustillo CH, Pascua N, Marquez C, Adornado H, Aguilos M (2022) Above-and-belowground carbon stocks in two contrasting peatlands in The Philippines. Forests 13(2):303
122. Bardgett RD, Freeman C, Ostle NJ (2008) Microbial contributions to climate change through carbon cycle feedbacks. ISME J 2(8):805–814
123. Mondal S, Pramanik K, Pal P, Mitra S, Ghosh SK, Mondal T et al (2023) Multifaceted roles of root exudates in light of plant-microbe interaction. In: Unravelling plant-microbe synergy. Academic Press, pp 49–76
124. Carneiro B, Cardoso P, Figueira E, Lopes I, Venâncio C (2023) Forward-looking on new microbial consortia: combination of rot fungi and rhizobacteria on plant growth-promoting abilities. Applied Soil Ecology: A Section of Agriculture, Ecosystems & Environment 182(104689):104689. https://doi.org/10.1016/j.apsoil.2022.104689
125. Odoh CK, Sam K, Zabbey N, Eze CN, Nwankwegu AS, Laku C, Dumpe BB (2020) Microbial consortium as biofertilizers for crops growing under the extreme habitats. In: Plant microbiomes for sustainable agriculture. Springer, Cham, pp 381–424
126. Verma RK, Sachan M, Sharma S (2020) Microbial consortia: emerging conglomerate for better and superior sustainable agricultural practices. In: The plant microbiome in sustainable agriculture. John Wiley & Sons, Hoboken, pp 141–155
127. Ríos-Ruiz WF, Torres-Chávez EE, Torres-Delgado J, Rojas-García JC, Bedmar EJ, Valdez-Nuñez RA (2020) Inoculation of bacterial consortium increases rice yield (Oryza sativa L.) reducing applications of nitrogen fertilizer in san Martin region, Peru. Rhizosphere 14:100200
128. Yadav A, Yadav K (2022) Comparative growth analysis of cumin plants (Cuminum cyminum L.) treated with glycerol and talc based phosphate solubilizing bacterial consortia. Journal of applied and natural. Science 14(4):1374–1386. https://doi.org/10.31018/jans.v14i4.3859
129. Kaur T, Devi R, Kumar S, Sheikh I, Kour D, Yadav AN (2022) Microbial consortium with nitrogen fixing and mineral solubilizing attributes for growth of barley (Hordeum vulgare L.). Heliyon 8(4):e09326. https://doi.org/10.1016/j.heliyon.2022.e09326
130. Tyagi J, Mishra A, Kumari S, Singh S, Agarwal H, Pudake RN, Varma A, Joshi NC (2023) Deploying a microbial consortium of Serendipita indica, Rhizophagus intraradices, and Azotobacter chroococcum to boost drought tolerance in maize. Environ Exp Bot 206: 105142. https://doi.org/10.1016/j.envexpbot.2022.105142
131. Kapadia C, Sayyed RZ, El Enshasy HA, Vaidya H, Sharma D, Patel N, Malek RA, Syed A, Elgorban AM, Ahmad K, Zuan ATK (2021) Halotolerant microbial consortia for sustainable mitigation of salinity stress, growth promotion, and mineral uptake in tomato plants and soil nutrient enrichment. Sustainability 13(15):8369. https://doi.org/10.3390/su13158369
132. Majumder S, Datta K, Datta SK (2019) Rice biofortification: high iron, zinc, and vitamin-a to fight against "hidden hunger". Agronomy 9(12):803
133. Bourassa MW, Osendarp SJM, Adu-Afarwuah S, Ahmed S, Ajello C, Bergeron G, Black R, Christian P, Cousens S, de Pee S, Dewey KG, Arifeen SE, Engle-Stone R, Fleet A, Gernand AD, Hoddinott J, Klemm R, Kraemer K, Kupka R, McLean E et al (2019) Review of the evidence regarding the use of antenatal multiple micronutrient supplementation in low- and middle-income countries. Ann N Y Acad Sci 1444(1):6–21. https://doi.org/10.1111/nyas.14121
134. United Nations System Standing Committee on Nutrition (UNSSCN) (2004) 5thReport on the world nutrition situation nutrition for improved development outcomes. SCN, Geneva
135. Upadhayay VK, Singh AV, Khan A, Pareek N (2021) Influence of zinc solubilizing bacterial co-inoculation with zinc oxide supplement on rice plant growth and Zn uptake. Pharm Innovat 10:113–116
136. Upadhayay VK, Singh AV, Khan A, Sharma A (2022d) Contemplating the role of zinc-solubilizing bacteria in crop biofortification: an approach for sustainable bioeconomy. Front Agron 4:903321. https://doi.org/10.3389/fagro.2022.903321
137. Upadhayay VK (2022) Bacterial based Biofortification. Vigyan Varta 3(4):99–101

138. Singh J, Singh AV, Upadhayay VK, Khan A (2020) Comparative evaluation of developed carrier based bio-formulations bearing multifarious PGP properties and their effect on shelf life under different storage conditions. Environ Ecol 38(1):96–103
139. Ali M, Ahmed I, Tariq H, Abbas S, Zia MH, Mumtaz A, Sharif M (2023) Growth improvement of wheat (Triticum aestivum) and zinc biofortification using potent zinc-solubilizing bacteria. Front Plant Sci 14:1140454. https://doi.org/10.3389/fpls.2023.1140454
140. Yuan Y, Liu D, Huang X, Wang S, Qiu R, Zhang Z, Ming J (2023) Effect of Enterobacter sp. EG16 on selenium biofortification and speciation in pak choi (Brassica rapa ssp. chinensis). Sci Hortic 310:111723. https://doi.org/10.1016/j.scienta.2022.111723
141. Marfetán JA, Gallo AL, Farias ME, Vélez ML, Pescuma M, Ordoñez OF (2023) Exiguobacterium sp. as a bioinoculant for plant-growth promotion and selenium biofortification strategies in horticultural plants. World J Microbiol Biotechnol 39(5):134. https://doi.org/10.1007/s11274-023-03571-x
142. Ahmad M, Hussain A, Dar A, Luqman M, Ditta A, Iqbal Z, Ahmad HT, Nazli F, Soufan W, Almutairi K, Sabagh AE (2023) Combating iron and zinc malnutrition through mineral biofortification in maize through plant growth promoting bacillus and Paenibacillus species. Front Plant Sci 13:1094551. https://doi.org/10.3389/fpls.2022.1094551
143. Sarwar S, Khaliq A, Yousra M, Sultan T (2022) Iron biofortification potential of siderophore producing rhizobacterial strains for improving growth, yield and iron contents of groundnut. J Plant Nutr 45(15):2332–2347. https://doi.org/10.1080/01904167.2022.2063733
144. Sun Z, Yue Z, Liu H, Ma K, Li C (2021) Microbial-assisted wheat iron biofortification using endophytic Bacillus altitudinis WR10. Front Nutr 8:704030. https://doi.org/10.3389/fnut.2021.704030
145. Kushwaha P, Srivastava R, Pandiyan K, Singh A, Chakdar H, Kashyap PL, Bhardwaj AK, Murugan K, Karthikeyan N, Bagul SY, Srivastava AK, Saxena AK (2021) Enhancement in plant growth and zinc biofortification of chickpea (Cicer arietinum L.) by Bacillus altitudinis. Journal of. Soil Sci Plant Nutr 21(2):922–935. https://doi.org/10.1007/s42729-021-00411-5
146. Mansani M, Polapally R, Subramaniam G, Vadlamudi S, Sayyed RZ, Hesham Ali EE, Maizatulakmal Y, Ali Tan KZ, Hazem SK, Bee H (2021) Co-inoculation of *bacillus* spp. for growth promotion and iron fortification in sorghum. Sustainability 13(21):12091. https://doi.org/10.3390/su132112091
147. Shakeel M, Hafeez FY, Malik IR, Farid A, Ullah H, Ahmed I et al (2023) Serratia marcescens strain FA-4 enhances zinc content in rice grains by activating the zinc translocating enzymes. SABRAO J Breed Genet 55(2):495–507
148. Ullah A, Farooq M, Hussain M (2020) Improving the productivity, profitability and grain quality of kabuli chickpea with co-application of zinc and endophyte bacteria Enterobacter sp. MN17 Archiv Fur Acker- Und Pflanzenbau Und Bodenkunde 66(7):897–912. https://doi.org/10.1080/03650340.2019.1644501

Chapter 11
Bioactive Potential of Actinomycetes in Agriculture Sector

Arun Kumar Rai

Abstract Agriculture sector has been under tremendous pressure for merely producing enough food products to fulfil the considerable demand of progressively expanding population in this planet. Although chemical inputs in productive agriculture are imperative for proper plant growth and good yield, negative impacts of such compounds have undoubtedly increased the apparent importance of environmentally friendly pesticides and fertilizers of plants or microbial origin. Microorganisms efficiently perform a significant role in improving plant health, increasing plant yield, effective promotion of plant growth under stressful and adverse conditions, mitigating plant diseases and inducing plant defences. Impressive array of specific functionalities from the beneficial microorganism could significantly increase the necessary sustenance of plants under hostile and adverse conditions as well as increase crop yields with least adverse effect to human health and the environment. Extensive exploration of beneficial microorganisms from poorly explored habitats is still in its nascent stage, and the active search for better microbial isolates from exotic habitats with prospective source of bioproducts could divulge some novel and promising isolates with unique functionalities suitable for use in sustainable agriculture practices.

Keywords Agriculturally important microorganisms · Biofertilizers · Actinomycetes · Endophytes

11.1 Introduction

Ever-growing population across the globe has resulted in increase in considerable demand for food which has led to severe depletion of nutrition in the soil thereby limiting the factors for decreasing agricultural produce. As per published reports, around 70% of the abundant produce from the productive agriculture could be

A. K. Rai (✉)
Department of Botany, Sikkim University, Gangtok, Sikkim, India
e-mail: akrai@cus.ac.in

R. Soni et al. (eds.), *Microbial Bioactive Compounds*,
https://doi.org/10.1007/978-3-031-40082-7_11

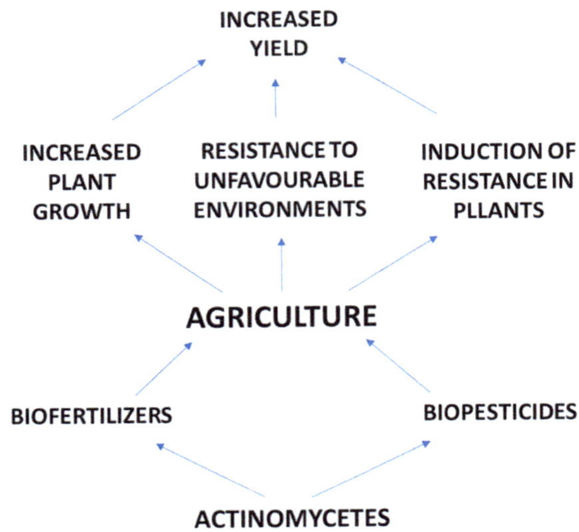

Fig. 11.1 Fundamental aspects of actinomycetes suitable for the possible enhancement of agriculture produce

progressively lost to pests without the extensive use of pesticides [1]. Pesticides are known to protect agricultural plants from phytopathogens such as bacteria, fungi, mites and insects [2]. This in turn results in the extensive use of considerable amount of fertilizers and pesticides which has been known to be harmful towards humans as well as the local environment. Biofertilizers and biopesticides derived from beneficial microorganisms have been thoughtfully considered as a viable alternative to naturally replenish the depleted soil fertility. Among beneficial microorganisms, actinomycetes have been known to secrete impressive array of bioactive metabolites with plant growth promoting functionalities (Fig. 11.1). Actinomycetes are gram-positive bacteria bearing similarities with mycelium of fungi [3]. These groups of specific microorganisms naturally possess plethora of desired functionalities such as secretion of beneficial enzymes, viz., proteases, chitinase and cellulase; potential bio-metabolites; decomposition of complex polysaccharides such as lignin, pectin, cellulose, etc.; and antimicrobial metabolites with antagonistic properties against suitable phytopathogens for sustainable agriculture practices [4]. Some of the actinomycetes belonging to the genus *Streptomyces*, *Actinoplanes* and *Micromonospora* have been known to considerably increase yield in agriculture sector [5, 6].

11.2 Plant Growth Promotion

Actinomycetes, especially *Streptomyces* spp., are able to colonize the rhizosphere as well as rhizoplane region of the plants efficiently. Distinct strains belonging to this recognized genus are equally able to colonize internal tissues of the plants therefore establishing an endophytic relationship with the host plant [7]. As per Olanrewaju and Babalola [8], beneficial properties of actinomycetes could possibly be due to following attributes, viz., gene expression controlled through the complex mechanism of quorum sensing, production of plant hormones, chitinases, lipases, synthesis of amino acids, secretion of antimicrobial compounds, etc. Actinomycetes obtained from the rhizosphere of the olive tree demonstrated plant growth-promoting traits such as the production of indole acetic acid, ammonia, solubilization of inorganic phosphates and siderophores [9].

In one study, it was reported that the actinomycete strains in study were able to produce ammonia and solubilize phosphate as well as effectively colonize the roots of *Solanum lycopersicum* which as observed through scanning electron microscopy thus indicating their potential as prospective biofertilizer [10]. Chloropyrroles, isolated from an actinomycete strain *Catellatospora* sp., and a trehalose, Trehangelin E obtained from the extract of *Polymorphospora* sp., were found to promote elongation of roots in the germinated seeds of lettuce [11, 12].

11.2.1 Plant Growth Promotion Under Salt-Stressed Environment

Growth promotion with the help of actinomycetes in durum wheat plants under salt stress was studied by Djebaili et al. [13, 14]. In the experiment, different strains of actinomycetes were able to solubilize phosphates and produce IAA, hydrocyanic acids and ammonia under different concentration of sodium chloride. The strains were able to produce aminocyclopropane-1carboxylate deaminase activity and alpha-ketobutyric acids. The study inferred that the salt-tolerant actinomycetes could promote healthy plant growth and alleviate the stress of high concentration of salt on wheat plants. In one study, inoculation of seedlings of tomato with a strain of *Streptomyces* sp. resulted in increase in the weight, length of roots, number of roots, enrichment in pathways responsible for the processing of proteins by endoplasmic reticulum and enhancement of biochemical pathways thus indicating the possibility of using the strain as a biofertilizer [15].

11.2.2 Plant Growth Promotion Under Drought

Actinomycete strains have been assessed for their potential as a plant growth promoter under drought condition. Maize plants inoculated with the strains of actinomycetes, i.e. *Arthrobacter arilaitensis* and *Streptomyces pseudovenezuelae*, were able to promote proper growth of the plants as well as demonstrated a significant increase in other growth parameters thus indicating their potential as bioinoculants [16]. In one study, treatment of wheat plants with *Streptomyces pactum* Act 12 enhanced adjustment of osmotic potential and antioxidant capacity with the help of accumulation of abscisic acid and regulation of genes related to resistance of drought thereby helping in resisting the stress on plants due to drought [17].

11.3 Biocontrol Attributes of Actinomycetes

Actinomycetes are known to secrete an array of hydrolysing enzymes that contribute to biocontrol aspect of the microorganisms. In one study, high levels of glucanases, proteases and chitinases were produced by *Arthrobacter humicola*, and moderate levels of amylases and pectinases were produced by *Streptomyces atratus* [18].

Some of the traits of actinomycete strains responsible for biocontrol of plant pathogens as indicated by Olanrewaju and Babalola [8] are as follows:

(a) Production of volatile organic compounds
(b) Secretion of antimicrobial metabolites
(c) Production of siderophores
(d) Production of growth regulators of plants

Actinomycete strains, viz., *Nocardiopsis aegyptica* and *Streptomyces lycopersicum*, secreted diffusible and volatile compounds with antifungal properties thus highlighting their potential as a biocontrol agent in a consortium [13]. Application of strains of *Streptomyces* spp. was able to control the effects of southern blight as well as root rot disease in *Aconitum carmichaelii* plants. Promotion of plant growth and control of disease was also visible even after stopping of application of the inoculum. Increase in the abundance of beneficial microorganisms in rhizosphere region was also observed [19]. An actinomycete strain *Streptomyces aureoverticillatus* was found to exhibit antibacterial activity against a bacterial phytopathogen *Ralstonia solanacearum* [20]. A strain of actinomycete, *Streptomyces griseorubiginosus*, demonstrated antifungal properties against *Fusarium oxysporum* responsible for causing wilt disease in banana plants. The strain was also able to produce plant growth-promoting properties [21]. A strain of actinomycetes, *Micromonospora* sp., was found to possess nematocidal potential. A compound, benzenepropanoic, extracted from the strain was able to kill 99% of nematode *Meloidogyne incognita* as well as inhibit hatching of the eggs thus establishing as a promising biocontrol agent against the nematode [22]. One of the strains of *Streptomyces antibioticus* was also able to inhibit the hatching of eggs of *M. incognita* as well as increase in the mortality of juvenile nematodes [23]. A pinewood nematode *Bursaphelenchus xylophilus* was inhibited by *Streptomyces* sp. through the production of a nematocidal metabolite teleocidin [23].

1. Properties of actinomycetes with possible application in sustainable agriculture system

Beneficial actinomycetes have immense potential in the production of biofertilizer and biopesticide formulations (Table 11.1). Their role in promoting

Table 11.1 Different properties of actinomycete species suitable for agricultural application

Sl. No.	Actinomycetes	Properties	Reference
1	*Streptomyces* sp.	Antagonistic effect against *Botrytis cinerea* causing spot disease in faba bean (*Vicia fabae* L.)	El-Shatoury et al. [24]
2	*Streptomyces rochei* ANH	Biosorption of heavy metals, viz., Cr^{6+}, Cd^{2+} and Pb^{2+} from industrial effluents and improve water quality suitable for irrigation	Hamdan et al. [25]
3	*Streptomyces* sp.	A part of consortia to convert waste from husbandry into organic fertilizer	Amrullah et al. [26]
4	*Streptomyces tsukiyonensis*	Antagonistic effect against *Colletotrichum dematium* responsible for causing anthracnose in *Sarcandra glabra*	Song et al. [27]
5	*Streptomyces* sp.	Degradation of lignocellulosic residues by enzymatic hydrolysis during composting	Buzón-Durán et al. [28]
6	*Kribbella speibonae*	Production of siderophores	Acquah et al. [29]
7	*Streptomyces* sp.	Production of exotoxins with insecticidal properties against Diamondback moth, *Plutella xylostella* pest of cabbage and cauliflower	Srujana et al. [30]
8	*Brachybacterium phenoliresistens* and *Microbacterium* sp.	Antifungal properties against *Peronophythora litchi* and *Rhizoctonia solani*	Wu et al. [31]
9	*Streptomyces* spp.	Promotes degradation of lignocellulosic residues during the composting through enzymatic hydrolysis	Buzón-Durán et al. [28]
10	*Streptomyces* sp.	Antifungal action against *Thielaviopsis paradoxa* that causes root rot, fruit rot and bleeding disease of coconuts	M. M. et al. [32]

plant growth, maintaining plant health and secretion of suitable bioactive metabolites with immense potential in agriculture cannot be ruled out [33].

11.4 Future Scopes and Prospects

Biofertilizers have undoubtedly gained considerable attention in the recent years. The untapped microorganisms laden with diverse features have been popularly considered over the chemical fertilizers due to the negative impacts of hazardous chemicals resulting in various health issues and adversely affecting the environment. Biofertilizers are known to assist the active management of critical components of assimilated nutrients for structured farming. Among various strains of microorganisms in this planet, the beneficial roles of actinomycete cannot be overlooked. The massive population of actinomycetes in diversified soil types and their prominent roles in adequately maintaining fertility of soil and nature are of prime importance in agriculture sector. The proper management of soil ecosystem by the actinomycetes through the diversified attributes such as an ability of plant growth promotion, properly managing the optimal health and vigour of the plants, and secretion of agro-active compounds are significant contributors to agriculture. The remarkable ability of actinomycetes to amply compensate for hazardous and harmful chemical fertilizers and also to boost beneficial effects in plants highlights their prominent role in the vulnerable ecosystem. These actinomycetes are universally recognized to exert substantial beneficial effects in soil and hence provide a substantial way to proportionately increase crop yield in the foreseeable future. Some of the strains of actinomycetes were at par with the chemical fertilizers when evaluated for their plant growth promoting attributes [34].

11.5 Conclusion

Beneficial microorganisms such as bacteria, along with actinomycetes, and fungi have been known to efficiently manage soil-borne phytopathogens. Some of the strains belonging to actinobacteria, viz., *Streptomyces*, *Nocardia*, *Frankia*, *Micromonospora* and *Amycolatopsis*, have been reported to adequately support the plants to absorb required nutrition from the surrounding soil as well as help in the control of pathogens.

The continuous exploration for plant growth promoting characteristics and antimicrobial properties remains a need of the hour to promote sustainable agriculture practices.

Extensive exploration and continuous evaluation of prospective actinomycetes laden with beneficial properties suitable for sustainable agriculture system should be a continuous process to replenish the existing consortia with the more efficient and effective ones.

References

1. Popp J, Pető K, Nagy J (2013) Pesticide productivity and food security. A review. Agron Sustain Dev 33:243–255. https://doi.org/10.1007/s13593-012-0105-x
2. Li S, Yang B, Tan G-Y et al (2021) Polyketide pesticides from actinomycetes. Curr Opin Biotechnol 69:299–307. https://doi.org/10.1016/j.copbio.2021.05.006
3. da Silva GC, Kitano IT, de Ribeiro IAF, Lacava PT (2022) The potential use of Actinomycetes as microbial inoculants and biopesticides in agriculture. Front Soil Sci 2. https://doi.org/10.3389/fsoil.2022.833181
4. Javed Z, Tripathi GD, Mishra M, Dashora K (2021) Actinomycetes – the microbial machinery for the organic-cycling, plant growth, and sustainable soil health. Biocatal Agric Biotechnol 31:101893. https://doi.org/10.1016/j.bcab.2020.101893

5. Das P, Singh SK, Singh P et al (2021) Actinomycetes as biostimulants and their application in agricultural practices. In: Microbiome stimulants for crops. Elsevier, pp 267–282
6. Rai AK, Sunar K, Sharma H (2021) Agriculturally important microorganism: understanding the functionality and mechanisms for sustainable farming. In: Soni R, Suyal DC, Prachi B, Reeta G (eds) Microbiological activity for soil and plant health management. Springer Singapore, Singapore, pp 35–64
7. de Sousa JAJ, Olivares FL (2016) Plant growth promotion by streptomycetes: ecophysiology, mechanisms and applications. Chem Biol Technol Agric 3:24. https://doi.org/10.1186/s40538-016-0073-5
8. Olanrewaju OS, Babalola OO (2019) Streptomyces: implications and interactions in plant growth promotion. Appl Microbiol Biotechnol 103:1179–1188. https://doi.org/10.1007/s00253-018-09577-y
9. Dede A, Güven K, Şahin N (2020) Isolation, plant growth-promoting traits, antagonistic effects on clinical and plant pathogenic organisms and identification of actinomycetes from olive rhizosphere. Microb Pathog 143:104134. https://doi.org/10.1016/j.micpath.2020.104134
10. Djebaili R, Pellegrini M, Smati M et al (2020) Actinomycete strains isolated from saline soils: plant-growth-promoting traits and inoculation effects on Solanum lycopersicum. Sustainability 12:4617. https://doi.org/10.3390/su12114617
11. Liu C, Yamamura H, Hayakawa M et al (2022) Plant growth-promoting and antimicrobial chloropyrroles from a rare actinomycete of the genus Catellatospora. J Antibiot (Tokyo) 75: 655–661. https://doi.org/10.1038/s41429-022-00567-x
12. Lu S, Harunari E, Oku N, Igarashi Y (2022) Trehangelin E, a bisacyl trehalose with plant growth promoting activity from a rare actinomycete Polymorphospora sp. RD064483. J Antibiot (Tokyo) 75:296–300. https://doi.org/10.1038/s41429-022-00519-5
13. Djebaili R, Pellegrini M, Ercole C et al (2021a) Biocontrol of soil-borne pathogens of *Solanum lycopersicum* L. and *Daucus carota* L. by plant growth-promoting actinomycetes: in vitro and in planta antagonistic activity. Pathogens 10:10
14. Djebaili R, Pellegrini M, Rossi M et al (2021b) Characterization of plant growth-promoting traits and inoculation effects on Triticum durum of Actinomycetes isolates under salt stress conditions. Soil Syst 5:26. https://doi.org/10.3390/soilsystems5020026
15. Gong Y, Chen L-J, Pan S-Y et al (2020) Antifungal potential evaluation and alleviation of salt stress in tomato seedlings by a halotolerant plant growth-promoting actinomycete *Streptomyces* sp. KLBMP5084. Rhizosphere 16:100262. https://doi.org/10.1016/j.rhisph.2020.100262
16. Chukwuneme CF, Babalola OO, Kutu FR, Ojuederie OB (2020) Characterization of actinomycetes isolates for plant growth promoting traits and their effects on drought tolerance in maize. J Plant Interact 15:93–105. https://doi.org/10.1080/17429145.2020.1752833
17. Li H, Guo Q, Jing Y et al (2020a) Application of Streptomyces pactum Act12 enhances drought resistance in wheat. J Plant Growth Regul 39:122–132. https://doi.org/10.1007/s00344-019-09968-z
18. Elshafie HS, Camele I (2022) Rhizospheric Actinomycetes revealed antifungal and plant-growth-promoting activities under controlled environment. Plants 11:1872. https://doi.org/10.3390/plants11141872
19. Li Y, Guo Q, He F et al (2020b) Biocontrol of root diseases and growth promotion of the tuberous plant Aconitum carmichaelii induced by Actinomycetes are related to shifts in the rhizosphere microbiota. Microb Ecol 79:134–147. https://doi.org/10.1007/s00248-019-01388-6
20. Zhao J, Han L, Yu M et al (2019) Characterization of Streptomyces sporangiiformans sp. nov., a novel soil Actinomycete with antibacterial activity against Ralstonia solanacearum. Microorganisms 7:360. https://doi.org/10.3390/microorganisms7090360
21. Lai B, Lin M, Wu Z, Wang J (2020) Identification of an antagonistic actinomycetes and its biological control effect on banana fusarium wilt. J South Agric 51:836–843. https://doi.org/10.3969/j.issn.2095-1191.2020.04.014

22. Ran Y, Zhang Y, Wang X, Li G (2022) Nematicidal metabolites from the Actinomycete Micromonospora sp. WH06. Microorganisms 10:2274. https://doi.org/10.3390/microorganisms10112274
23. Raja VV, Dhanasekaran D (2022) An in vitro Nematicidal activity of Actinobacteria: juvenile mortality and egg hatching inhibition. In: Dharumadurai D (ed) Methods in Actinobacteriology. Springer US, New York, NY, pp 615–619
24. El-Shatoury SA, Ameen F, Moussa H et al (2020) Biocontrol of chocolate spot disease (Botrytis cinerea) in faba bean using endophytic actinomycetes Streptomyces: a field study to compare application techniques. PeerJ 8:e8582
25. Hamdan AM, Abd-El-Mageed H, Ghanem N (2021) Biological treatment of hazardous heavy metals by *Streptomyces rochei* ANH for sustainable water management in agriculture. Sci Rep 11:9314. https://doi.org/10.1038/s41598-021-88843-y
26. Amrullah S, Amin M, Ali M (2021) Converting husbandry waste into liquid organic fertilizer using probiotic consortiums (Lactobacillus sp., Rhodopseudomonas sp., Actinomycetes sp., Streptomyces sp.). IOP Conf Ser Earth Environ Sci 679:012001. https://doi.org/10.1088/1755-1315/679/1/012001
27. Song L, Jiang N, Wei S et al (2020) Isolation, screening, and identification of Actinomycetes with antifungal and enzyme activity assays against Colletotrichum dematium of Sarcandra glabra. Mycobiology 48:37–43. https://doi.org/10.1080/12298093.2020.1716604
28. Buzón-Durán L, Pérez-Lebeña E, Martín-Gil J et al (2020) Applications of Streptomyces spp. enhanced compost in sustainable agriculture. In: Meghvansi MK, Varma A (eds) BT - Biology of composts. Springer International Publishing, Cham, pp 257–291
29. Acquah KS, Beukes DR, Warner DF et al (2020) Novel South African rare Actinomycete Kribbella speibonae strain SK5: a prolific producer of Hydroxamate Siderophores including new Dehydroxylated congeners. Molecules 25:2979. https://doi.org/10.3390/molecules25132979
30. Srujana Y, Hugar PS, Krishnaraj PU (2022) Exploring the entomotoxic ability of actinomycetes against Diamondback moth, Plutella xylostella L. (Lepidoptera: Plutellidae). Int J Trop Insect Sci 42:2559–2566. https://doi.org/10.1007/s42690-022-00783-w
31. Wu Y, Liu Y, Yu J et al (2022) Observation of the antimicrobial activities of two Actinomycetes in the harvester Ant Messor orientalis. Insects 13:691. https://doi.org/10.3390/insects13080691
32. Nisha MN, Kumar SS, Ibomcha N et al (2022) In vitro study on Actinomycetes extracts against the stem bleeding disease of coconut. CORD 38:1–7. https://doi.org/10.37833/cord.v38i.431
33. Shanthi V (2021) Actinomycetes: implications and prospects in sustainable agriculture. In: Biofertilizers. Wiley, pp 335–370
34. Asadu CO, Ike IS, Onu CE et al (2020) Investigation of the influence of biofertilizer synthesized using microbial inoculums on the growth performance of two agricultural crops. Biotechnol Reports 27:e00493. https://doi.org/10.1016/j.btre.2020.e00493

Chapter 12
Environmental Sustainability Through Microbes and Their Metabolites

Safina Ismail, Kalp Das, Deep Chandra Suyal, and Ravindra Soni

Abstract Environmental sustainability refers to the responsible management of natural resources and ecosystems to ensure their long-term health and resilience. Microbes, which are microscopic organisms such as bacteria, fungi, and algae, play a crucial role in maintaining environmental sustainability through their metabolic activities. Ecologically sustainable agricultural practices are essential to ensure food security. Agrochemicals can be replaced in the production of food by microorganism-based inoculants that improve nutrient uptake, encourage crop development, or shield plants against pests and diseases. Effective agriculturally beneficial microbes (microbial inoculants) are potentially playing a role in sustainable crop production due to their immense plant growth-promoting attributes, better adaptability to survival under stresses and other uses that result in attenuating the pesticides/fertilizers use in agriculture. By fixing N_2, solubilizing K and P, releasing soil trace elements, secreting exopolysaccharides, converting organic matter into usable nutrients, increasing soil water-holding capacity and strengthening soil health overall, effective microbes aid in crop development and welfare. In order to promote plant development, such microbes secrete biocontrol agents and improve drought tolerance, and they also produce bioactive substances like vitamins, hormones, and enzymes. In addition to their positive role in stimulating plant growth and development, microorganisms possess the ability to clean contaminated sites from accumulated pesticides, heavy metals, polyaromatic hydrocarbons, and other industrial effluents. Microbes can synthesize biopolymers such as polyhydroxyalkanoates (PHAs) and polylactic acid (PLA) through their metabolic pathways. These

S. Ismail · R. Soni (✉)
Department of Agricultural Microbiology, Indira Gandhi Krishi Vishwavidyalaya, Raipur, Chhattisgarh, India

K. Das
Department of Vegetable Science, Punjab Agricultural University, Ludhiana, Punjab, India

D. C. Suyal
Vidyadayini Institute of Science, Management and Technology, Bhopal, Madhya Pradesh, India

R. Soni et al. (eds.), *Microbial Bioactive Compounds*,
https://doi.org/10.1007/978-3-031-40082-7_12

biopolymers have properties similar to conventional plastics but are biodegradable and derived from renewable sources. Moreover, employing potentially effective microorganisms can greatly contribute in environmental sustainability.

Keywords Metabolites · Bioremediation · Wastewater · Sustainability

12.1 Introduction

In recent decades, the concept of environmental sustainability has gained increasing attention due to the growing concerns about the impact of human activities on the environment. According to the Intergovernmental Panel on Climate Change (IPCC), global greenhouse gas emissions have increased by about 70% between 1970 and 2004, with the majority of these emissions coming from the burning of fossil fuels [1]. These emissions are causing significant changes in the Earth's climate system, including rising temperatures, changing precipitation patterns, and more frequent and severe weather events, which in turn are affecting ecosystems and human well-being [2]. Microbes, which are tiny organisms that are ubiquitous in nature, play a critical role in maintaining environmental sustainability through their various metabolic processes. For example, some microbes can break down pollutants such as pesticides and heavy metals, while others can fix atmospheric nitrogen and recycle nutrients in soil, water, and other ecosystems. Additionally, microbial metabolites, which are compounds produced by microbes, have been shown to have various applications in promoting environmental sustainability, including biodegradable plastics, biofuels, and biofertilizers [3–5].

The purpose of this chapter is to provide an overview of the role of microbes and their metabolites in environmental sustainability. This paper will explore the various types of microbes, the metabolites they produce, and the applications of these metabolites in promoting environmental sustainability. Additionally, this chapter will highlight some of the challenges and limitations associated with the use of microbial metabolites and provide recommendations for future research. Overall, this chapter aims to contribute to our understanding of the potential of microbial metabolites in promoting environmental sustainability.

12.2 Microbes and Their Metabolites

Microbes are diverse organisms that can be found in virtually all environments, including soil, water, air, and inside other organisms. The three main types of microbes are bacteria, fungi, and viruses. Bacteria are single-celled organisms that can be either beneficial or harmful to the environment and human health. Fungi, which are typically multicellular, can also be beneficial or harmful and are important decomposers in ecosystems. Viruses, which are not considered living organisms, can infect both bacteria and other living organisms and can have significant impacts on ecosystems [6].

12.2.1 Microbial Metabolites

Microbial metabolites are compounds that are produced by microbes as a result of their metabolic processes. These metabolites have diverse chemical structures and biological activities and have been found to have various applications in promoting environmental sustainability. For example, some microbial metabolites have been shown to have antimicrobial, antioxidant, and anticancer properties, while others have been used in the production of biodegradable plastics, biofuels, and biofertilizers [3, 7].

12.3 Applications of Microbial Metabolites

12.3.1 Biodegradable Plastics

One of the most promising applications of microbial metabolites is in the production of biodegradable plastics. Biodegradable plastics are a type of plastic that can be broken down by microbes into natural compounds, such as carbon dioxide and water. This is in contrast to traditional plastics, which can persist in the environment for hundreds of years and have significant impacts on ecosystems. Microbial metabolites such as polyhydroxyalkanoates (PHAs) have been shown to be effective in the production of biodegradable plastics and have the potential to significantly reduce the environmental impact of plastic waste [4, 5].

12.3.2 Biofuels

Microbial metabolites have also been used in the production of biofuels, which are fuels that are derived from renewable biomass sources. One such metabolite is ethanol, which is produced by certain types of bacteria and yeast during fermentation. Ethanol has been used as a biofuel for many years and has the potential to significantly reduce greenhouse gas emissions when used as a replacement for traditional fossil fuels [6].

12.3.3 Biofertilizers

Finally, microbial metabolites have been used in the production of biofertilizers, which are fertilizers that are derived from natural sources such as plants, animals, and microbes. Microbial metabolites such as indole acetic acid (IAA) have been shown to stimulate plant growth and improve soil fertility, making them a promising alternative to traditional chemical fertilizers [8].

12.3.4 Microbes and Their Role in Environmental Sustainability

12.3.4.1 Soil Health

- **Soil Microbes**

Soil microbes play a vital role in maintaining soil health and fertility. These microorganisms are involved in various processes such as nutrient cycling, decomposition of organic matter, and the formation of soil aggregates. They also help in the production of plant growth-promoting substances, which can improve crop yields and reduce the need for chemical fertilizers. In addition, soil microbes can help to mitigate climate change by storing carbon in the soil, thereby reducing atmospheric carbon dioxide levels [9].

- **Bioremediation**

Microbes can also be used for bioremediation, which is the process of using living organisms to remove or detoxify pollutants from the environment [10]. For example, some bacteria are capable of breaking down toxic compounds such as petroleum hydrocarbons and heavy metals and can be used to clean up contaminated soil and water [11]. This approach is more environmentally friendly and cost-effective than traditional methods such as excavation and disposal [12].

12.3.4.2 Water Quality

- **Wastewater Treatment**

Microbes are also important in the treatment of wastewater. Many microorganisms are capable of breaking down organic matter and converting it into harmless substances such as water and carbon dioxide. This process, known as biological wastewater treatment, is widely used in municipal and industrial wastewater treatment plants. By using microbes to treat wastewater, we can reduce the amount of pollutants that are discharged into waterways and improve overall water quality [13].

- **Algal Blooms**

Microbes also play a role in preventing harmful algal blooms (HABs) in waterways. HABs are caused by the rapid growth of certain types of algae, which can deplete oxygen levels in the water and release toxins that can be harmful to aquatic life and human health. Some types of bacteria, known as probiotics, can help to control the growth of harmful algae by outcompeting them for resources and producing compounds that inhibit their growth [14].

- **Air Quality**

Microbes can also have a positive impact on air quality. For example, some bacteria are capable of breaking down volatile organic compounds (VOCs), which are a major contributor to air pollution. By using microbes to remove VOCs from the air, we can improve air quality and reduce the risk of respiratory illnesses [15].

12.4 Environmental Sustainability Through Microbial Metabolites

12.4.1 Bioactive Compounds

- **Antibiotics**

Microbes produce a wide range of bioactive compounds, many of which have important applications in medicine, agriculture, and other fields. Antibiotics, for example, are natural products produced by bacteria and fungi that can kill or inhibit the growth of other microorganisms. These compounds have been used for decades to treat bacterial infections in humans and animals and are also used in agriculture as growth promoters and to prevent diseases in livestock. However, the overuse and misuse of antibiotics have led to the emergence of antibiotic-resistant bacteria, which is a growing public health concern [16].

- **Enzymes**

Microbes also produce a wide range of enzymes that have important applications in industry and biotechnology. For example, some bacteria produce enzymes that can break down plant fibres and convert them into biofuels and other value-added products. Other enzymes produced by microbes are used in the production of food, textiles, and paper, among other applications [17].

12.4.2 Bioplastics

- **Polyhydroxyalkanoates**

Microbes also produce bioplastics, which are biodegradable plastics made from renewable resources such as plant sugars or waste biomass. Polyhydroxyalkanoates (PHAs) are a type of bioplastic produced by many types of bacteria that can be used to replace conventional plastics in a variety of applications. PHAs have several advantages over conventional plastics, including biodegradability, renewability, and reduced dependence on fossil fuels [18].

12.4.3 Biofertilizers

- **Plant Growth-Promoting Substances**

Microbial metabolites can also be used as biofertilizers, which are products that contain living microorganisms or their metabolites and are used to enhance plant growth and productivity. Some bacteria produce plant growth-promoting substances such as indole acetic acid (IAA), which can stimulate root growth and improve nutrient uptake in plants. These substances can also improve soil health by increasing microbial diversity and nutrient availability [19].

12.5 Challenges and Limitations

12.5.1 Regulatory Issues

- **Lack of Regulations for Microbial Metabolites**

The use of microbial metabolites in various industries has raised concerns about their safety and environmental impact. However, there is a lack of regulatory guidelines for the production and use of these metabolites. This has led to uncertainty about their safety and efficacy and has hindered their commercialization [20].

- **Intellectual Property Issues**

The development of microbial metabolites as commercial products can be hindered by intellectual property issues. The patenting of microbial strains and their metabolites can be complex, and disputes over patent ownership can arise. This can make it difficult for companies to invest in the development and commercialization of these products [20].

12.6 Technical Limitations

- **Yield Optimization**

The production of microbial metabolites can be limited by low yields, which can make their commercialization difficult. Optimization of fermentation conditions and strain engineering can help to increase yields, but this can be a time-consuming and expensive process [21].

- **Scale-Up**

The scale-up of microbial metabolite production from laboratory to industrial scale can be challenging. Fermentation conditions and downstream processing methods may need to be optimized for larger-scale production, and this can require significant investment in equipment and infrastructure [21].

12.7 Future Prospects

12.7.1 Industrial Applications

- **Agriculture**

Microbial metabolites have great potential in agriculture for improving soil health, plant growth, and nutrient uptake. Several metabolites, such as indole acetic acid, gibberellins, and siderophores, have been shown to promote plant growth and protect crops from pathogens [22].

- **Bioremediation**

Microbes and their metabolites can play a vital role in the bioremediation of contaminated soils and water bodies. Microbial enzymes, such as laccases, peroxidases, and cellulases, can break down pollutants into harmless compounds. Additionally, microbial biosurfactants can help to remove hydrophobic pollutants from the environment [23].

12.7.2 Advances in Technology

- **Synthetic Biology**

Advances in synthetic biology have opened up new possibilities for the production of microbial metabolites. Synthetic biology allows for the engineering of microbial strains to produce specific metabolites and has the potential to overcome the limitations of natural microbial strains [24].

- **Metagenomics**

Metagenomics is a powerful tool for the discovery of novel microbial metabolites. Metagenomic analysis of environmental samples can identify new microbial strains and their associated metabolites, which can then be isolated and studied [25, 26].

12.8 Conclusion

Microbes and their metabolites play a vital role in environmental sustainability. Their metabolites have applications in various industries, including agriculture, bioremediation, and medicine. Challenges and limitations, such as regulatory issues, technical limitations, and scale-up, need to be addressed for the commercialization of microbial metabolites. Further, advances in technology, such as synthetic biology and metagenomics, offer new opportunities for the discovery and production of microbial metabolites.

Continued research is needed to discover new microbial strains and their associated metabolites. Moreover, efforts should be made to optimize the production and commercialization of microbial metabolites. Regulatory guidelines should be established for the safe production and use of microbial metabolites. Furthermore, collaboration between industry, academia, and regulatory bodies is necessary for the successful development and commercialization of microbial metabolites.

References

1. IPCC (2007) Climate change 2007: the physical science basis. Cambridge University Press
2. IPCC (2014) Climate change 2014: impacts, adaptation, and vulnerability. Cambridge University Press
3. Das S, Lyla PS, Khan SA (2015) Microbial metabolites in nutrition, healthcare and environmental sustainability. J Appl Pharm Sci 5(11):001–010
4. Singh N, Rai V, Singh P (2019a) Biodegradable plastics: microbial metabolites, production strategies, and applications. Crit Rev Biotechnol 39(4):492–507
5. Singh P, Srivastava S, Singh SP (2019b) Microbial metabolites and their potential applications in biodegradable plastics. J Polym Environ 27(4):730–743
6. Janssens SB, Horemans B, Van Houdt R, Provoost A (2019) Microbial biotechnology for sustainable agriculture and environmental health. J Ind Microbiol Biotechnol 46(2):155–164
7. Sathishkumar M, Sneha K (2018) Microbial metabolites: a sustainable eco-friendly approach for crop protection, growth and yield enhancement. J Appl Microbiol 124(6):1369–1383
8. El-Naggar MY, Abdelwahed NAM (2020) Microbial metabolites and their applications in sustainable agriculture. Appl Soil Ecol 154:103592
9. Chaudhry V, Rehman A, Mishra A, Chauhan PS, Nautiyal CS (2019) Changes in microbial diversity and soil physicochemical properties in long-term field trials of inorganic and organic fertilization in rice-wheat cropping system. PLoS One 14(11):e0225052

10. Kumar P, Dash B, Suyal DC, Gupta SB, Singh AK, Chowdhury T, Soni R (2021) Characterization of arsenic-resistant Klebsiella pneumoniae RnASA11 from contaminated soil and water samples and its bioremediation potential. Curr Microbiol 78:3258–3267. https://doi.org/10.1007/s00284-021-02602-w
11. Dash B, Sahu N, Singh AK, Gupta SB, Soni R (2021) Arsenic efflux in Enterobacter cloacae RSN3 isolated from arsenic-rich soil. Folia Microbiol 66:189–196. https://doi.org/10.1007/s12223-020-00832-2
12. Bhatt P, Patel KS (2021) Microbial bioremediation: a sustainable approach for environmental cleanup. Environmen Sustain 4(2):201–215
13. Khan MZ, Ahmad I, Khan S (2018) Microbial diversity and functionality in wastewater treatment plants. Microb Biotechnol 11(5):813–828
14. Jain S, Mishra A, Kumar R (2019) Probiotics as potential biocontrol agents in the management of harmful algal blooms (HABs). Environ Sustain 2(2):137–147
15. Almeida OAC, de Araujo NO, Mulato ATN, Persinoti GF, Sforça ML, Calderan-Rodrigues MJ, Oliveira JVdC (2023) Bacterial volatile organic compounds (VOCs) promote growth and induce metabolic changes in rice. Front Plant Sci 13:1056082. https://doi.org/10.3389/fpls.2022.1056082
16. Ventola CL (2015) The antibiotic resistance crisis: part 1: causes and threats. Pharm Ther 40(4):277
17. Bhalla TC, Kumar P, Sharma P (2020) Microbial enzymes: industrial progress in 21st century. 3 Biotech 10(5):199
18. Majeed A, Naveed M, Ali Z, Ashraf M, Akram N (2021) Bioplastics: an eco-friendly substitute of petrochemical-based plastics. J Environ Manag 278:111532
19. Jha Y, Saraf M (2014) Plant growth promoting rhizobacteria (PGPR): a review. J Agric Res Dev 4(1):71–79
20. Lau MS, Chan CY, Zada S (2018) Regulatory considerations for microbial metabolites. J Ind Microbiol Biotechnol 45(5):315–330
21. Bhattacharjee A, Ghosh D, Pradhan B (2020) Microbial metabolites: current perspective and applications. J Biosci Bioeng 129(6):609–619
22. Dutta S, Shivaji S, Mandal SM (2020) Microbial metabolites in agriculture and environment: an overview. In: Microbial Metabolites in Plants, Progress and Prospects, pp 27–56
23. Mnif I, Ghribi D (2015) Microbial enzymes and their applications in hydrocarbon polluted environments. J Environ Manag 157:48–64
24. Medema MH, Fischbach MA (2015) Computational approaches to natural product discovery. Nat Chem Biol 11(9):639–648
25. Soni R, Suyal DC, Sahu B, Phulara SC (2021) Metagenomics. In: Verma A, Saini JK, Hesham AEL, Singh HB (eds) Phytomicrobiome interactions and sustainable agriculture. https://doi.org/10.1002/9781119644798.ch3
26. Charlop-Powers Z, Owen JG, Reddy BV (2015) Terrestrial metagenomics: challenges and opportunities for mining microbial biodiversity. Appl Environ Microbiol 81(20):6870–6877

Chapter 13
Induction of Stress Tolerance in Plants by Metabolic Secretions of Endophytes for Sustainable Development

Anand Kumar Chaubey, Vijay Sharma, Pawan Kumar Prajapati, Suraj Mishra, Rakesh Pandey, S. V. Dwivedi, Ajeet Singh, and Ravindra Soni

Abstract Endophytes are microbes that can survive inside of a plant's stable tissues without wreaking havoc on the host. Endophytes aid plant adaptation by conferring a variety of determining effects that can counteract the harmful impacts of abiotic or biotic stressors. As a result, there is significant potential for long-term agricultural output if endophytic bacteria are used to increase crop performance under stress circumstances including low temperatures, high salt, low humidity, and heavy metal contamination. In order to benefit from symbiotically conferred resistance to abiotic stress, at least two routes must activate host stress response systems soon after stress exposure. That way, plants can prevent or lessen the impact of the stress on their systems. Endophytes increase a plant's resilience to stress through biochemical processes, such as the activation of biomolecules and plant stress genes. Endophytes are essential to sustainable agriculture due to their many beneficial impacts on the host plant. These effects include the regulation of phytohormone signalling, metabolic activity, and plant defence response pathways.

Keywords Endophytes · Metabolites · Abiotic stress · Phytohormones

---

A. K. Chaubey (✉) · S. Mishra
Department of Soil Science and Agricultural Chemistry, College of Agriculture, Banda University of Agriculture and Technology, Banda, India

V. Sharma · P. K. Prajapati
Department of Genetics and Plant Breeding, College of Agriculture, Banda University of Agriculture and Technology, Banda, India

R. Pandey
Department of Entomology, College of Agriculture, Banda University of Agriculture and Technology, Banda, India

S. V. Dwivedi · A. Singh
Department of Vegetable Science, College of Horticulture, Banda University of Agriculture and Technology, Banda, India

R. Soni
Department of Agricultural Microbiology, Indira Gandhi Krishi Vishwavidyalaya, Raipur, India

R. Soni et al. (eds.), *Microbial Bioactive Compounds*,
https://doi.org/10.1007/978-3-031-40082-7_13

13.1 Introduction

De Barry [1] first used the term endophytes, which literally means "in the plant" (edon = within, phyte = plant). This phrase has a wide range of applications, encompassing not only the algae that may live within the algae but also the bacteria, fungus, plants, and insects that may live within them [2]. Consequently, endophytes are symbiotic bacteria that promote plant growth and live within plant tissues. It also plays beneficial roles related with plant responses under conditions of biotic and abiotic stress without being responsible for any disease symptoms on the host plant. There are a variety of methods by which endophytes influence plant growth in response to abiotic stimuli like salt, heat, temperature, heavy metal toxicity, and nutritional stress. These microbial communities generate numerous secondary active chemicals that shield plants from insect and fungal diseases, hence promoting plant growth. In addition, they can produce extracellular enzymes that promote endophyte colonisation of the host plant. Endophytic microbes play a significant role in helping plants adapt to stress and environmental variables that limit growth and output. Symbiotic connections between plants and microbes enable them thrive in harsh environments by facilitating mutually beneficial changes in both partners' rates of evolution and fitness.

It has been shown that endophytic microbes like bacteria, actinomycetes, and fungi create a protective microbial "nest" around their plant hosts, making them more resistant to frost and other environmental stresses [3, 4]. The absence of endophytes in a plant is not a natural condition [5]. Without endophytes, plants can't fight off infections and will easily succumb to environmental stresses [6]. Endophytic microbes, particularly fungi like *Sebacina vermifera* and *Piriformospora indica* and other *Colletotrichum* and *Penicillium* species, differentiate under unfavourable conditions to have more potent effects on plant growth [7, 8].

Generally speaking, the fungus, bacteria, and nematodes that aren't directly hazardous to plants aren't nearly as harmful when they're among plants that have plant growth-promoting microorganisms (PGPM) attached to them. Phytohormones are produced as a result of the primary effects of PGPM. Modifications in chemical or physical plant defence strategies, known as mediated systemic resistance (ISR), may also be affected by PGPM [9]. Under a variety of adverse environmental conditions, PGPM has consistently expanded, proving its usefulness to plant life. Increasing tolerance to environmental stimuli like sun, drought, salinity, cold, and heavy metals has been proved time and time again to be one of the key activities of plant growth-promoting fungus (PGPF) [10–12]. Abscisic acid (ABA)-independent or abscisic acid-dependent pathways convey osmotic stress from salinity and drought [13], and low ABA development levels have been achieved through fungal activity [14, 15]. Treatment with endophytic *Penicillium* spp. brought about water balance in plants, as reported by Miransari [16], so plants didn't have to try very hard to synthesize ABA and shield the progress of stressed cells.

13.2 Role of Endophytes in Abiotic Stress Management

Suboptimal to supra-optimal temperatures, soil pH imbalance (from acidity to salinity), soil moisture deficits and surpluses, heavy metal toxicity, ultraviolet radiation, and many other environmental factors have all been stressors for plants ever since they first appeared [17]. Due to their brief life span, endophytes are able to quickly adapt to their environment and impart a wide variety of stresses on to their host plant [18]. In reaction to abiotic stress, endophytes either (i) activate the host plant's response system or (ii) produce compounds that are toxic to the stress [19]. In the following paragraphs, we will discuss the mechanisms involved in coping with abiotic stresses in greater depth.

Extraction of endophytic fungi from medicinal plants by Chathurdevi and Gowrie [20] revealed that these fungi release extracellular enzymes that aided the plants' growth when subjected to abiotic stress. More than 50 unique endophytic fungal strains rich in enzymes like laccase, amylase, pectinase, cellulase, lipid hydrolase, and proteinase were isolated and identified by Sunitha et al. [21]. Breakdown enzymes for 1-aminocyclopropane-1-carboxylate (ACC) Bacterial endophytes have been studied in relation to the enzymes amylase, deaminase, esterase, pectinase, cellulases, lipids, protease, phytase, asparaginase, and xylanase [22–25]. A group of researchers led by Vijayalakshmi [26] has recently isolated bacterial endophytes from medicinal plants. These endophytes secrete extracellular enzymes such as amylase, protease, and cellulase.

The potential for several different types of endophytic bacteria with molecular weights between 400 and 1500 daltons to create siderophores has been studied [27]. In addition to catecholate and salicylate, bacteria can also produce hydroxamate and carboxylate as siderophores. Streptomyces, Pseudonocardia, Actinopolyspora, Nocardia, Salinispora, Micromonospora, Actinomadura, and Kibdelosporangium are all examples of endophytic actinobacteria that create siderophores [27–29]. To regulate plant growth and confer disease resistance, plants rely on endophytic actinobacteria, which use an exomechanism to produce siderophores [30].

The phytohormone salicylic acid (SA) plays an important role in a wide range of processes, including development of the plant's root system, germination of seeds, induction of flowering, closure of the plant's stomata, and resistance to abiotic and biotic stress. Produced by bacterial endophytes, SA promotes development and protects plants against pathogens such as fungus, making them more resilient to drought [31].

13.3 Endophytes in Biotic Stress Management

Generally speaking, endophytic bacteria are considered to be effective biocontrol agents. Endophytic fungi are extremely important for both grasses and conifers in mitigating the damage caused by insect herbivores. *Bacillus subtilis*, an endophytic bacterium isolated from *Speranskia tuberculata* (Bge.) Baill, has been shown to have antagonistic effect against *Botrytis cinerea*, a fungus responsible for the spoilage of tomato fruits during storage [32]. Poplar canker was the subject of a biocontrol study in which novel endophytes including *Burkholderia pyrrocinia* JK-SH007 and *Bacillus cepacia* were employed [33].

Recombinant endophytic strains, which may be found in many plants, can be used to create a wide variety of anti-pest proteins, which can then be used to combat a wide range of plant pests. Hassan et al. [34] biocontrolled *Culex pipiens* and *Musca domestica* with copper nanoparticles made by the endophyte *Streptomyces capillispiralis* Ca-1. The endophytic actinomycetes *Streptomyces zaomyceticus* Oc-5 and *Streptomyces pseudogriseolus* Acv-11, found in the plant *Oxalis corniculata* L., synthesised copper oxide nanoparticles with antimicrobial activity against four phytopathogenic fungi: *Phoma destructiva*, *Alternaria alternaria*, *Fusarium oxysporum*, and *Curvularia lunata* [35].

In asymptomatic colonisation, the host and the endophyte maintain a dynamic equilibrium in which their antagonistic interactions are roughly balanced. Although only a fraction of endophytes are thought to be dormant pathogens, all endophytes investigated so far have developed the exoenzymes essential to infect and colonise the host [36–39]. Almost all of them can produce compounds that are poisonous to plants (phytotoxins) [2, 40]. Hosts, just like they would in reaction to pathogens, can develop preformed and induced defensive metabolites [2, 41–45]. If the virulence of the fungus and the defence mechanisms of the plant are about equal, there will be no visible signals of danger.

The dynamics of an antagonistic relationship can shift depending on the host and endophyte's tolerance to biotic and abiotic environmental conditions and the state of health of both parties. Many endophytes, for instance, can infect as a pathogen, colonise cryptically, and sporulate as either a pathogen or a saprophyte, making them masters of phenotypic plasticity. Therefore, diversity is required to act as a check on this trend; if this is the case, then endophytic interactions are creative and can drive evolutionary change, with symbioses having the potential to develop into both highly specialised mutualisms and parasitisms or forms of exploitation [46].

13.4 Signalling During Abiotic and Biotic Stresses

Plants have a number of built-in systems that allow them to sense stress signals and continue to grow even in adverse conditions. Information is routinely relayed across pathways and signal molecules/cofactors in the signalling response to any stressor, biotic event abiotic [47]. Reactive oxygen species (ROS) such as NO_2, Ca^{2+}, inositol phosphates, and systemin have a function in signalling as well, complementing the effect of phytohormones. Drought causes osmotic stress, while salt stress causes ionic stress [48]. ROS production has been proposed as a critical mechanism for responding to biotic and abiotic stresses. New research reveals that Ca^{2+} and NO have a large impact on hormone signalling, which is important in stress response pathway crosstalk. Plant defence, ABA-dependent stomata movement, and drought stress responses depend on nitric oxide and Ca^{2+} signalling [49]. MAPK/MPK cascades regulate proliferation, cell differentiation, cell death, development, and stress responses. The mitogen-activated protein kinase (MAPK) cascade drives cellular responses to biotic and abiotic stresses. Cellular responses to biotic and abiotic stressors depend on the MAPK cascade. Plants produce heat shock proteins to avoid protein denaturation and maintain protein homeostasis in severe temperatures [50].

Fungal endophytes have been discovered to contain compounds that counteract the effects of flavonoids, phenols, terpenoids, alkaloids, saponin, nematode polysaccharides, and tannins [51, 52]. Treatment of diseases caused by a wide range of pathogens may be possible in future, with the help of bioactive compounds synthesised by endophytic actinomycetes [53]. In addition to their biocontrol actions, endophytic bacteria also have favourable impacts on abiotic stress.

13.5 Induced Systemic Resistance (ISR)

Plants' natural defences are boosted by endophytic bacteria, making them more resilient to disease. "Induced systematic resistance" (ISR) describes this phenomenon [54]. Endophytic microbes colonise plants by escaping defence responses, as seen in *Bacillus* and *Pseudomonas* [45]. ISR can be activated by several bacterial agents, including salicylic acid, antibiotics, siderophores, N-acyl-homoserine lactones, jasmonic acid, volatiles (such acetoin), and lipopolysaccharides [55]. ISR was associated with the development of defences and immunity against herbivorous insects and diseases. Many types of endophytic bacteria have had their ISR triggered by salicylic acid, but it is also known that the plant hormones ethylene (ET) and jasmonic acid (JA) play crucial regulatory roles in the signalling pathways involved in ISR induction [56]. ISR induction by the endophytic bacterium *Pseudomonas fluorescens* 89B-61 protects cucumbers from the disease anthracnose [45]. Changes in the native endophytic population were associated with improved plant resistance to *Pectobacterium atrosepticum* when the endophytic bacterium *Methylobacterium* sp. IMBG290 was present in potato soil. Correlations between changes in the endophytic community and resistance to disease show the crucial role this population plays in preventing illness [57]. Endophytic fungi have been involved in defence mechanisms via ISR induction to a lesser degree than endophytic bacteria [58]. The ability of endophytic fungi to create metabolites has been linked to both herbivore control and disease prevention. There are numerous different types of metabolites, including alkaloids, steroids, terpenoids, peptides, polyketones, flavonoids, chlorinated compounds, phenols, and quinols [59, 60]. Fungal endophytes cause localised illness in their hosts but have been connected to the discovery of compounds with antibacterial, antiviral, insecticidal, and antifungal activities due to their horizontal spread [61].

13.6 Abiotic Stress Alleviation by Microbial Endophytes

Different extreme situations, such as environmental pressures and strains induced by living communities, limit plant growth and development.

Plants are able to withstand abiotic stress in two ways: (i) by immediately activating response systems after being stressed [62] and (ii) by generating biochemical compounds that act as antistress agents, which are then metabolised by endophytes [40]. Both up- and down regulation of several stress-inducible genes were reduced in pepper plants after inoculation with the endophyte *Arthrobacter* sp. and *Bacillus* sp. Cucumber plants exposed to salt chloride and dehydration stress benefited greatly from inoculation with *Phoma glomerata* and *Penicillium* sp., as measured by increased nutrient absorption (particularly magnesium, potassium, and calcium), increased plant biomass, and enhanced growth metrics, as well as decreased sodium toxicity [63]. Bailey et al. [64] concluded that *Trichoderma* sp. isolated from *Theobroma cacao* improved the cocoa plant's tolerance to abiotic stress, particularly drought, via modifying gene expression.

Plants grown from *Kalmia latifolia* L. tissue cultures were found to be more resilient to drought when inoculated with the endophytic fungus *Streptomyces padanus* AOK-30, according to research by Hasegawa et al. [65]. Under drought stress, sugars and amino acids were shown to be considerably higher in endophyte-colonised plants compared to non-colonised plants [66]. Due to a complex symbiotic interaction, plants with a drought-tolerant phenotype are able to produce more sugar and amino acids, both of which are signs of higher osmolytic activity [67]. Plants that have been colonised by endophytes are more tolerant to environmental challenges such low water availability, high temperatures, and high salt concentrations [7]. A boost in antioxidant activity, as discovered by Chugh et al. [68], was determined to be the cause of accelerated seedling development in response to dryness.

There was also evidence that endophyte colonisation under low-water conditions led to increases in biomass, proline concentrations, and relative water content [69, 70].

13.7 Drought Stress

As an abiotic stressor, drought is particularly detrimental to plant development and output. Due to root water restriction or excessive transpiration, plants experience drought [71]. Most plant species, especially those adapted to temperate climes, experience diurnal water stress during the middle of the day, even when soil moisture levels are within normal range. Growth is stunted due to the temporary drought stress [72]. Drought causes an increase in reactive oxygen species generation, lower germination rates, and membrane disruption [73]. The principal causes of osmotic stress in plants were, in addition, prolonged periods of dryness and high salinity. Cells are affected by osmotic stress, which dryness causes; ionic or ion-toxicity signs appear in high salinity [48]. The effects of drought stress, such as stunted growth and leaf senescence in the shoot system, are counteracted by osmotic stress, induced by salinity [74].

Plants with symbiotic relationships (like rice, tomatoes, dune grass, and panic grass) produce greater biomass with less water.

Endophyte-associated plants may be more resistant to drought because they accumulate more solutes in their tissues than noninfected plants. This could be the result of a slower transpiration rate, a thicker cuticle layer, or reduced leaf conductivity [75]. Changes in a plant's structure, genetic make-up, and metabolic processes may all contribute to its resilience in the face of water stress. Yet, in response to water scarcity, plants primarily increase ABA production and/or decrease ABA breakdown [76]. It is often believed that ABA acts as a signal in drought-stricken plants, primarily regulating transpiration and stomatal closure to decrease water loss [77]. Some evidence also suggests that ABA helps plants develop more extensive root systems, which improves their capacity to take in water [78].

Full scan mass spectrometry was used to isolate ABA from *Azospirillum brasilense* Sp 245 cells that had been grown at a heightened rate due to chemical stimulation. Increased amounts of ABA were seen in *Azospirillum brasilense* Sp 245-infected *Arabidopsis thaliana* seedlings, and the addition of sodium chloride to the growth medium increased the rate of bacterial ABA synthesis [79].

13.8 Salinity Stress

Soil salinisation, caused by the build-up of water-soluble salts, is a problem for farmers all over the world and endangers ecosystem vitality, food supplies, and economic development. Initially, salt has a chilling effect on the distribution and metabolism of soil microorganisms and other creatures that make their home in the soil. It first reduces crop yields and then, in its later phases, completely wipes out the local flora, turning once-productive land into a barren wasteland [80, 81]. Soil is regarded to be saline if its electrical conductivity (EC) in the root zone is greater than 4 dS m^{-1} at 25 ^{0}Celsius and an exchangeable sodium concentration of 15% (almost 40 mMNaCl). Most plant yields are diminished by this degree of salinity. In the world, high salinity affects 20% of all crop land and 33% of irrigated agriculture [82]. By 2050, half of the world's arable land would be affected by salinity, according to a 2017 assessment of the literature by Machado, Rui M.A., and Serralheiro, Ricardo P. They also found that salt accumulation degrade 10 million ha of agriculture land every year. This harm can be hastened by aggressive groundwater use, a rise in the use of low-quality water in agriculture, and climate change.

13.9 Effect of Soil Salinity on Plants

The production of agricultural crops, especially vegetable crops, which have a low tolerance sensitivity to soil saline, is drastically reduced due to soil salinity (Table 13.1). In general, Compared to field crops, vegetable crops offer a better yield per acre under irrigated conditions. Vegetable crops needed higher total water application rates and more frequent irrigation than other agronomic crops. Despite the need for more fertilisers and irrigation, vegetable crop production still takes place in dry and semi-arid regions where rainfall is scarce and temperatures are high. It's common knowledge that vegetables are an excellent source of many different vitamins and minerals, as well as dietary fibre. Salty soil possess a challenges for plants as it hinders their metabolic function, making it difficult for them to thrive. Effects on reproductive development, such as the lengthening of stamen filaments,

Table 13.1 Soil salinity (ECe) tolerance in different crops

Rating group	Crop/Vegetable	Tolerance based on	Soil threshold (dSm^{-1}) ECe	Reference
Tolerant	Barley	Grain Yield	8.0	Maas and Grattan [83]
Tolerant	Canola/Rape seed	Seed yield	9.7	Francois [84, 85]
Tolerant	Cotton	Seed cotton yield	7.7	Maas and Grattan [83]
Tolerant	Rye	Grain yield	11.4	Maas and Grattan [83]
Moderately tolerant	Sorghum	Grain yield	6.8	Maas and Grattan [83]
Moderately tolerant	Wheat	Grain yield	6.0	Maas and Grattan [83]
Moderately tolerant	Sunflower	Seed yield	4.8	Francois [86]
Moderately tolerant	Red beet	Storage root	4.0	Machado and Serralheiro [87]
Moderately sensitive	Onion seed	Seed yield	1.0	Mangal et al. [88]
Moderately sensitive	Eggplant	Fruit yield	1.1	Machado and Serralheiro [87]
Moderately sensitive	Garlic	Bulb yield	3.9	Francois [84, 85]
Moderately sensitive	Potato	Tuber yield	1.7	Machado and Serralheiro [87]
Sensitive	Mung been	Seed yield	1.8	Minhas [89]
Sensitive	Onion bulb	Bulb yield	1.2	Maas and Grattan [83]
Sensitive	Rice	Gran Yield	3.0	Venkateswarlu et al. [90]
Sensitive	Spinach	Top fresh weight	2.0	Machado and Serralheiro [87]

ECe—electrical conductivity (EC) of saturated paste extract of soil the suppression of microsporogenesis, the aborting of ovules, the senescence of fertilised embryos, and the enhancement of cell death in different tissues, can be mediated by high concentrations of K^+, which in turn affect mitosis and meiosis of nucleic acid. When K^+ is replaced by Na^+ in these processes, soil salinity causes ion toxicity. Protein conformational changes are also produced by Cl^- and Na^+. Loss of turgor, cellular dehydration, and cell death can all result from the osmotic stress caused by high soil salt levels. Metabolic imbalance, brought on by osmotic stress and ion toxicity, results in oxidative stress. As salts are also nutrients for plants, too much salt in the soil can disrupt the plant's nutritional balance or prevent it from absorbing key minerals (nitrogen, phosphorus, potassium, iron, and zinc). Salinity decreases photosynthesis by lowering photosystem II capability, chlorophyll content, leaf area, and stomatal conductance in photosynthesis.

Growth is stunted as a result of the high salinity [91]. In addition, cyclin-dependent kinase activity is decreased because of the post translational inhibition that occurs during periods of high salt [92].

13.10 Salinity Stress Alleviation by Microbial Endophytes

Over 20% of farmable soil is at risk from salt right now, and experts predict that by 2050, half of all prime farmland will be under salinity stress. Plant-associated microorganisms use a wide variety of metabolic and genetic methods to better adapt to abiotic and biotic stress. Endophytic bacteria not only react to root-secreted signal molecules but also produce their own signalling molecules, all of which have positive impacts on plant health, such as enhanced root growth, resistance or tolerance to biotic and abiotic challenges, and general plant health [9]. Endophytic fungi *Yarrowia lipolytica* controlled the production of proline in salt stressed maize plant [93]. In another study, Abdelaziz et al. [94] observed that endophytic fungi *Piriformospora indica* caused considerable reduction in shoot proline content in *Solanum lycopersicum* under salinity stress. *Piriformospora indica* also responsible for significant increase in shoot proline in *Trichoderma harzianum* salt stressed plant [95].

13.11 Primary Benefits of Endophytes in Reducing the Negative Effects of Salinity on Plants

13.11.1 Plant Antioxidant Status

Numerous organisms in the microbial world exhibit comparable responses to oxidative stress. That ROS production in plants is mediated by endophytic fungus which was discovered by Hamilton and colleagues in 2012 [96]. Previous research has established a connection between the suppression of antioxidant enzymes and salt tolerance in plants [97]. There are many enzymes in the body that can neutralise reactive oxygen species, including superoxide dismutases (SOD), glutathione reductases (GR), dehydroascorbate reductases (DHAR), catalases (CAT), ascorbate or thiol-dependent peroxidases (APX), and mono-dehydroascorbate reductases (MDHAR) [98]. APX, SOD, and CAT are all direct or indirect participants in the detoxification of reactive oxygen species (ROS). In a 500 mmol NaCl solution, the nonsymbiotic plant *Leymus mollis* (dune grass) shrivels, dries out within in 7 days, and dies after 14 days [99]. After being exposed to 500 mmol NaCl, *Fusarium culmorum*-infected plants became active for 14 days. Barley's salt tolerance is improved by the endophyte *Piriformospora indica ulmus*, which also increases the grain's antioxidant levels [100].

13.11.2 ACC Deaminase

ACC deaminase, produced by endophytic bacteria, is essential to plant growth and stress tolerance but useless to the bacteria [101]. ACC deaminase breaks ACC (1-aminocyclopropane-1-carboxylate) into 2-oxobutanoate and ammonia, lowering ethylene levels and blocking plant ethylene signalling [102]. Ethylene's fundamental involvement in bacterial colonisation of plant tissues affects seed germination and plant responses to various stresses [103].

Over production of ethylene in plants as a response to stress can be harmful to their health and growth [104]. The ACC deaminase enzyme does more than just help plants deal with stress; it also encourages the colonisation of the plant by microorganisms known as endophytes. Silencing the ACC deaminase gene in *Burkholderia phytofirmans* PsJN may prevent a bacterial infection that causes canola seedlings to fail to develop strong roots [105]. Branch invasion by endophytic bacteria has been observed in prior investigations of cut flowers and blocking ACC deaminase helped keep flowers from getting old too quickly [106].

13.11.3 Phytohormone Production

Endophytes produce auxins, most notably indole-3-acetic acid (IAA) that can significantly increase plant growth [107]. Auxins, which counteract the effects of ethylene, are crucial for root growth and development. Endophytic regulation of auxin production in halophytic plants, thus, has the potential to be an important technique for granting salt resistance. There were two groups of bacteria that were found to produce IAA: (i) salinity-tolerant rhizobacteria (*Halomonas* sp., *Arthrobacter* sp., *Pseudomonas mendocina*, *Bacillus pumilus*, and *Nitrinicolalacis aponensis*) and (ii) microorganisms such as *Serratia*, *Bacillus*, *Vibrio*, *Brevundimonas*, and *Oceanobacillus* [108–110]. There were ABA, gibberellins, and IAA generated by the halophytic *Prosopis strombulifera* [111]. Plants produce more of the growth hormone abscisic acid (ABA) when they are under stress.

ABA is primarily responsible for regulating water balance and osmotic stress tolerance in plants [112]. Wheat plants that were grown in salty soil benefited from the presence of IAA-producing rhizobacteria [108]. It is unknown if mycorrhizal or endophytic root fungus get salt tolerance from phytohormones [113].

13.11.4 Nitrogen Fixation

Endophytes help their host plants in many ways, including by preventing disease, creating beneficial hormones, increasing the availability of nutrients, and fixing nitrogen. These mechanisms also contribute to endophytes' buffering effect when the host plant is exposed to unfavourable ecological conditions [113]. Nitrogen could be fixed by a wide range of root endophytes (e.g., *Azoarcus* spp., *Acetobacter diazotrophicus*, and *Herbaspirillum* spp.). Host plant fitness is increased through nitrogen fixation, especially in low-nitrogen conditions. In cases when only a little amount of fixed nitrogen is found in a single species, it is important to determine if this nitrogen is meant to meet the needs of the microbes in the soil or those of the host plants. Poplar trees' endophytic bacteria *Paenibacillus* P22 contributed to the host plant's total nitrogen pool and triggered metabolic shifts [114].

13.11.5 Compatible Solutes

Osmotic pressure results from the accumulation of Na^+ and Cl^- ions in the vacuole of a plant cell. To counteract this force, organelles and the cytoplasm must collect (even at high concentrations) organic solutes that are metabolically compatible. Most commonly found sugars, amino acids, and amino acids are glycine betaine, proline, and proline [115].

Increased salt tolerance in plants colonised by endophytes has been examined, and proline amino acid has been of particular interest because it has been hypothesised that organic solute accumulation is a critical mechanism for halophytic plants to offset osmotic pressure [116]. Proline accumulation appears to be an outcome rather than a cause of salt tolerance, despite contradictory findings about the role of mycorrhizal fungus [117].

Betaines and carbohydrates can also control osmosis. Elevated sugar and betaine levels in mycorrhizal plants have been linked to a potential involvement in salt tolerance [118]. *Pseudomonas pseudoalcaligenes* is an endophyte that increased rice's salt tolerance by encouraging the formation of glycine betaine-like molecules [119].

13.11.6 Temperature Stress

High temperatures have a lethal effect on plants because they cause the proteins inside the cells to get denaturised and agglomerate, ultimately killing the plants. Metabolism slows down as a result of low temperatures because of their effect on enzyme activity, macromolecule interactions, protein structure alterations, and modulation of membrane characteristics [120].

Extreme heat is rarely reported, despite its negative effects, which are frequently linked to a lack of water. The bacterium *Burkholderia phytofirmans* increases cold hardiness in plants [121]. Due to *Curvularia protuberata* and its thermal endurance mycovirus *Curvularia* (CThTV), the grass *Dichanthelium lanuginosum* was able to live in Yellowstone National Park, where soil temperatures ranged from 38 °C to 65 °C [122].

Wheat's endurance to high temperatures has been improved by the presence of fungal endophytes, which has led to higher crop yields and improved germination rates in following generations [123].

Endophyte composition may be affected by a variety of environmental factors, including but not limited to temperature, humidity, and latitude. Lower annual precipitation and higher latitudes favour *Paenibacillus* strains in sweet root (*Osmorhiza depauperata*) endophytes, while greater annual precipitation and lower latitudes favour *Sinorhizobium meliloti* and *Agrobacterium tumefaciens* [124].

Ascorbate and glutathione are oxidised to reduced forms, and lipid peroxidation is reduced in endophyte-colonised plants, which makes them more resistant to temperature and salt stress, as found by Matsouri et al. [125]. By increasing its resistance to cold, endophytes increase a plant's chance of survival. Accumulated phenolic compounds, proline, and starch are downregulated, and cellular damage and photosynthetic activity are elevated in response to cold stress.

Endophytes have a protective effect on wheat development during drought stress due to their positive effect on metabolic balance [70].

13.11.7 Heavy Metal Stress

Heavy metal toxicity is a major abiotic stressor that is responsible for the loss of anywhere from 25% to 80% of many types of farmed crops. Because of toxicity from manganese and aluminium and a lack of potassium, magnesium, phosphorus, and calcium in acidic soils, agricultural output and soil fertility are negatively affected [126]. Exposure to heavy metals significantly slows a plant's root system and is also toxic to plant tissue [126]. Heavy metal toxicity in acidic soils is problematic because it interferes with several vital physiological and biochemical processes, such as nutrition intake, protein and nitrogen metabolism, photosynthesis, and respiration [127].

It is well known that the availability of cations to plants is affected by the immobilisation and mobilisation of metal cations by bacterial endophytes [128]. Higher activity of the antioxidant enzymes was found in Cd-stressed soil when the dark septate endophyte (DSE) *Exophiala pisciphila* was combined with the root of Zea mays [129]. In 2016, Wang et al. DSE-inoculated plants subjected to high amounts of Cd showed upregulation of genes related in Cd detoxification, transport, and absorption, while ZIP was downregulated. Plant ethylene levels are affected by heavy metal tolerance, and *Gigaspora* and *Pseudomonas* can directly affect ethylene levels by varying the amount of 1-aminocyclopropane-1-carboxylate (ACC) [130].

13.11.8 Nutrient Stress

For growth, development, and reproduction, plants need mineral nutrients, light, water, and carbon. Examples of abiotic conditions that can cause damage to plants include hunger and nutrient deficits [131].

Endophytes provide their hosts with both micro and macronutrients.

Amino acids can be synthesised by plants, thanks to nitrogen-fixing bacteria that can metabolise plant root exudates. Growth-promoting gibberellins (GAs), phosphate solubilisation, cytokinins, indole-3-acetic acid (IAA), and siderophore synthesis, as well as important vitamins, are all produced by endophytes and used by the host plant [119]. The solubilisation of phosphate in wheat and rice was found to be enhanced by gibberellic acid generated by *Pseudomonas* sp., according to research by Choi et al. [132]. Zinc uptake in wheat plants can be improved using either *Azotobacter chroococcum* or *Piriformospora indica* [133].

Endophytes have been shown to aid in the biological breakdown of dead host plants. Endophytes colonise plants at first and then they actively work against saprophytic bacteria, speeding up the degradation of plant matter [134–138]). Another study showed that all endophytes can break down lignin, cellulose, and hemicelluloses, which help nutrient cycling [17].

13.12 Role of Microbial Metabolites in Stress Mitigation of Plants

When plants are attacked by the different microbial species belonging to different microbiomes, i.e. rhizosphere microbiome, epiphytic microbiome, endophytic microbiome, seed microbiome, core microbiome, etc., they show differential responses which alter their resistance mechanism against the stresses prevailing. Interaction of microbial metabolites to the plant system leads to the synthesis of different important secondary metabolites, e.g. phenolics, alkaloids, steroids, and flavonoids which are positively correlated with stress resistance in plants [137]. Generally, it is observed that microbe induced production of secondary metabolites that helps in abiotic stress mitigation [138]. Reactive oxygen species level was found to be reduced in crop plants like wheat, soy bean, and peanut due to the activity of metabolites released from different strains of *Pseudomonas* spp. reported by Shaik et al. [139], Kang et al. [140], and Sharma et al. [141]. In another study, by Ghosh and co-workers [142] also observed amelioration of osmotic stresses in *Arabidopsis thaliana* by exopolysaccharides released from *Bacillus* spp., whereas Liu et al. [143], Balsanelli et al. [144], and Mahmood et al. [145] found mitigation of salt stress by IAA, SmR1, and exopolysaccharides released from *Klebsiella oxytoca* Rs-5, *Herbaspirillum seropedicae*, and *Enterobacter cloacae* P6, respectively. Adjustment in production of metabolites for the sake of adaptation to the changing environment is also observed [146], in addition to the enhancement in uptake of plant nutrients, and the formation of soil humus [147]. However, it is also reported by Burkhead et al. [148], Haas and Defago [149], and Sankari et al. [150] that biotic stress can be mitigated by microbial species. The adverse impact of stem rot of chickpea, foot rot of tomato, early blight of tomato, head blight of wheat, and black scurf of potato were observed to be minimised by phenylpropanoid, harzianic acid, siderophores, bacillomycin D, and surfactin by Sathya et al. [151], Manganiello et al. [152], Verma et al. [153], Gu et al. [154], and Kong et al. [155], respectively. Some microbial metabolites which are useful in plant stress mitigation are given in Table 13.2.

Table 13.2 Metabolites from different endophytes

S. No.	Microbe	Metabolite	Stress	Reference
1	*Azospirillum* spp.	IAA/IBA	Inhibition of uptake of nutrients like nitrogen and phosphorus	Malhotra and Srivastava [156]
2	*Pseudomonas putida*	Pyoverdine	Fusarium wilt	Hass and Defago [149]
3	*Pseudomonas fluorescence*	Proline	Salinity stress in *Vicia faba*	Metwali et al. [157]
4	*Bacillus subtilis*	Surfactin	Damping off of cole crops	Kong et al. [155]
5	*Trichoderma koningii*	Koninginin C	Take all disease of wheat	Vinale et al. [158]
6	*Azospirillum brasilense*	Cadaverine	Osmotic stress in rice	Cassan et al. [159]
7	*Pseudomonas aeruginosa*	Glycine betaine	Drought stress in *Vigna radiata*	Sarma et al. [160]
8	*Rhizopus arrhizus*	Raphorin	Fe deficiency in solanaceous crops	Shenker et al. [161]
9	*Streptomyces acidiscabies*	Coelichelin	Nickel stress in *Vigna unguiculata*	Sathya et al. [151]
10	*Azotobacter chroococcum*	Exopolysaccharide	Environmental stress in *Vicia faba*	El-Ghany and Attia [162]
11	*Sphingomonas* sp.	Gibberellic acid	Salinity stress in Tomato	Halo et al. [163]
12	*Streptomyces platensis*	Phenylethyl alcohol	Seedling blight of rice	Wan et al. [164]
13	*Bradyrhizobium* sp.	Nitrogenase	Water stress in Cowpea	Fugyeuredi et al. [165]

13.13 Conclusions

Considering the role of microbes, it can be possible to make a holistic approach which can mitigate the different types of stress in association with other mechanism of mitigation, i.e. avoidance and escape mechanism, genotypic tolerance mechanism, etc. Ultimately, these mechanisms will result in a sustainable and eco-friendly mitigation of the prevailing stress and also make possible the adaptation of crop plants to the changing environment. However, this approach is emerging as a revolution in sustainable development, and it still requires thorough study of mechanism of release and action of the secondary metabolite and the signalling crosstalk in plant-microbiome interactions to make it more effective and reliable. The focus should also be given to understand the genetic controls on plant secondary metabolites and their adjustment according to the changing microbiomes and environmental conditions.

References

1. De Barry A (1866) Morphologie und Physiologie der Pilze, Flechten, und Myxomyceten. Hofmeister's handbook of physiological botany, vol II, Leipzig
2. Schulz B, Boyle C (2006) What are endophytes? In: Schulz B, Boyle C, Sieber TN (eds) Soil biology Vol 9 microbial root endophytes. Springer-Verlag, Berlin Heidelberg, pp 1–13
3. Passari AK, Mishra VK, Singh G, Singh P, Kumar B, Gupta VK, Sarma RK, Saikia R, Donovan AO, Singh BP (2017) Insights into the functionality of endophytic actinobacteria with a focus on their biosynthetic potential and secondary metabolites production. Sci Rep 7: 11809. https://doi.org/10.1038/s41598-017-12235-4
4. Zhao K, Penttinen P, Guan T, Xiao J, Chen Q, Xu J et al (2011) The diversity and antimicrobial activity of endophytic actinomycetes isolated from medicinal plants in Panxi Plateau China. Curr Microbiol 62:182–190
5. Partida-Martínez LP, Heil M (2011) The microbe-free plant: fact or artifact? Front Plant Sci 29(2):100. https://doi.org/10.3389/fpls.2011.00100.eCollection
6. Timmusk S, Paalme V, Pavlicek T, Bergquist J, Vangala A, Danilas T, Nevo E (2011) Bacterial distribution in the rhizosphere of wild barley under contrasting microclimates. PLoS One 6:e17968
7. Redman RS, Kim YO, Woodward CJ, Greer C, Espino L, Doty SL, Rodriguez RJ (2011) Increased fitness of rice plants to abiotic stress via habitat adapted symbiosis: a strategy for mitigating impacts of climate change. PLoS One 6:e14823
8. Waller F, Achatz B, Baltruschat H, Fodor J, Becker K, Fischer M, Heier T, Hückelhoven R, Neumann C, von Wettstein D, Franken P, Kogel KH (2005) The endophytic fungus *Piriformospora indica* reprograms barley to salt stress tolerance, disease resistance, and higher yield. Proc Natl Acad Sci U S A 102:13386–13391
9. Eid M, Salim SS, Hassan SE-D, Ismail MA, Fouda A (2019) Microbiome in plant health and disease. Springer Nature Singapore, pp 119–143. https://doi.org/10.1007/978-981-13-8495-0_6
10. Khan AL, Hamayun M, Hussain J, Kang SM, Lee IJ (2012) The newly isolated endophytic fungus *Paraconiothyrium* sp. LK1 produces ascotoxin. Molecules 17:1103–1112
11. Chaubey AK, Singh HP, Sahi UP (2015) Mycorrhizal strains efficacy in soybean. Agric Sci Digest 35(3):183–186

12. Chaubey AK, Singh HP, Singh SR, Shahi UP (2008) Impact of paper mill effluent irrigation on arbuscular mycorrhizal fungi in wheat (*Triticum aestivum*). J Mycol Plant Pathol 38(2):333–336
13. Cao MJ, Wang Z, Zhao Q, Mao JL, Speiser A, Wirtz M, Hell R, Zhu JK, Xiang CB (2014) Sulfate availability affects ABA levels and germination response to ABA and salt stress in Arabidopsis thaliana. Plant J 77:604–615
14. Jahromi F, Aroca R, Porcel R, Ruiz-Lozano JM (2008) Influence of salinity on the in vitro development of *Glomus intraradices* and on the *in vivo* physiological and molecular responses of mycorrhizal lettuce plants. Microb Ecol 55:45–53
15. Khan AL, Waqas M, Hussain J, Al-Harrasi A, Lee IJ (2014) Fungal endophyte *Penicillium janthinellum*LK5 can reduce cadmium toxicity in *Solanum lycopersicum* (Sitiens and Rhe). Biol Fertil Soils 50:75–85
16. Miransari M (2012) Role of phytohormone signaling during stress. In: Ahmad P, Prasad MNV (eds) Environmental adaptations and stress tolerance of plants in the era of climate change. Springer, New York, pp 381–393
17. He M, Cheng-Qiang H, Nai-Zheng D (2018) Abiotic stresses: general defenses of land plants and chances for engineering multistress tolerance. Front Plant Sci 9:1771. https://doi.org/10.3389/fpls.2018.01771
18. Singh S, Kumar V, Dhanjal DS, Sidhu GK, Datta S, Kumar S, Singh J (2020) Endophytic microbes in abiotic stress management. In: Kumar A, Singh VK (eds) Microbial endophytes: prospects for sustainable agriculture. Woodhead Publishing, pp 91–123
19. Lata R, Chowdhury S, Gond SK, White Jr JF (2018) Induction of abiotic stress tolerance in plants by endophytic microbes. Lett Appl Microbiol 66(4):268–276
20. Chathurdevi G, Gowrie SU (2016) Endophytic fungi isolated from medicinal plant—a source of potential bioactive metabolites. Int J Curr Pharm Res 8:50–56
21. Sunitha VH, Devi DN, Srinivas C (2013) Extracellular enzymatic activity of endophytic fungal strains isolated from medicinal plants. WJAS 9:1–9
22. Akinsanya MA, Ting A, Goh JK, Lim SP (2016) Biodiversity, enzymatic and antimicrobial activities of bacterial endophytes in selected local medicinal plants. J Biomed Pharm Res 5(1):16
23. Carrim AJ, Barbosa EC, Vieira JD (2006) Enzymatic activity of endophytic bacterial isolates of Jacaranda decurrens Cham (Carobinha-do-campo). Braz Arch Biol Technol 49:353–359
24. Fouda AH, Hassan SE, Eid AM, Ewais EE (2015) Biotechnological applications of fungal endophytes associated with medicinal plant *Asclepias sinaica*(Bioss.). Ann Agric Sci 60:95–104
25. Gupta V, Trivedi N, Kumar M, Reddy CR, Jha B (2013) Purification and characterization of exob-Agarase from an endophytic marine bacterium and its catalytic potential in bioconversion of red algal cell wall polysaccharides into galactans. Biomass Bioenergy 28:290–298
26. Vijayalakshmi R, Kairunnisa K, Sivvaswamy SN, Dharan SS, Natarajan S (2016) Enzyme production and antimicrobial activity of endophytic bacteria isolated from medicinal plants. Indian J Sci Technol 9(14):23–32
27. Kannahi M, Senbagam N (2014) Studies on siderophore production by microbial isolates obtained from rhizosphere soil and its antibacterial activity. J Chem Pharm Res 6(4):1142–1145
28. Bhosale HJ, Kadam TA (2015) Generic diversity and a comparative account on plant growth promoting characteristics of actinobacteria in roots and rhizosphere of *Saccharum officinarum*. Int J Curr Microbiol App Sci 4:230–244
29. Gangwar M, Dogra S, Sharma N (2011) Antagonistic bioactivity of endophytic actinobacteria isolated from medicinal plants. J Adv Lab Res Biol 2(4):1–4
30. Rungin S, Indanand C, Suttiviriya P, Kruasuwan W, Jaemsaeng R, Thamchaipenet A (2012) Plant growth enhancing effects by a siderophore producing endophytic streptomycete isolated from a Thai jasmine rice plant (*Oryza sativa* L. cv. KDML105). Antonie Van Leeuwenhoek 102:463–472

31. Klessig DF, Tian M, Choi HW (2016) Multiple targets of salicylic acid and its derivatives in plants and animals. Front Immunol 7:206. https://doi.org/10.3389/fimmu.2016.00206
32. Parker MA (1995) Plant fitness variation caused by different mutualist genotypes. Ecology 76(5):1525–1535
33. Ren JH, Ye JR, Liu H, Xu XL, Wu XQ (2011) Isolation and characterization of a new *Burkholderia pyrrocinia* strain JK SH007 as a potential biocontrol agent. World J Microbiol Biotechnol 27(9):2203–2215
34. Hassan SE, Salem SS, Fouda A, Awad MA, El-Gamal MS, Abdo AM (2018) New approach for antimicrobial activity and bio-control of various pathogens by biosynthesized copper nanoparticles using endophytic actinomycetes. J Radiat Res Appl Sci 11:262–270
35. Hassan SE, Fouda A, Radwan AA, Salem SS, Barghoth MG, Awad MA, El-Gamal MS, Abdo AM (2019) Endophytic actinomycetes *Streptomyces* spp. mediated biosynthesis of copper oxide nanoparticles as a promising tool for biotechnological applications. J Biol Inorg Chem 24(3):377–393
36. Ahlich-Schlegel K (1997) Vorkommen und Charakterisierung von dunklen, septiertenHyphomyceten (DSH) in Geholzwurzeln. Dissertation, EidgenossischeTechnische Hochschule, Zurich, Switzerland Petrini O, Sieber T, Toti L, Viret O (1992) Ecology, metabolite production, and substrate utilization in endophytic fungi. Nat Toxins 1:185–196
37. Sieber TN, Sieber-Canavesi F, Petrini O, Ekramoddoullah AK, Dorworth CE (1991) Characterization of Canadian and European *Melanconium* from some *Alnus* by morphological cultural and biochemical studies. Can J Bot 69:2170–2176
38. Boyle C, Gotz M, Dammann-Tugend U, Schulz B (2001) Endophyte–host interactions III. Local vs. systemic colonisation. Symbiosis 31:259–281
39. Lumyong S, Lumyong P, McKenzie EHC, Hyde K (2002) Enzymatic activity of endophytic fungi of six native seedling species from Doi Suthep-PuiNathional Park, Thailand. Can J Microbiol 48:1109–1112
40. Schulz B, Boyle C, Draeger S, R€ommert AK, Krohn K (2002) Endophytic fungi: a source of novel biologically active secondary metabolites. Mycol Res 109:996–1004
41. Schulz B, Rommert A-K, Dammann U, Aust H-J, Strack D (1999) The endophyte-host interaction: a balanced antagonism. Mycol Res 103:1275–1283
42. Yates I, Bacon CW, Hinton DM (1997) Effects of endophytic infection by Fusarium moniliforme on corn growth and cellular morphology. Plant Dis 81:723–728
43. Mucciarelli M, Scannerini S, Bertea C, Maffei M (2003) In vitro and in vivo peppermint (Mentha piperita) growth promotion by nonmycorrhizal fungal colonisation. New Phytol 158: 579–591
44. Narisawa K, Usuki F, Hashiba T (2004) Control of Verticillium yellows in Chinese cabbage by the dark septate endophytic fungus LtVB3. Phytopathology 94:412–418
45. Kloepper JW, Ryu CM (2006) Bacterial endophytes as elicitors of induced systemic resistance. In: Schulz BJE, Boyle CJC, Sieber TN (eds) Microbial root endophytes. Springer, Berlin, pp 33–52. https://doi.org/10.1007/3-540-33526-9_3
46. Imaizumi-Anraku H, Takeda N, Charpentier M, Perry J, Miwa H, Umehara Y, Kouchi H, Murakami Y, Mulder L, Vickers K, Pike J, Downie JA, Wang T, Sato S, Asamizu EE, Tabata S, Yoshikawa M, Murooka Y, Wu G, Kawaguchi M, Kawasaki S, Parniske M, Hayashi M (2005) Plastid proteins crucial for symbiotic fungal and bacterial entry into plant roots. Nature 433:527–531
47. Dombrowski JE (2003) Salt stress activation of wound-related genes in tomato plants. Plant Physiol 132:2098–2107
48. Zhu JK (2002) Salt and drought stress signal transduction in plants. Annu Rev Plant Biol 53: 247–273
49. Mauch-Mani B, Mauch F (2005) The role of abscisic acid in plant–pathogen interactions. Curr Opin Cell Biol 8:409–414
50. Scharf KD, Berberich T, Ebersberger I, Nover L (2012) The plant heat stress transcription factor (Hsf) family: structure, function and evolution. Biochim Biophys Acta 1819:104–119

51. Bogner CW, Kamdem RST, Sichtermann G, Mattheaus C, Heolscher D, Popp J, Proksch P, Grundler FMW, Schouten A (2017) Bioactive secondary metabolites with multiple activities from a fungal endophyte. Microb Biotechnol 10:175–188. https://doi.org/10.1111/1751-7915.12467
52. Liu G, Lai D, Liu ZQ, Zhou L, Liu LZ (2016) Identification of nematicidal constituents of *Notopterygium incisum* rhizomes against *Bursaphelenchus xylophilus* and *Meloidogyne incognita*. Molecules 21:1276. https://doi.org/10.3390/molecules21101276
53. Prashith-Kekuda TR (2016) Isolation, characterization and antimicrobial potential of endophytic actinomycetes. Int J Curr Microbiol App Sci 5:100–116. https://doi.org/10.20546/ijcmas.2016.507.008
54. Zamioudis C, Pieterse CM (2012) Modulation of host immunity by beneficial microbes. Mol Plant-Microbe Interact 25(2):139–150
55. Bordiec S, Paquis S, Lacroix H, Dhondt S, Ait Barka E, Kauffmann S, Jeandet P, Mazeyrat-Gourbeyre F, Clement C, Baillieul F, Dorey S (2011) Comparative analysis of defence responses induced by the endophytic plant growthpromoting rhizobacterium *Burkholderia phytofirmans* strain PsJN and the non-host bacterium *Pseudomonas syringae* pv. pisi in grapevine cell suspensions. J Exp Bot 62(2):595–603
56. Pieterse CM, Van der Does D, Zamioudis C, Leon-Reyes A, Van Wees SC (2012) Hormonal modulation of plant immunity. Annu Rev Cell Dev Biol 28:489–521
57. Pavlo A, Leonid O, Iryna Z, Natalia K, Maria PA (2011) Endophytic bacteria enhancing growth and disease resistance of potato (*Solanum tuberosum* L.). Biol Control 56:43–49
58. Bae H, Roberts DP, Lim HS, Strem MD, Park SC, Ryu CM, Melnick RL, Bailey BA (2011) Endophytic *Trichoderma* isolates from tropical environments delay disease onset and induce resistance against *Phytophthora capsici* in hot pepper using multiple mechanisms. Mol Plant-Microbe Interact 24(3):336–351. https://doi.org/10.1094/mpmi-09-10-0221
59. Higginbotham SJ, Arnold AE, Ibañez A, Spadafora C, Coley PD, Kursar TA (2013) Bioactivity of fungal endophytes as a function of endophyte taxonomy and the taxonomy and distribution of their host plants. PLoS One 8:e73192
60. Tejesvi MV, Segura DR, Schnorr KM, Sandvang D, Mattila S, Olsen PB, Neve S, Kruse T, Kristensen HH, Pirttilä AM (2013) An antimicrobial peptide from endophytic *Fusarium tricinctum* of *Rhododendron tomentosum* Harmaja. Fungal Divers 60:153–159
61. Gunatilaka AAL (2006) Natural products from plant-associated microorganisms: distribution, structural diversity, bioactivity, and implications of their occurrence. J Nat Prod 69:509–526
62. Meena KK, Sorty AM, Bitla UM, Choudhary K, Gupta P, Pareek A, Singh DP, Prabha R, Sahu PK, Gupta VK, Singh HB (2017) Abiotic stress responses and microbe-mediated mitigation in plants: the omics strategies. Front Plant Sci 8:172. https://doi.org/10.3389/fpls.2017.00172
63. Waqas M, Khan AL, Kamran M, Hamayun M, Kang SM, Kim YH, Lee IJ (2012) Endophytic fungi produce gibberellins and indoleacetic acid and promotes host-plant growth during stress. Molecules 17:10754–10773. https://doi.org/10.3390/molecules170910754
64. Bailey BA, Bae H, Strem MD, Roberts DP, Thomas SE, Crozier J, Samuels GJ, Choi IY, Holmes KA (2006) Fungal and plant gene expression during the colonization of cacao seedlings by endophytic isolates of four *Trichoderma* species. Planta 224:1449–1464
65. Hasegawa S, Meguro A, Toyoda K, Nishimura T, Kunoh H (2005) Drought tolerance of tissue-cultured seedlings of mountain laurel (*Kalmia latifolia* L.) induced by an endophytic actinomycete II. Acceleration of callose accumulation and lignification. Actinomycetologica 19(1):13–17. https://doi.org/10.3209/saj.19.13
66. Bae H, Sicher RC, Kim MS, Kim SH, Strem MD, Melnick RL, Bailey BA (2009) The beneficial endophyte *Trichoderma hamatum* isolate DIS 219b promotes growth and delays the onset of the drought response in *Theobroma cacao*. J Exp Bot 60(11):3279–3295. https://doi.org/10.1016/j.envexpbot.2017.08.008
67. Shinozaki K, Yamaguchi-Shinozaki K (2007) Gene networks involved in drought stress response and tolerance. J Exp Bot 58:221–227. https://doi.org/10.1093/jxb/erl164

68. Chugh V, Kaur D, Purwar S, Kaushik P, Sharma V, Kumar H, Rai A, Singh CM, Kamaluddin, Dubey RB (2022 Dec 28) Applications of molecular markers for developing abiotic-stress-resilient oilseed crops. Life 13(1):88. https://doi.org/10.3390/life13010088
69. Zhang X, Li C, Nan Z (2010) Effects of cadmium stress on growth and anti-oxidative systems in *Achnatherum inebrians* symbiotic with *Neotyphodium gansuense*. J Hazard Mater 175(1–3):703–709
70. Naveed M, Mitter B, Reichenauer TG, Wieczorek K, Sessitsch A (2014) Increased drought stress resilience of maize through endophytic colonization by *Burkholderia phytofirmans* PsJN and Enterobacter sp. FD17. Environ Exp Bot 97:30–39. https://doi.org/10.1016/j.envexpbot.2013.09.014
71. Anjum SA, Xie XY, Wang LC, Saleem MF, Man C, Lei W (2011) Morphological, physiological and biochemical responses of plants to drought stress. Afr J Agric Res 6:2026–2032
72. Tardieu F, Granier C, Muller B (1999) Modelling leaf expansion in a fluctuating environment: are changes in specific leaf area a consequence of changes in expansion rate? New Phytol 143:33–43
73. Sharma V, Singh CM, Chugh V, Prajapati PK, Mishra A, Kaushik P, Dhanda PS, Yadav A (2023) Morpho-physiological and biochemical responses of field pea genotypes under terminal heat stress. Plan Theory 12(2):256. https://doi.org/10.3390/plants12020256
74. Munns R (2002) Comparative physiology of salt and water stress. Plant Cell Environ 25:239–250. https://doi.org/10.1046/j.0016-8025.2001.00808.x
75. Malinowski DP, Belesky DP (2000) Adaptations of endophyte-infected cool-season grasses to environmental stresses: mechanisms of drought and mineral stress tolerance. Crop Sci 40 (4):923–940. https://doi.org/10.2135/cropsci2000.404923x
76. Bray EA (2002) Abscisic acid regulation of gene expression during waterdeficit stress in the era of the Arabidopsis genome. Plant Cell Environ 5:153–161. https://doi.org/10.1046/j.1365-3040.2002.00746.x
77. Zhang SO, Outlaw Jr WH (2001) Abscisic acid introduced into transpiration stream accumulates in the guard cell apoplast and causes stomatal closure. Plant Cell Environ 24:1045–1054. https://doi.org/10.1046/j.1365-3040.2001.00755.x
78. De Smet I, Signora L, Beeckman T, Inzé D, Foyer CH, Zhang H (2003) An abscisic acid-sensitive checkpoint in lateral root development of Arabidopsis. Plant J 33(3):543–555
79. Cohen AC, Bottini R, Piccoli PN (2008) *Azospirillum brasilense* Sp 245 produces ABA in chemically-defined culture medium and increases ABA content in *Arabidopsis* plants. Plant Growth Regul 54:97–103. https://doi.org/10.1007/s10725-007-9232-9
80. Jones A, Panagos P, Barcelo S, Bouraoui F, Bosco C, Dewitte O, Gardi C, Hervás J, Hiederer R, Jeffery S (2012) The state of soil in Europe - a contribution of the JRC to the European Environment Agency's Environment State and Outlook R- SOER 2010
81. Sahab S, Suhani I, Srivastava V, Chauhan PS, Singh RP, Prasad V (2021) Potential risk assessment of soil salinity to agroecosystem sustainability: current status and management strategies. Sci Total Environ 764:144164
82. Shrivastava P, Kumar R (2015) Soil salinity: a serious environmental issue and plant growth promoting bacteria as one of the tools for its alleviation. Saudi J Biol Sci 22(2):123–131. https://doi.org/10.1016/j.sjbs.2014.12.001
83. Maas EV, Grattan SR (1999) Crop yields as affected by salinity. Agric Drain 38:55–108
84. Francois LE (1994a) Growth, seed yield, and oil content of canola grown under saline conditions. Agron J 86(2):233–237
85. Francois LE (1994b) Yield and quality response of salt-stressed garlic. HortScience 29 (11):1314–1317
86. Francois LE (1996) Salinity effects on four sunflower hybrids. Agron J 88(2):215–219
87. Machado RMA, Serralheiro RP (2017) Soil salinity: effect on vegetable crop growth. Management practices to prevent and mitigate soil salinization. Horticulturae 3(2):30. https://doi.org/10.3390/horticulturae3020030

88. Mangal JL, Lal S, Hooda PS (1989) Salt tolerance of the onion seed crop. J Hortic Sci 64 (4):475–477
89. Minhas PS, Sharma DR, Khosla BK (1990) Mungbean response to irrigation with waters of different salinities. Irrig Sci 11:57–62. https://doi.org/10.1007/BF00189996
90. Venkateswarlu J, Ramesam M, Murali Mohan Rao GV (1972) Salt tolerance in rice varieties. J Ind Soc Soil Sci 20:169–113
91. Ashraf M (2004) Some important physiological selection criteria for salt tolerance in plants. Flora 199:361–376
92. Seckin B, Sekmen AH, Turkan I (2009) An enhancing effect of exogenous mannitol on the antioxidant enzyme activities in roots of wheat under salt stress. J Plant Growth Regul 28:12–20
93. Gul Jan F, Hamayun M, Hussain A, Jan G, Iqbal A, Khan A, Lee IJ (2019) An endophytic isolate of the fungus *Yarrowia lipolytica* produces metabolites that ameliorate the negative impact of salt stress on the physiology of maize. BMC Microbiol 19(1):1–10
94. Abdelaziz ME, Abdelsattar M, Abdeldaym EA, Atia MA, Mahmoud AWM, Saad MM, Hirt H (2019) *Piriformospora indica* alters Na+/K+ homeostasis, antioxidant enzymes and LeNHX1 expression of greenhouse tomato grown under salt stress. Sci Hortic 256:108532
95. Li L, Li L, Wang X, Zhu P, Wu H, Qi S (2017) Plant growth-promoting endophyte *Piriformospora indica* alleviates salinity stress in *Medicago truncatula*. Plant Physiol Biochem 119:211–223
96. Hamilton CE, Gundel PE, Helander M, Saikkonen K (2012) Endophytic mediation of reactive oxygen species and antioxidant activity in plants: a review. Fungal Divers 54(1):1–10. https://doi.org/10.1007/s13225-012-0158-9
97. Sekmen AH, Trkan I, Takio S (2007) Differential responses of antioxidative enzymes and lipid peroxidation to salt stress in salt-tolerant *Plantago maritima* and salt-sensitive *Plantago media*. Physiol Plant 131:399–411
98. Rouhier N, San Koh C, Gelhaye E, Corbier C, Favier F, Didierjean C, Jacquot JP (2008) Redox based anti- oxidant systems in plants: biochemical and structural analyses. Biochim Biophys Acta 1780:1249–1260
99. Rodriguez RJ, Henson J, Van Volkenburgh E, Hoy M, Wright L, Beckwith F, Kim YO, Redman RS (2008) Stress tolerance in plants via habitat-adapted symbiosis. ISME 2:404–416. https://doi.org/10.1038/ismej.2007.106
100. Baltruschat H, Fodor J, Harrach BD, Niemczyk E, Barna B, Gullner G, Janeczko A, Kogel KH, Schäfer P, Schwarczinger I, Zuccaro A (2008) Salt tolerance of barley induced by the root endophyte *Piriformospora indica* is associated with a strong increase in antioxidants. New Phytol 180(2):501–510. https://doi.org/10.1111/j.1469-8137.2008.02583.x
101. Glick BR (2014) Bacteria with ACC deaminase can promote plant growth and help to feed the world. Microbiol Res 169(1):30–39. https://doi.org/10.1016/j.micres.2013.09.009
102. Glick BR, Penrose DM, Li J (1998) A model for the lowering of plant ethylene concentrations by plant growth-promoting bacteria. J Theor Biol 190:63–68
103. Iniguez AL, Dong Y, Carter HD, Ahmer BM, Stone JM, Triplett EW (2005) Regulation of enteric endophytic bacterial colonization by plant defenses. Mol Plant-Microbe Interact 18 (2):169–178. https://doi.org/10.1094/MPMI-18-0169
104. Czarny JC, Grichko VP, Glick BR (2006) Genetic modulation of ethylene biosynthesis and signaling in plants. Biotechnol Adv 24(4):410–419. https://doi.org/10.1016/j.biotechadv.2006.01.003
105. Sun Y, Cheng Z, Glick BR (2009) The presence of a 1-aminocyclopropane-1-carboxylate (ACC) deaminase deletion mutation alters the physiology of the endophytic plant growth-promoting bacterium *Burkholderia phytofirmans* PsJN. FEMS Microbiol Lett 296(1):131–136. https://doi.org/10.1111/j.1574-6968.2009.01625.x
106. Ali S, Charles TC, Glick BR (2012) Delay of flower senescence by bacterial endophytes expressing 1-aminocyclopropane-1-carboxylate deaminase. J Appl Microbiol 113(5):1139–1144. https://doi.org/10.1111/j.1365-2672.2012.05409.x

107. Witzel K, Gwinn-Giglio M, Nadendla S, Shefchek K, Ruppel S (2012) Genome sequence of Enterobacter radicincitans DSM16656(T), a plant growth-promoting endophyte. J Bacteriol 194:5469. https://doi.org/10.1128/JB.01193-12
108. Tiwari S, Singh P, Tiwari R, Meena KK, Yandigeri M, Singh DP, Arora DK (2011) Salt tolerant rhizobacteria mediated induced tolerance in wheat (*Triticum aestivum*) and chemical diversity in rhizosphere enhance plant growth. Biol Fertil Soils 47(8):907–916. https://doi.org/10.1007/s00374-011-0598-5
109. Suzuki S, He YX, Oyaizu H (2003) Indole-3-acetic acid production in *Pseudomonas fluorescens* HP72 and its association with suppression of creeping bentgrass brown patch. Curr Microbiol 47:138–143. https://doi.org/10.1007/s00284-002-3968-2
110. Navarro L, Dunoyer P, Jay F, Arnold B, Dharmasiri N, Estelle M, Voinnet O, Jones JD (2006) A plant miRNA contributes to antibacterial resistance by repressing auxin signaling. Science 312(5772):436–439
111. Piccoli P, Travaglia C, Cohen A, Sosa L, Cornejo P, Masuelli R, Bottini R (2011) An endophytic bacterium isolated from roots of the halophyte *Prosopis strombulifera* produces ABA, IAA, gibberellins A(1) and A(3) and jasmonic acid in chemically defined culture medium. Plant Growth Regul 64(2):207–210. https://doi.org/10.1007/s10725-010-9536-z
112. Tuteja N (2007) Abscisic acid and abiotic stress signaling. Plant Signal Behav 2(3):135–138. https://doi.org/10.4161/psb.2.3.4156
113. Ruppel S, Franken P, Witzel K (2013) Properties of the halophyte microbiome and their implications for plant salt tolerance. Funct Plant Biol 40(9):940–951. https://doi.org/10.1071/FP12355
114. Hardoim PR, Van Overbeek LS, Berg G, Pirttilä AM, Compant S, Campisano A, Döring M, Sessitsch A (2015) The hidden world within plants: ecological and evolutionary considerations for defining functioning of microbial endophytes. Microbiol Mol Biol Rev 79(3):293–320. https://doi.org/10.1128/mmbr.00050-14
115. Munns R, Tester M (2008) Mechanisms of salinity tolerance. Annu Rev Plant Biol 59:651–681
116. Flowers TJ, Colmer TD (2008) Salinity tolerance in halophytes. New Phytol:945–963
117. Ruiz-Lozano JM, Porcel R, Bárzana G, Azcón R, Aroca R (2012) Contribution of arbuscular mycorrhizal symbiosis to plant drought tolerance: state of the art. Plant responses to drought stress: from morphological to molecular features. In: Plant responses to drought stress: from morphological to molecular features. Springer, Heidelberg, pp 335–362. https://doi.org/10.1007/978-3-642-32653-0_13
118. Manchanda G, Garg N (2011) Alleviation of salt-induced ionic, osmotic and oxidative stresses in *Cajanus cajan* nodules by AM inoculation. Plant Biosyst 145(1):88–97
119. Jha Y, Subramanian R, Patel S (2011) Combination of endophytic and rhizospheric plant growth promoting rhizobacteria in *Oryza sativa* shows higher accumulation of osmoprotectant against saline stress. Acta Physiol Plant 33:797–802. https://doi.org/10.1007/s11738-010-0604-9
120. Andreas T, Christophe C, Essaïd AB (2012) Physiological and molecular changes in plants at low temperatures. Planta 235:1091–1105. https://doi.org/10.1007/s00425-012-1641-y
121. Ait Barka E, Nowak J, Clément C (2006) Enhancement of chilling resistance of inoculated grapevine plantlets with a plant growth-promoting rhizobacterium, *Burkholderia phytofirmans* strain PsJN. Appl Environ Microbiol 72(11):7246–7252
122. Redman RS, Sheehan KB, Stout RG, Rodriguez RJ, Henson JM (2002) Thermotolerance generated by plant/fungal symbiosis. Science 298:1581–1581
123. Hubbard M, Germida JJ, Vujanovic V (2014) Fungal endophytes enhance wheat heat and drought tolerance in terms of grain yield and second-generation seed viability. J Appl Microbiol 116:109–122
124. Li L, Sinkko H, Montonen L, Wei G, Lindstrom K, Rasanen LA (2012) Biogeography of symbiotic and other endophytic bacteria isolated from medicinal *Glycyrrhiza* species in China. FEMS Microbiol Ecol 79:46–68

125. Matsouri F, Björkman T, Harman GE (2010) Seed treatment with *Trichoderma harzianum* alleviates biotic, abiotic, and physiological stresses in germinating seeds and seedlings. Biol Control 100:1213–1221
126. Singh D, Singh NP, Chauhan SK, Singh P (2011) Developing aluminium tolerant crop plants using biotechnological tools. Curr Sci 100(12):1807–1814
127. Zhang X, Ervin EH, Evanylo GK, Haering KC (2009) Impact of biosolids on hormone metabolism in drought-stressed tall fescue. Crop Sci 49:1893–1901
128. Pandey PK, Singh S, Singh AK, Samanta R, Yadav RNS, Singh MC (2016) Inside the plant: bacterial endophytes and abiotic stress alleviation. J Appl Nat Sci 8(4):1899–1904
129. Wang XQ, Liu AX, Guerrero A, Liu J, Yu XQ, Deng P, Ma L, Baird SM, Smith L, Li XD, Lu SE (2016) Occidiofungin is an important component responsible for the antifungal activity of *Burkholderia pyrrocinia* strain Lyc2. J Appl Microbiol 120:607–618. https://doi.org/10.1111/jam.13036
130. Friesen ML, Porter SS, Stark SC, von Wettberg EJ, Sachs JL, Martinez Romero E (2011) Microbially mediated plant functional traits. Annu Rev Ecol Evol Syst 42:23–46
131. Chaves MM, Oliveira MM (2004) Mechanisms underlying plant resilience to water deficits: prospects for water-saving agriculture. J Exp Bot 55:2365–2384
132. Choi O, Kim J, Kim JG, Jeong Y, Moon JS, Park CS, Hwang I (2008) Pyrroloquinoline quinone is a plant growth promotion factor produced by Pseudomonas fluorescens B16. Plant Physiol 146:657–668
133. Abadi VAJM, Sepehri M (2015) Effect of Piriformospora indica and Azotobacter chroococcum on mitigation of zinc deficiency stress in wheat (*Triticum aestivum* L). Symbiosis 69:9–19
134. Pang Z, Chen J, Wang T, Gao C, Li Z, Guo L, Xu J, Cheng Y (2021) Linking plant secondary metabolites and plant microbiomes: a review. Front Plant Sci 12:621276
135. Sasse J, Martinoia E, Northen T (2018) Feed your friends: do plant exudates shape the root microbiome? Trends Plant Sci 23:25–41. https://doi.org/10.1016/j.tplants.2017.09.003
136. Terekhova VA, Semenova TA (2005) The structure of micromycete communities and their synecologic interactions with basidiomycetes during plant debris decomposition. Microbiology 74:91–96
137. Kessler A, Kalske A (2018) Plant secondary metabolite diversity and species interactions. Annu Rev Ecol Evol Syst 49:115–138. https://doi.org/10.1146/annurevecolsys-110617-062406
138. de Vries FT, Griffiths RI, Knight CG, Nicolitch O, Williams A (2020) Harnessing rhizosphere microbiomes for drought-resilient crop production. Science 368:270. https://doi.org/10.1126/science.aaz5192
139. Shaik ZA, Sandhya V, Grover M, Linga VR, Bandi V (2011) Effect of inoculation with a thermotolerant plant growth promoting Pseudomonas putida strain AKMP7 on growth of wheat (Triticum spp.) under heat stress. J Plant Interact 6:239–246
140. Kang MS, Radhakrishnan R, Khan LA, Kim JM, Kim JM, Park JM, Kim RB, Shin HD, Lee JI (2014) Gibberellin secreting rhizobacterium, Pseudomonas putida H-2-3 modulates the hormonal and stress physiology of soybean to improve the plant growth under saline and drought conditions. Plant Physiol Biochem 84:115–124
141. Sharma S, Kulkarni J, Jha B (2016) Halotolerant rhizobacteria promote growth and enhance salinity tolerance in peanut. Front Microbiol 7:1600
142. Ghosh D, Gupta A, Mohapatra S (2019) A comparative analysis of exopolysaccharide and phytohormone secretions by four drought-tolerant rhizobacterial strains and their impact on osmotic-stress mitigation in Arabidopsis thaliana. World J Microbiol Biotechnol 35:90
143. Liu Y, Shi Z, Yao L, Yue H, Li H, Li C (2013) Effect of IAA produced by Klebsiella oxytoca Rs-5 on cotton growth under salt stress. J Gen Appl Microbiol 59:59–65
144. Balsanelli E, de Baura VA, Fdo P, de Souza EM, Monteiro RA (2014) Exopolysaccharide biosynthesis enables mature biofilm formation on abiotic surfaces by Herbaspirillum seropedicae. PLoS One 9:110392

145. Mahmood S, Daur I, Al-Solaimani GS, Ahmad S, Madkour HM, Yasir M, Hirt H, Ali S, Ali Z (2016) Plant growth promoting rhizobacteria and silicon synergistically enhance salinity tolerance of mung bean. Front Plant Sci 7:876
146. Bont Z, Züst T, Arce CCM, Huber M, Erb M (2020) Heritable variation in root secondary metabolites is associated with recent climate. J Ecol 108:2611–2624. https://doi.org/10.1111/1365-2745.13441
147. Sokol NW, Kuebbing SE, Karlsen-Ayala E, Bradford MA (2019) Evidence for the primacy of living root inputs, not root or shoot litter, in forming soil organic carbon. New Phytol 221:233–246. https://doi.org/10.1111/nph.15361
148. Burkhead KD, Schisler AD, Slininger JP (1994) Pyrrolnitrin production by biological control agent Pseudomonas cepacia B37w in culture and in colonized wounds of potatoes. Appl Environ Microbiol 60:2031–2039
149. Haas D, Defago G (2005) Biological control of soilborne pathogens by fluorescent pseudomonads. Nat Rev Microbiol 3:307
150. Sankari U, Dinakar S, Sekar C (2011) Dual effect of Azospirillum exopolysaccharides (EPS) on the enhancement of plant growth and biocontrol of blast (Pyricularia oryzae) disease in upland rice (var. ASD-19). J Phytology 3:16–19
151. Sathya A, Vijayabharathi R, Gopalakrishnan S (2017) Plant growth-promoting actinobacteria: a new strategy for enhancing sustainable production and protection of grain legumes. Biotech 7:102
152. Manganiello G, Sacco A, Ercolano MR, Vinale F, Lanzuise S, Pascale A, Napolitano M, Lombardi N, Lorito M, Woo SL (2018) Modulation of tomato response to Rhizoctonia solani by Trichoderma harzianum and its secondary metabolite harzianic acid. Front Microbiol 9:9
153. Verma VC, Singh SK, Prakash S (2011) Research paper bio-control and plant growth promotion potential of siderophore producing endophytic Streptomyces from Azadirachta indica A. Juss. J Basic Microbiol 51(5):550–556
154. Gu Q, Yang Y, Yuan Q, Shi G, Wu L, Lou Z, Huo R, Wu H, Borriss R, Gao X (2017) Bacillomycin D produced by Bacillus amyloliquefaciens is involved in the antagonistic interaction with the plant-pathogenic fungus Fusarium graminearum. Appl Environ Microbiol 83(19):e01075–e01017
155. Kong AHG, Kim JC, Choi GJ, Lee KY, Kim HJ, Hwang EC, Moon BJ, Lee SW (2010) Production of surfactin and iturin by Bacillus licheniformis N1 responsible for plant disease control. Plant Pathol J 26(2):170–177
156. Malhotra M, Srivastava S (2009) Stress-responsive indole-3-acetic acid biosynthesis by Azospirillum brasilense SM and its ability to modulate plant growth. Eur J Soil Biol 45(1): 73–80
157. Metwali E, Abdelmoneim TS, Bakheit MA, Kadasa NMS (2015) Alleviation of salinity stress in faba bean (Vicia faba L.) plants by inoculation with plant growth promoting rhizobacteria (PGPR). Plant Omics 8(5):449–460
158. Vinale F, Sivasithamparam K, Ghisalberti EL, Woo SL, Nigro M, Marra R, Lombardi N, Pascale A, Ruocco M, Lanzuise S, Manganiello G, Lorito M (2014) Trichoderma secondary metabolites active on plants and fungal pathogens. Open Mycol J 8:127–139
159. Cassa'n F, Maiale S, Masciarelli O, Vidal A, Luna V, Ruiz O (2009) Cadaverine production by Azospirillumbrasilense and its possible role in plant growth promotion and osmotic stress mitigation. Eur J Soil Biol 45:12–19
160. Sarma RK, Gogoi A, Dehury B, Debnath R, Bora TC, Saikia R (2014) Community profiling of culturable fluorescent pseudomonads in the rhizosphere of green gram (Vigna radiata L.). PLoS One 9(10):e108378
161. Shenker M, Oliver I, Helmann M, Hadar Y, Chen Y (1992) Utilization by tomatoes of iron mediated by a siderophore produced by Rhizopus arrhizus. J Plant Nutr 15(10):2173–2182
162. El-Ghany MFA, Attia M (2020) Effect of exopolysaccharide-producing bacteria and melatonin on faba bean production in saline and non-saline. Agronomy 10:316

163. Halo AB, Khan LA, Waqas M, AlHarrasi A, Hussain J, Ali L, Adnan M, Lee JI (2015) Endophytic bacteria (Sphingomonas sp. LK11) and gibberellin can improve Solanum lycopersicum growth and oxidative stress under salinity. J Plant Interact 10(1):117–125
164. Wan W, Li C, Zhang J, Jiang D, Huan HC (2008) Effect of volatile substances of Streptomyces platensis F-1 on control of plant fungal diseases. Biol Control 46:552–559
165. Fugyeuredi MVB, Vilhar JJ, Burity HA, de Franca FP (1999) Alleviation of water stress effects in cowpea by Bradyrhizobium spp. inoculation. Plant Soil 207:67–75

Chapter 14
Importance of Antagonistic Activities of Microbes and Their Metabolites

Parth Choudhary, Manu Pant, and Kumud Pant

Abstract Microbial antagonism is a complex phenomenon that involves various interactions among microbial communities. The importance of antagonistic activities of microbes and their metabolites in various fields such as biopreservation, biological control, and bioremediation cannot be overstated. This chapter provides an overview of the various types of antagonistic activities of microbes, including nutrient competition, production of inhibitory compounds, and modulation of host immunity. Additionally, this chapter discusses the applications of microbial antagonistic activities in food and beverage preservation, biological control of plant diseases, and production of antibiotics and probiotics. Despite the significant potential of microbial antagonistic activities, challenges in the discovery and development of antagonistic compounds still exist, including limited knowledge of microbial communities and limited screening methods. Future research efforts should aim to address these challenges and expand our understanding of the complex interactions among microbial communities. This chapter emphasizes the importance of continued research and development in the field of microbial antagonism to improve the health and well-being of humans and the environment.

Keywords Antagonism · Microbiome · Interactions · Metabolite · Probiotic

14.1 Introduction

Microbial interactions refer to the complex relationships and interactions between microorganisms, including bacteria, fungi, viruses, and other microorganisms [1]. These interactions can be either synergistic, where the microorganisms work together and benefit each other, or antagonistic, where one microorganism negatively affects another. The study of microbial interactions has led to important

P. Choudhary · M. Pant · K. Pant (✉)
Department of Biotechnology, Graphic Era Deemed to be University, Dehradun, Uttarakhand, India
e-mail: choudharyparth21@gmail.com; Manupant@geu.ac.in; kumud.pant@geu.ac.in

R. Soni et al. (eds.), *Microbial Bioactive Compounds*,
https://doi.org/10.1007/978-3-031-40082-7_14

discoveries and innovations in various fields, including medicine, agriculture, biotechnology, and ecology.

Microbial interactions play a critical role in the development and maintenance of the human microbiome, which is the collective community of microorganisms that reside on and inside the human body [2]. The microbiome is involved in essential physiological processes, including digestion, metabolism, and immune system function. In agriculture, microbial interactions are important for maintaining healthy soil and promoting plant growth [3]. Some microorganisms help to break down organic matter and make nutrients available to plants, while others can protect plants from pests and diseases. Also, microbial interactions have led to the development of a range of products and processes. For example, the discovery of naturally occurring antibiotics produced by microorganisms has revolutionized medicine, and bioremediation technologies use microorganisms to clean up environmental pollutants [4]. Microbial interactions also play a role in the development of new vaccines and drugs. For ecology, microbial interactions are essential for the functioning of ecosystems. Microorganisms play a key role in the cycling of nutrients, decomposition of organic matter, and maintenance of soil health. These interactions help to sustain the balance and health of ecosystems and are critical for the long-term health of the planet.

Microbial antagonism is a phenomenon in which microorganisms interact with each other in a way that negatively impacts the growth or survival of one or more of the microorganisms involved [2]. This can occur through a variety of mechanisms, such as competition for resources, production of toxic compounds, or stimulation of host immune responses [5]. By understanding the mechanisms and implications of microbial antagonism, scientists and researchers can develop new strategies to harness the beneficial interactions between microorganisms and improve our understanding of complex microbial ecosystems.

14.2 Types of Microbial Antagonistic Activities

A few antagonistic activities of microbes and their metabolites include:

14.2.1 Production of Bacteriocins

Bacteriocins are antimicrobial peptides or proteins produced by bacteria that can inhibit the growth of other bacteria [4]. These are produced by a wide range of bacteria, including lactic acid bacteria, *Bacillus* species, and *Enterococcus* species [4, 5]. The use of bacteriocins as antimicrobial agents has received increasing attention in recent years, as they have the potential to be used as alternatives to traditional antibiotics in the prevention and treatment of bacterial infections [4].

14.2.2 Production of Antibiotics

Antibiotics are antimicrobial compounds produced by microorganisms, including bacteria and fungi, that can kill or inhibit the growth of other microorganisms. There are several types of antibiotics, each with a unique mechanism of action and spectrum of activity. These are produced by a wide range of microorganisms, including *Streptomyces* and *Penicillium* species. The discovery and development of antibiotics have revolutionized the field of medicine, providing effective treatments for bacterial infections. However, the overuse and misuse of antibiotics have led to the emergence of antibiotic-resistant bacteria, highlighting the need for new antimicrobial strategies [2].

The production of bacteriocins and antibiotics is often controlled by quorum sensing and can be influenced by environmental factors such as pH, temperature, and nutrient availability [6].

14.2.3 Production of Quorum-Sensing Molecules

Quorum sensing (QS) is a process by which bacteria communicate with each other using chemical signals, known as autoinducers, to coordinate their behavior and gene expression. In addition to regulating bacterial behavior, QS can also play a role in microbial antagonism. Some bacteria produce quorum-sensing molecules (QSMs) that can interfere with the QS of other bacteria, leading to a disruption of their behavior and virulence. There are different types of QSMs produced by bacteria, including acyl-homoserine lactones (AHLs), autoinducer-2 (AI-2), and peptides. The production of QSMs can have different effects on bacterial populations. For example, some QSMs can inhibit the growth of competing bacteria or prevent biofilm formation, while others can induce the expression of virulence factors and promote bacterial pathogenesis. By targeting the QS system of pathogenic bacteria, it may be possible to disrupt their behavior and virulence, without necessarily killing them.

This approach could help to reduce the selective pressure for the development of antibiotic resistance as well as limit the damage to beneficial microbial communities. One example of a bacterium that produces QSMs with antagonistic activity is *Pseudomonas aeruginosa*, which produces AHLs that can inhibit the QS of other bacteria. Another example is *Bacillus thuringiensis*, which produces cyclic lipopeptides that can inhibit the growth of other bacteria and fungi [7].

14.3 Mechanisms of Microbial Antagonism

Microbial antagonism is a fundamental aspect of microbial ecology, and the mechanisms by which microorganisms interact with each other can have profound effects on the structure and function of microbial communities. These interactions can occur through a variety of mechanisms, including competitive exclusion, nutrient competition, production of inhibitory compounds, and modulation of host immunity [8].

14.3.1 Competitive Exclusion

One of the most common mechanisms of microbial antagonism is competitive exclusion, which occurs when one microbe is able to outcompete another for a particular resource. By competing for resources, microbes are able to maintain balance in their environment and prevent the overgrowth of any one particular species [8]. This is particularly important in environments such as the gut, where the presence of pathogenic bacteria can have serious health consequences that make it helpful in making probiotics. Competitive exclusion can occur through a number of different mechanisms:

- Some microbes are able to produce antimicrobial compounds, such as bacteriocins or antibiotics, that can kill or inhibit the growth of other microbes in their environment [4].
- Some microbes have mechanisms for acquiring nutrients more efficiently, such as the ability to break down complex organic compounds into simpler forms that can be readily absorbed [9].
- The production of biofilms, which are complex communities of microbes that adhere to surfaces and form protective matrices that can prevent other microbes from colonizing the same surface [10].
- Some microbes are able to outcompete others through physical interactions, such as through the use of pili or other structures that allow them to attach to surfaces or other microbes [8].

14.3.2 Nutrient Competition

One of the most important mechanisms of microbial antagonism is nutrient competition, which occurs when microbes compete for limited sources of nutrients, such as carbon, nitrogen, and phosphorus. Despite the importance of nutrient competition in microbial communities, much remains to be understood about the mechanisms by which microbes compete for resources. Ongoing research is focused on understanding the complex interactions between different microbes and the role that nutrient competition plays in shaping microbial populations [9]. Nutrient competition can occur through a number of different mechanisms:

- Some microbes may be able to secrete enzymes that break down complex organic molecules into simpler forms that can be readily absorbed.
- Some are able to absorb nutrients more efficiently than others or may have specialized mechanisms for acquiring specific nutrients that are scarce in their environment.
- In addition to direct competition for nutrients, some microbes are able to produce secondary metabolites that can inhibit the growth of other microbes in their environment. These secondary metabolites may act as antibiotics, bacteriocins, or other types of antimicrobial compounds that interfere with the growth and reproduction of competing microbes [9].

14.3.3 Production of Inhibitory Compounds

One mechanism of microbial antagonism is the production of inhibitory compounds, including antibiotics, bacteriocins, and other types of antimicrobial agents [2] as discussed earlier in the chapter. The production of inhibitory compounds is a critical mechanism of microbial antagonism that plays a crucial role in maintaining microbial diversity and preventing the overgrowth of pathogenic microorganisms. The wide range of inhibitory compounds produced by microbes, including antibiotics, bacteriocins, and quorum-sensing inhibitors, highlights the complexity of microbial interactions and the potential for developing new strategies for managing microbial populations in a range of environments [2].

14.3.4 Modulation of Host Immunity

Microbes can interact with host immune cells and alter their function, leading to changes in the host's immune response that can have both beneficial and detrimental effects. The modulation of host immunity is a complex process that involves the interplay of multiple factors, including microbial and host factors [8]. Microbes can modulate host immunity in several ways:

- Some microbes can directly interact with host immune cells and activate or suppress specific immune responses. This can be achieved through the secretion of microbial factors, such as lipopolysaccharides or flagellin, that can bind to receptors on host immune cells and trigger specific signaling pathways.
- In addition to direct interactions with immune cells, microbes can also modulate host immunity by altering the composition of the microbiota. The microbiota is a complex community of microorganisms that resides within the host and plays an essential role in shaping the host's immune system. Microbes can influence the

composition of the microbiota by producing metabolites that can either promote or inhibit the growth of specific microorganisms.

By modulating host immunity, microbes can exert both beneficial and detrimental effects on host health. For example, some microbes can promote immune tolerance, reducing the risk of inflammatory diseases such as allergies or autoimmune disorders. Other microbes can enhance the host's immune response, providing protection against infectious diseases [8].

14.4 Applications of Microbial Antagonistic Activities

Microbial antagonistic activities have broad applications in various fields and can be used to promote beneficial interactions between microorganisms, prevent the growth of harmful pathogens, and produce bioactive compounds with diverse biological activities. Further research is needed to better understand the mechanisms underlying these interactions and to develop new strategies for harnessing the potential of microbial antagonistic activities [11].

14.4.1 Biopreservation of Food and Beverages

An important application of microbial antagonistic activities is in the biopreservation of food and beverages. Biopreservation refers to the use of microbial antagonists to control the growth of spoilage and pathogenic microorganisms in food and beverages and improving the quality and shelf life of food and beverage products; microbial antagonists can help to ensure the safety and availability of nutritious and delicious food and beverage products for consumers.

Microbial antagonists, such as lactic acid bacteria and bacteriocin-producing strains, such as *Listeria monocytogenes* and *Staphylococcus aureus*, can be added to food and beverage products to prevent the growth of harmful microorganisms and extend their shelf life. In addition to inhibiting the growth of spoilage and pathogenic microorganisms, microbial antagonists can also improve the sensory and nutritional quality of food and beverage products [11].

14.4.2 Biological Control of Plant Diseases

Another important application of microbial antagonistic activities is in the biological control of plant diseases. Biological control refers to the use of microorganisms to suppress plant pathogens and protect crops from disease. The use of microbial antagonistic activities for biological control of plant diseases offers a promising and eco-friendly alternative to traditional chemical-based methods. By reducing the need for synthetic pesticides and fungicides, microbial antagonists can help to minimize the environmental impact of crop protection while promoting sustainable agriculture [10].

Microbial antagonists, such as bacteria and fungi, can be used as biocontrol agents to suppress the growth and spread of plant pathogens in the soil and on plant surfaces. For example, certain strains of *Bacillus* and *Pseudomonas* have been shown to produce antimicrobial compounds that inhibit the growth of plant pathogens, such as *Fusarium* and *Phytophthora*. In addition to producing antimicrobial compounds, microbial antagonists can also compete with plant pathogens for nutrients and space on plant surfaces, thereby reducing their ability to infect plants. Moreover, some microbial antagonists can stimulate the plant's immune system, which enhances its resistance to pathogen infection [10].

14.4.3 Production of Antibiotics and Probiotics

Microbial antagonistic activities are also widely used in the production of antibiotics and probiotics. Antibiotics are a class of compounds produced by microorganisms that inhibit or kill the growth of other microorganisms. They are widely used in the treatment of bacterial infections in humans, animals, and plants. The production of antibiotics relies on the ability of microorganisms to produce and secrete inhibitory compounds that target specific bacterial pathogens. Probiotics, on the other hand, are live microorganisms that confer a health benefit to the host when administered in adequate amounts. They are commonly used as dietary supplements to promote gut health and prevent or treat various diseases. The production of probiotics relies on the ability of microorganisms to compete with and inhibit the growth of harmful bacteria in the gut [8].

14.4.4 Bioremediation and Wastewater Treatment

Microbial antagonistic activities have important applications in bioremediation and wastewater treatment. The ability of some microorganisms to degrade environmental pollutants can be enhanced by the production of inhibitory compounds that target and eliminate harmful microorganisms in the contaminated site. Additionally, the use of microbial consortia with antagonistic activities can help to create a balanced ecosystem and reduce the need for chemical interventions. In wastewater treatment, the use of microbial antagonistic activities can help to control pathogenic bacteria and prevent the spread of waterborne diseases. Moreover, it can aid in the removal of organic pollutants and nitrogen compounds, which can contribute to eutrophication and other environmental problems [12]. Overall, the application of microbial antagonistic activities in bioremediation and wastewater treatment provides a promising avenue for sustainable environmental management.

14.5 Challenges in the Discovery and Development of Antagonistic Compounds

Despite the promising applications of microbial antagonistic activities, there are several challenges in the discovery and development of antagonistic compounds. One of the primary challenges is the identification of potential antagonists, as many microbial interactions are complex and difficult to study. Moreover, many of the compounds produced by microorganisms may have multiple biological activities, making it difficult to identify the specific mechanism of antagonism. Additionally, the production of antagonistic compounds may be affected by environmental factors, which can limit their efficacy or production levels [13]. Regulatory requirements and safety concerns can present additional obstacles in the development and commercialization of antagonistic compounds [5].

14.5.1 Limited Knowledge of Microbial Communities

Many microorganisms exist in complex and diverse ecosystems, and their interactions with other microorganisms are often poorly understood. As a result:

- Identifying potential antagonistic compounds can be difficult and may require extensive study of the microbial community and its interactions.
- The composition and diversity of microbial communities can vary significantly depending on environmental factors, making it challenging to identify and isolate specific microorganisms or compounds [6].
- The study of microbial communities requires advanced techniques such as metagenomics, which can be expensive and time-consuming [12].

To overcome these challenges, researchers are increasingly turning to advanced technologies such as high-throughput screening and synthetic biology. However, these technologies are still in the early stages of development, and their widespread use in the discovery and development of antagonistic compounds is limited [14].

14.5.2 Limited Screening Methods

Limited screening methods pose a significant challenge in the discovery and development of antagonistic compounds.

- Traditional screening methods based on culturing individual microbial strains in vitro may not accurately represent the complex interactions that occur within natural microbial communities.
- Many microorganisms are unculturable or difficult to culture, further limiting the pool of potential antagonistic compounds that can be identified through traditional methods [3].

Advances in high-throughput sequencing and other omics technologies have allowed for a more comprehensive understanding of microbial communities and their interactions, but the translation of this knowledge into the identification of specific antagonistic compounds remains a challenge. Therefore, the development of new screening methods that can effectively capture the complexity of microbial interactions is critical for the discovery and development of novel antagonistic compounds [9].

14.5.3 Need for Better Understanding of Molecular Mechanisms

The discovery and development of antagonistic compounds also face the challenge of limited understanding of the molecular mechanisms underlying microbial interactions. While high-throughput sequencing and other omics technologies have provided a wealth of information about the genetic and metabolic potential of microbial communities [7], the specific mechanisms by which microorganisms interact with each other and with their environment remain poorly understood. This lack of knowledge can make it difficult to predict which microbes are likely to produce effective antagonistic compounds or how these compounds may function to inhibit the growth or activity of target organisms. Therefore, there is a need for more research to better understand the molecular mechanisms underlying microbial interactions and the production of antagonistic compounds. Such knowledge will aid in the development of more effective screening methods and the design of new compounds with improved efficacy and specificity.

14.6 Future Directions

Continued emphasis should be laid on understanding the molecular mechanisms underlying these interactions, as well as the development of more effective screening methods for identifying novel antagonistic compounds. Advances in omics technologies, such as metagenomics and metatranscriptomics, may provide new insights into the interactions between microorganisms in natural communities and allow for the discovery of new antagonistic compounds [15]. Additionally, the use of synthetic biology and genetic engineering to manipulate microbial communities and optimize the production of specific antagonistic compounds shows promise for the development of more targeted and effective biopreservation and biological control strategies [16]. Finally, further exploration of the potential therapeutic applications of microbial antagonistic compounds, including their use as antibiotics and immunomodulators, may lead to new treatments for a range of human diseases.

14.7 Conclusion

The antagonistic activities of microbes and their metabolites have immense importance in various fields including food preservation, plant disease control, and bioremediation. The potential for the discovery and development of novel antagonistic compounds remains high, but significant challenges such as limited screening methods and a lack of understanding of molecular mechanisms must be addressed. Continued research and development in the field is crucial to unlock the full potential of microbial antagonism and its applications. The future holds great promise for the use of microbial antagonistic activities in promoting human health and sustainable agriculture practices.

References

1. Özer ED, Yildirim M (2021) Chemical modification of Nisin with homocysteine thiolactone and its antibacterial activity. Available at https://doi.org/10.21203/rs.3.rs-649358/v1
2. Peterson SB, Bertolli SK, Mougous JD (2020) The central role of interbacterial antagonism in bacterial life. Curr Biol 30(19):R1203–R1214
3. Berg G, Kusstatscher P, Abdelfattah A, Cernava T, Smalla K (2021) Microbiome modulation—toward a better understanding of plant microbiome response to microbial inoculants. Front Microbiol 12:650610
4. Bashir A, Ali K (2021) Antagonistic activity of Bacteriocins produced by lactobacillus isolates against multidrug resistant pathogens. J Pharm Res Int 33:76–87
5. Elnahal ASM, El-Saadony MT, Saad AM et al (2022) The use of microbial inoculants for biological control, plant growth promotion, and sustainable agriculture: a review. Eur J Plant Pathol 162:759–792. https://doi.org/10.1007/s10658-021-02393-7
6. Zhao X, Lin C, Tan J et al (2023) Changes of rhizosphere microbiome and metabolites in Meloidogyne incognita infested soil. Plant Soil 483:331–353. https://doi.org/10.1007/s11104-022-05742-5
7. Prazdnova EV, Gorovtsov AV, Vasilchenko NG, Kulikov MP, Statsenko VN, Bogdanova AA et al (2022) Quorum-sensing inhibition by gram-positive bacteria. Microorganisms 10(2):350
8. Tariq M, Khan A, Asif M, Khan F, Ansari T, Shariq M, Siddiqui MA (2020) Biological control: a sustainable and practical approach for plant disease management. Acta Agric Scand B Soil Plant Sci 70(6):507–524
9. Kumar V, Ahluwalia V, Saran S, Kumar J, Patel AK, Singhania RR (2021) Recent developments on solid-state fermentation for production of microbial secondary metabolites: challenges and solutions. Bioresour Technol 323:124566

10. Cesa-Luna C, Baez A, Quintero-Hernández V, Cruz-Enríquez JDL, Castañeda-Antonio MD, Muñoz-Rojas J (2020) The importance of antimicrobial compounds produced by beneficial bacteria on the biocontrol of phytopathogens. Acta Biológica Colombiana 25(1):140–154
11. Rathod NB, Phadke GG, Tabanelli G, Mane A, Ranveer RC, Pagarkar A, Ozogul F (2021) Recent advances in bio-preservatives impacts of lactic acid bacteria and their metabolites on aquatic food products. Food Biosci 44:101440
12. Villela HD, Vilela CL, Assis JM, Varona N, Burke C, Coil DA et al (2019) Prospecting microbial strains for bioremediation and probiotics development for metaorganism research and preservation. J Vis Exp 152:e60238
13. Li M (2008) Structure-based discovery and experimental verification of novel AI-2 quorum sensing inhibitors against Vibrio harveyi. ChemMedChem 3:1242
14. Butt UD, Lin N, Akhter N, Siddiqui T, Li S, Wu B (2021) Overview of the latest developments in the role of probiotics, prebiotics and synbiotics in shrimp aquaculture. Fish Shellfish Immunol 114:263–281
15. Singh J, Faull JL (2020) Antagonism and biological control. In: Biocontrol of plant diseases. CRC Press, pp 167–177
16. Fijan S, Frauwallner A, Langerholc T, Krebs B, ter Haar née Younes JA, Heschl A et al (2019) Efficacy of using probiotics with antagonistic activity against pathogens of wound infections: an integrative review of literature. BioMed Research International

Chapter 15
Microbial Community Dynamics of Antarctica: Their Ecological Potential and Industrial Importance

Amir Khan, Arjita Punetha, Bharti Kukreti, Raj Shekhar Sharma, Divyansh Panthari, Neetika Naudiyal, Vinita Gouri, Harminder Singh Baweja, and Ajay Veer Singh

Abstract Antarctica is covered with ice and therefore the coldest, driest and least populated continent of the Earth. Apart from the challenges of sustaining the life in such environment, the microbial diversity of archaea, bacteria and fungi can thrive here by maintaining the membrane fluidity, producing the antifreeze proteins, cold-shock proteins, cryoprotectants, osmolytes, antioxidants, cold active enzymes, alteration in DNA, etc. and thereby adapted for the cold environments. In Antarctica, these microbes are the main basis for the biogeocycling of nutrients in extreme environments. The enzymes produced by these psychrophilic organisms are cold

---

Amir Khan and Arjita Punetha contributed equally and share first authorship.

A. Khan · B. Kukreti · A. V. Singh (✉)
Biofortification Laboratory, Department of Microbiology, College of Basic Sciences and Humanities, Govind Ballabh Pant University of Agriculture and Technology, Pantnagar, Uttarakhand, India

A. Punetha
Doon University, Dehradun, Uttarakhand, India

R. S. Sharma
Department of Microbiology, School of Basic and Applied Science, Sri Guru Ram Rai University, Dehradun, India

D. Panthari
Department of Botany, School of Basic and Applied Science, Sri Guru Ram Rai University, Dehradun, India

N. Naudiyal
Department of Biochemistry, College of Basic Sciences and Humanities, Govind Ballabh Pant University of Agriculture and Technology, Pantnagar, Uttarakhand, India

V. Gouri
Cell and Molecular Biology Laboratory, Department of Zoology, Soban Singh Jeena University, Nainital, Uttarakhand, India

H. S. Baweja
Department of Horticulture and Food Processing, Chaubatia, Uttarakhand, India

R. Soni et al. (eds.), *Microbial Bioactive Compounds*,
https://doi.org/10.1007/978-3-031-40082-7_15

active and therefore gained the spotlight owing to their significance in environment and research. The behaviour of these cold-active enzymes has a wide and potentially biotechnological applications in fields as wide as the detergent, textile and food industries, medical and pharmaceutical preparations, bioremediation, etc. Therefore, the current manuscript illustrates the native microbial diversity and their significance and application aspect in environment and industry.

Keywords Antarctica · Biogeocycling · Cold environments · Cold active enzyme · Microbial diversity

15.1 Introduction

Antarctica is located close to the South Pole. The entire continent covers an area of 14 million km^2 of which only 0.3% is not covered in ice. A massive ice sheet covering almost the whole continent has more than 26,106 km^3 of ice, an amount which if calculated in terms of sea-level rise will be equivalent to about 58 m [1]. Average thickness of the ice sheet of the continent is about 2.1 km. The four main geographical regions of Antarctica are West and East Antarctica, the Antarctic Peninsula and the Sub-Antarctic region. Antarctica and its residents on daily basis are subjected to extremely harsh climatic conditions such as low humidity, frequent freeze-thaw and wet-dry cycles, low temperatures, variable UV radiation and strong desiccating winds [2]. Extreme environmental conditions result in formation of oligotrophic environments in which only exceedingly particular organisms can thrive. The primary basis of ecosystem processes in Antarctica is formed by the microbial autotrophs, which are crucial to the processes of primary colonization and stabilization, thereby paving the way to secondary colonization and succession by other microbiota, plants and metazoans. Advancement in molecular techniques and their increased use for the study of microbes dwelling in Antarctic region have revealed the presence of a very rich microbial diversity, including viral diversity [3]. Numerous microbes have successfully colonized the harsh environment of Antarctica, and the richness of these microorganisms is now well known. The dynamics of microbial communities and their function in the Antarctic at present and in the upcoming future have been revealed by recent metagenomic data. Sadaiappan et al. [4] conducted Qiime analysis and revealed that the reads were signified by the presence of archaea and bacteria. The reads altogether represented microbes belong to 412 genera, 86 classes and 38 phyla. With 96.8% of the total readings, bacteria were found to have a larger diversity than archaea, which had only 1.7%. A total of 33 phyla of bacteria with the dominance of proteobacteria (87%) were reported, which was found to be followed by phylum Bacteroidetes (4.2%) and Firmicutes (1.70%). Major types of the proteobacteria reported were Gammaproteobacteria (58%), Alphaproteobacteria (29%), and Deltaproteobacteria (0.4%). Acidobacteria, Actinobacteria, Bacteroidetes, Chloroflexi, Cyanobacteria,

Deinococcus-Thermus, Firmicutes, Nitrospirae, Planctomycetes, Proteobacteria, and Verrucomycetes made up the maximum variety of the bacteria and 1.5% of the total reads remained unclassified. Archaeal diversity was composed of five major phyla, namely, Crenarchaeota, Diapherotrites, Euryarchaeota, Hadesarchaeaota and Thaumarchaeota. These are typically psychrophilic or psychrotolerant bacteria, which have been evolved a variety of adaptations to withstand the harsh effects of such cold surroundings [5]. These bacteria use a variety of techniques to thrive in such extreme environments, including maintenance of membrane fluidity, producing antifreeze proteins, cryoprotectants, cold-shock proteins, cold active enzymes, etc. Enzymes produced by these microorganisms are being increasingly reported as essential components responsible for their adaptation. The enzymes produced by these psychrophilic organisms are constantly in the spotlight owing to their significance in both basic and applied research. The properties of cold-adapted enzymes offer a wide range of potentially beneficial biotechnological applications in fields as wide as the detergent, textile and food industries, medical and pharmaceutical preparations, bioremediation, etc. The search for cold-adapted enzymes with potential utility in biotechnological sectors is consequently receiving more attention. The current manuscript discusses the diversity of bacteria found in the Antarctic, their coping mechanisms and potential ecological and industrial applications.

15.2 Challenges for Life in Antarctica

The environmental and geographical conditions of various locations impart various direct and indirect stresses and thereby play an important role in evolution of species as well as their distribution. The Earth harbours almost around 10 million species which are the result of long-term evolution in various ecosystems from high mountains to deep oceans and from tropics to the arctic tundra. Antarctica is covered by ice sheet that can be up to 3 km thick and surrounded by an annual ice pack, and surprisingly, these ice sheets represent 97% of the Earth's ice and 70% of its freshwater. Life in the largest freezer of Earth, i.e. Antarctica, is subjected to dangerous environmental stresses, which is more than any other desert or terrestrial ecosystem. Various chemical and physical gradients along with harsh climatic conditions such as very low temperature, humidity, low availability of water, high salinity, low carbon and nutrient concentration, inexhaustible freeze-thaw cycle, low annual precipitation, strong winds causing abrasions, high sublimation, seasonal day length variations, evaporation, limited nutrition and high exposure to ultraviolet radiations are few to be named [6]. The combination and interactions of these environmental stresses act as limiting factors for the survival of living organisms thus determining the taxonomical diversity of the continent. The soil of this continent is usually considered biologically depauperate or very simple in this term. The diversity of photosynthetic autotrophs is only covered by lower plants such as algae, mosses and lichens. Due to low diversity of flora, the terrestrial faunal diversity is also low due to lack of sufficient food and mostly consists of simple communities of invertebrates, nematodes, springtails, mites, tardigrades and rotifers [7].

The food web of the Antarctic continent is thought to be simplest on the planet, yet we lack fundamental understanding between the relationship of abiotic factors and biological diversity. Various studies have revealed that the chemistry of the soil is one of the most primary drivers for the establishment of soil biota than any other environmental factor. Magalhaes et al. [7] showed that the soil with lower conductivity and high C/N ratio were found to have higher diversity of microbial biota. The study also indicated that as the pH of the soil increases, the abundance of the cyanobacteria also increases, and this correlation is also relevant in alpine environment. The microbiota of soil also plays an important role in recycling of organic and inorganic nutrients and decomposition. Low level of microbiota insures flow of nutrients in low diversity of food web and vice versa. Similar correlation of soil chemistry is observed with diversity of invertebrates, lichens and algae, i.e. low salt concentration, higher pH and high C/N ratio [8]. The studies have shown that soil moisture ranging from 0.7 to 11.3% acted as an important determinant for the presence of species in the Antarctic continent [7]. The other studies have indicated that composition of organic matter, soil salinity and availability of liquid water also acts as strong limiting factors for biological colonization. Magalhaes et al. [7] also found that both the bacteria and cyanobacteria respond to physical and chemical parameters in almost similar ways, but the abundance and diversity of bacteria were found to be greater than cyanobacteria which could imply that bacteria possess high tolerance levels to the environmental factors than the cyanobacteria. It is often observed that primary colonizer of ice-free areas are bacteria, algae and cyanobacteria which bind the soil before the development of higher plants such as mosses and lichens on the substratum. Convey et al. [9] hypothesized that the primary colonizers including mosses and lichens are subjected to environmental stress to much greater extent than higher colonizer as they are unprotected from greater water retention and buffering of temperature and soil pH. The studies have also revealed that the presence of invertebrates, lichens and algae were found to be more abundant in the areas where the diversity of bacteria and cyanobacteria was more, indicating the importance of well-developed microbial biota for the development of complex multi-tropical community.

Terrestrial ecosystem of Antarctica covers a little proportion (0.32%) of the whole continent's area. Except high cliffs and exposed ridges of mountains, most of the habitats also suffer extended seasonal snowfall or ice cover which helps in protection of biotic components from extremely low temperatures and windy abrasions. But the same ice cover unavoidably limits biological activity for a long period of time. The little possibility of biological activity under ice is also limited to seasonal duration [9]. Some researchers have found that even in the ice-free areas, the diversity and abundance of living organisms were irregular. This makes it unclear to intemperate that these irregularities are due to edaphic factors, topology and vegetation or due to microclimatic conditions. Studies conducted in the ice-free areas showed seasonal variation in the flora and fauna of the region with respect to abiotic factors. For example, in summers, the moisture content of the ice-free area increases due to the increase in the temperature, the abundance and diversity of the microbes, and photosynthetic autotrophs, and microfaunal species increases drastically. Similarly, the coastal areas also experience higher temperature and thus higher water availability which results in more suitable habitats and hence higher production of biomass.

The presence of human beings in Antarctica is very low due to harsh conditions, but still this continent has not escaped the negative effects of anthropogenic activities worldwide. The global climatic changes have also shown deleterious effects on the physical, chemical and biological components of this continent. The most obvious and known effect of anthropogenic activities on Antarctica is the creation of ozone hole which had serious consequences to biological systems including humans. However in case of marine biota, the negative effects of the global warming are more prominent. The marine life of Antarctica usually thrives at temperature range of -1.9 °C to -0.5 °C. Various studies have been conducted on various marine animals of Antarctica in order to understand the effect of elevated temperature on them. The most obvious result of these studies was that most of the Antarctic marine species are very poor in surviving elevated temperatures. As the temperature increases, the levels of available oxygen decrease, and concentration of oxygen increases in water which creates a problem for numerous species due to increased cell damage from reactive oxygen species. The other negative impact of increased temperature is reduced oxygen availability and increased metabolic rates which would in turn affect the energy production. Apart from global activities affecting various aspects of the continent, local activities such as fishing, shipping traffic, tourism, pollutants released by research stations of various countries like the USA, Argentina, China, Russia, etc. are also the cause of great concern [10]. The contaminants released by different anthropogenic activities include leachate from historic disposal sites, petrochemicals from fuel spillage, organic enrichments, disposed chemicals through sewage systems and other harmful gases like SO_2, NO_2, CO, CO_2, etc. During the early periods of fishing, almost all the types of finfishes and their predators were fished causing a sharp decline in their number over a period of time. Shipping traffic is another major problem which is used for many purposes. These ships not only contribute to air and water pollution by secreting poisonous gases, sewage waste and polluted liquids but also create noise pollution which creates disturbance for the marine life.

Tourism in Antarctica has extended for its exclusive "expedition cruises", and it is expected to increase even more in the near future. This huge number could be a potential threat to the wildlife as the increasing pollution, risks of invasive species and disruption of scientific research activities. This ever-growing tourism with the landing of the tourists greatly disturbs the fragile polar system and has great negative potential towards both terrestrial and marine habitats. Major source of marine pollution is sewage outfalls, abandoned dump sites accidental oil spills and exhaust emissions. These contaminants are persistent as they continue to leach for a very long durations into the marine environment majorly effecting benthic communities. Apart from these contaminants, Padeiro et al. [11] reported high levels of heavy metals such as Zn, Pb, Cd, Cr and Ni in the soil samples of Antarctica due to human activities. Ogaki et al. [12] also reported the effect of heavy metal on the fungal diversity of different lakes of Antarctica. They observed that lakes which were in close proximity to human had high sedimentation levels of heavy metals and reduced diversity and richness of fungi. In contrast, low levels of sedimented heavy metals and high diversity and richness of the same were reported in the lakes which were far away from human impact.

15.3 Diversity of Microorganisms Found in Antarctic

As mentioned in the above section, abiotic factors are the major determining factor of biotic diversity in an ecosystem. The Antarctic region has vast variation in the abiotic factors; hence, a huge diversified microbial community has been developed, which has been reported by various scientists. Some of them are discussed below:

(a) **Fungal Diversity**

In Antarctica, microorganism-dominated food chains are mainly present in different pristine ecosystems. Concerning to fungal diversity, two basic forms, i.e. yeast and filamentous fungal forms of fungi, are mainly found in Antarctica, which represent colonies with different colours and morphologies [13]. These colonies display a very high level of genetic plasticity, which in turn allows them to survive under very unfavourable conditions of low temperatures, freeze-thaw cycles, variable pH levels, high UV irradiation, strong winds, dehydration, minimal nutrient concentration and osmotic stress conditions. The Antarctic fungal community includes taxa belonging to the major fungal groups like Ascomycota, Zygomycota, Basidiomycota, Glomeromycota and Zygomycota. In addition to these taxa, some assemblages also include Mycetozoa (slime moulds) and stramenopiles (oomycota). But, according to some recent taxonomic studies considering phylogenetic analysis as well as characterization of uncultivable taxa, this traditional taxonomic hierarchy is changing [14]. Tedersoo et al. [14] reported colonization of fungi belonging to 18 different phyla, i.e. *Ascomycota*, *Basidiobolomycota*, *Aphelidiomycota*, *Basidiomycota*, *Chytridiomycota*, *Calcarisporiellomycota*, *Entorrhizomycota*, *Entomophthoromycota*, *Glomeromycota*, *Muccoromycota*, *Kickxellomycota*, *Mortierellomycota*, *Monoblepharomycota*, *Neocallimastigomycota*, *Zoopagomycotina*, *Olpidiomycota* and *Rozaellomycota*. Bridge and Spooner [15] documented about the 1000 fungal species (without lichen forms) of Antarctic region and also concluded that true diversity of Antarctic fungi is far more than the recorded estimate.

Several recent studies have concluded that there is a vast difference between the fungal diversity of Antarctic peninsula and continental Antarctica environments because the harsh condition of Antarctic peninsula is milder than that of continental regions as well as the Antarctic peninsula has more life in the form of macroalgae, plants, vertebrates, invertebrates etc., which provides nutrients and organic matter to survive the fungi and thus creating various ecological niche and microenvironments for the fungal food web. In contrast, in Continental Antarctica, nutrients and organic matter are very limited due to the absence of life forms. Fungi in the region of continental Antarctica are majorly present in lichen symbiosis (symbiotically associated with lichens), while the presence of free living fungi in soils is poorly understood [13]. On the other hand, cosmopolitan psychrotolerant fungi are ectotypes with mesophilic psychrotolerant nature. Species like *Penicillium antarcticum*, *Penicillium chrysogenum*, *Antarctomyces psychrotrophicus*, *Antarctomyces pellizariae*, *Thelebolus* spp., *Metschnikowia australis*, *Mortierella antarctica*, etc. are considered as endemic psychrophilic species, and different species of *Aspergillus*, *Penicillium*, *Cladosporium*, *Colletotrichum* and *Rhodotorula* are known as cosmopolitan cold-tolerant taxa [16].

(b) **Bacterial, Cyanobacterial and Archaeal Diversity**

Antarctic, which is also known as driest, coldest continent, is separated from other continents via the Antarctic circumpolar current and the southern ocean. Massive ice sheets present here reflect 40–90% of incident solar radiation and cause a mass of cold dense air to accumulate on the polar plateau. In these very harsh conditions, only cold-adapted microorganisms like microorganisms, tardigrades, mites, tundra vegetation, penguins, seals and several types of algae can survive. In spite of these extreme conditions, prokaryotes and archaea dominate in the most Antarctic ecosystem and thus play major role in biogeochemical cycles, food webs and mineralization of pollutants. A major phylum of gram-negative bacteria, namely, Proteobacteria, is frequently found in Antarctic soil. Several other studies which are mainly focused on microbial diversity in Antarctic soil suggested Actinobacteria and Proteobacteria as the major phyla; on the other hand, the phyla Cyanobacteria and Firmicutes are common but less frequent [17].

Various new approaches like multi-omics approach reveal the several communities of prokaryotes present with in soil at Edmondson point (ice-free region on eastern slope at the base of Mount Melbourne, Victoria Land, Antarctica). The above region consists of Actinobacteria, Proteobacteria, Acidobacteria, Planctomycetes, Verrucomicrobia, Bacteroidetes and Chloroflexi. Bacteroidetes, Proteobacteria, Actinobacteria and Firmicutes are majorly reported through cultivation-based methods [18]. Some advanced analysis as well as 16S rRNA sequencing also reveals that the anaerobic, spore-forming Firmicutes are the most abundant group present in the rhizosphere of the Antarctic vascular plants [19]. During the summer season, snow melts and Antarctic soil becomes free and receives nutrients from the ocean as well as animal life (i.e. penguin guano), which enrich the soil with phosphorus and nitrogen, hence create an environment that supports abundant microbial growth. O'Brien et al. [20] reported that the ornithogenic soil samples of Antarctica support high bacterial yield counts (6–9 log CFU g^{-1}) with respect to other samples. Some cold-adapted *Burkholderia* species were also discovered in the coastal region of Ross sea in Antarctica.

With the help of applied metatranscriptomics, metaproteomics and metagenomics to permafrost, thermokarst soil, it can be concluded that Actinobacterial lineages are the most active as well as numerically dominated members of the prokaryotic community in the seasonally thawed soil [21]. Actinobacteria also showed the flexible nature towards both short- and long-term changes towards environmental conditions. On the other hand, Acidobacteria has been found very common to a vast range of Antarctic and Arctic soil biotopes. Despite the fact that Actinobacteria, Acidobacteria and Proteobacteria are numerically most abundant, cyanobacteria also plays an impressive role in terms of significant colonist in the cold soil. Cyanobacteria like *Nostoc commune* are majorly present in both Antarctic and Arctic soil and hence drive several important functional processes related to the nitrogen and carbon cycling. Increasing soil temperature due to global warming may extend microbial growth period. New meta-analysis reveals that the warming of cold soil directly increases the abundance of microbes, which indirectly impact the stored carbon. Thus, there is an urgent need of more extensive datasets on the effect of different climatic conditions on microbial community's functional processes and composition.

15.4 Survival Strategy of Antarctica Bacteria

Microbes residing in the Antarctica region face numerous challenges. Among them, cold temperature conditions are the most and consistent stress. In order to grow in such environmental conditions, microbes have developed some survival strategies. Out of which some are discussed below:

15.4.1 Production of Cryoprotectant and Osmolytes

The microbes in Antarctica are exposed to cold environment, which accelerates the osmotic damage and dehydration resulting in disruption of the functioning of cells and their survival by deactivating the enzymes and causing several negative effects. But to avert this cold aggregation of protein, there are many psychrophiles that are reported to secrete various ice-nucleating proteins, extracellular polymeric substances (EPS), compatible solutes and bio-surfactants. All these exopolymeric substances or osmolytes help to maintain optimum membrane fluidity and prevent cell shrinkage by simply accumulating within the cell and suppressing the solutes to freeze inside the cell membrane that protects the cell from cryo-injuries [22]. During exposure of the cells to cryoprotectants, ice disrupts the cell membranes mechanically, which is protected by secreting extracellular polymeric substances (EPS) within the cell; similarly, a sea ice bacterium, namely, *Colwellia psychrerythraea*, secretes EPS that protect the cell membrane. Moreover, the severe damage caused by ice crystals is protected through biofilms that are formed by psychrophilic microbial cells. Not only this, biofilms are also necessary to acquire the nutrients within the channels. On growing *Mesorhizobium* sp. strain N33 at 4 °C, it was reported that threonine, valine and sarcosine were accumulated in a higher amount which is considered as a cryoprotectant for microbes [23]. Certain type of growth-promoting substance such as glycine, and betaine, also known as compatible solutes has the ability to protect bacteria at low temperature. These compatible solutes avoid the cold-induced cellular protein aggregation. Moreover, these osmolytes also influence the fluidity of the bacterial membrane adapted to the cold stress.

15.4.2 Maintain Membrane Fluidity

To overcome the adverse effects of harsh environments, modulation of the fluidity of the membrane and providing interface between the external and internal environment are a strategy adopted by microbes in the freezing environment. Activation of the membrane-associated sensor, upregulating membrane fluidity and increase in membrane rigidity are the major aspects of maintaining the membrane fluidity. At low temperature, the function and physical properties of membranes are affected, leading to decrease in the membrane fluidity, the beginning of gel-phase transitions and the loss of cell function. In accordance to the function of bacterial membrane, some significant adaptations are present in bacterial membrane including the degree of unsaturation, branched fatty acids chain and chain length. Cell membrane associated with the modification of lipid fatty acyl chains helps to maintain optimum membrane fluidity. Increase in the methyl branched fatty acid helps cold microbial cells to survive harsh environmental conditions. The presence of desaturase enzyme in cold-adapted bacteria converts saturated acyl fatty acids to unsaturated acyl chains by removing two hydrogen atoms and helps in cold-adapted bacterial survival. Apart from this, steric constraints reduce number of interaction in membrane as well as changing packing order which playing important role in membrane fluidity of psychrophiles. Increase in the degree of unsaturated fatty acid in *Leucosporidium*, *Candida* and *Torulopsis* has been reported by Nagy and Kerekes [24]. Additionally, the reported study shows that the level of wax ester synthase increases when *Psychrobacter arcticus* exposed to cold temperatures [25].

15.4.3 Antifreezing Protein

For cold environment, psychrophiles have antifreezing proteins (AFPs) as a survival strategy. By synthesizing such unique proteins, they prevent the growth and recrystallization of ice. Bacteria reported to have AFP activity are *Psychrobacter* sp., *Rhodococcus* sp., *Stenotrophomonas maltophilia*, *P. fluorescens*, *Marinomonas protea* and *Enterobacter agglomerans* [26]. Thermal hysteresis (TH) is the difference in freezing and melting point which results in the adsorption of AFP on crystal surface of the ice, which results in the ice growth on convex surface between the adjoining AFPs thereby decreases in freezing point. Isolated antifreeze protein (AFN) helps maintaining the frozen food cell structure, showing its high value for frozen food industry isolated from the Antarctic bacterial culture GU3.1.1 [27].

15.4.4 Alterations in DNA Replication

Temperature-induced conformational or physicochemical changes in proteins, RNA and DNA could form basis for sensing temperature. Compaction might be due to one or more reasons like (1) temperature-induced conformational change in the DNA, (2) alteration in the activity and/or amounts of the DNA proteins like DNA gyrase, histone-like proteins including H-NS and topoisomerase I and/or impaired synthesis of the proteins that influence condensation of nucleoid [28]. DNA conformation is essential for cold adaptation, and the expression of many genes is dependent on the DNA conformation that in turn depends on the temperature-dependent changes in DNA supercoiling. Demonstration of link between expression of DNA supercoiling and cold-inducible genes inhibits the negative supercoiling leading to inhibition of the cold-inducible gene expression. The nucleoid-associated proteins like H-NS and topoisomerase II and I regulate supercoiling in bacteria at low temperature and thereby control cold-inducible gene [29]. Under cold conditions, expression of such genes induces and the product takes part in the adaptation.

15.4.5 Antioxidant Production

Bacteria in Antarctica have to resist not only to cold temperatures but also other stress conditions like high UV radiation. These bacteria have developed different antioxidant systems to survive in this extreme environment by avoiding the oxidative stress caused by reactive oxygen species (ROS). Hydrogen peroxide (H_2O_2), anion superoxide (O^{2-}) and hydroxyl radicals (OH) are the most common species of ROS. Oxidative damage occurs in the microbial cell despite adverse conditions, but some psychrophilic bacteria have the ability to counter the situation and reduce oxidative stress through certain defence mechanisms [30]. Glutathione, vitamins and some pigments like carotenoids are the examples of nonenzymatic antioxidant defence, which are reported to present in high amounts in psychrophiles. Carotenoids are pigment antioxidants that neutralize free radicals through various reactions in the microbial cell. For example, the interaction of carotenoids with singlet oxygen ($^1O_2$) occurs via transferring excitation energy to the carotenoid or by chemical quenching of $^1O_2$. Carotenoids pigment can interact with oxygen radicals in three main ways: electron transfer, hydrogen abstraction and the addition of radical species [31]. Apart from this, various psychrophiles are reported to produce various antioxidant enzymes, which can neutralize the free oxygen radicals and thereby protect microbes from oxidative damage.

15.5 Ecological Significance of Microbes

The microbes are the basis of nutrient cycling and community development at any naive regions. The microbes that reside in the Antarctica region are also reported to play crucial roles in ecosystem.

15.5.1 Biogeochemical Cycling

Microorganisms are the major drivers of biogeochemical cycles on Earth. They have a significant impact on carbon (C) as well as breakdown of organic matter, nitrogen (N) efflux and phosphorus (P) mobilization in the environment [32, 33]. These processes may result in CO_2 elevation, greenhouse gas release, nutrient loading and water consumption [34, 35]. The Earth's biosphere is dominated by cold environments, and the cold biosphere is dominated by microbes. Microorganisms in cold Southern Ocean waters are recognized for having key roles in global biogeochemical cycles.

(i) **Carbon Cycle**

The carbon cycle involves several microbial processes such as CO_2 fixation, organic compound decomposition, mineralization to CO_2, oxidation of methane (methanotrophy) and methane production (methanogenesis). The latter is solely anaerobic process, while the other activities may occur under both aerobic and anaerobic conditions. In the cryosphere, microorganisms are believed to carry out biogeochemical cycles within soil snow, lakes, hills and both in supraglacial and subglacial environments [36, 37]. Supraglacial microbial communities (within the snowpack or within 0.1–3 mm dark granular aggregates, cryoconite debris) are known to cycle carbon through photosynthesis and respiration pathways. Antarctic cryoconite holes (CHs) are water-filled depressions upon the glacier surface which contain coating of cryoconite debris that constitute important features on glaciers and ice sheets. Once hydrologically connected, these microbially dominated mini-ecosystems supply nutrients and biota for downstream environments. Within cryoconite holes, and especially in cryoconite debris, diverse communities of eubacteria, archaebacteria and eukaryotes exist. These microbial communities play a crucial role in carbon cycling such as primary producers like cyanobacteria and non-cyanobacterial group which fix inorganic carbon via photosynthesis and provide nutrients for the heterotrophic fraction of the community. Previous morphological studies of the Antarctic region suggested the presence of several different species of cyanobacteria, including *Hormathonema* spp., *Gloeocapsa* spp., *Anabaena* spp., *Aphanocapsa* spp., *Lyngbya* spp., etc. [38]. Heterotrophic community was dominated by relatively few taxa of *Gammaproteobacteria*, *Sphingobacteria–Flavobacteria* and *Alphaproteobacteria* reported by Straza et al. [39] who demonstrated substrate utilization by these taxa. Several mechanisms of carbon dioxide fixation are known of which Calvin-Benson-Bassham (CBB) cycle is the most important autotrophic pathway. The first rate-limiting step is catalysed by the ribulose1,5-biphosphate carboxylase/oxygenase (RubisCO) enzyme. Four different natural types of RuBisCO are known: type I which is encoded by *cbbL* genes which have been found within green-like autotrophic bacterial groups, green algae, cyanobacteria and representatives of some Alpha-, Beta- and Gammaproteobacteria and red-like autotrophic bacterial groups, including non-green algae and representatives of some Alpha- and Betaproteobacteria and autotrophic eukaryotes. Type II (*cbbM* gene) is found in purple non-sulphur bacteria, aerobic and facultative anaerobic chemoautotrophic bacteria and dinoflagellates. RuBisCO type III has only been found in Archaea. Finally, type IV RuBisCO is designated as RuBisCO-like and is considered not to be involved in the Calvin cycle [40].

However, complexity of genes and processes involved in the carbon cycling makes it difficult to understand how these complex processes could be affected by global change. Another carbon cycle is methane cycle which is probably one of the most studied carbon-related processes and has been shown to occur widely in Antarctic soils. Production of methane from organic matter is a multistep process, in which methanogenesis, carried out by microorganisms from the Archaea domain, is the final step. Oxidation of methane is a bacterial process, while carbon dioxide fixation can be carried out by chemolithotrophic bacteria and by oxygenic and anoxygenic photosynthetic organisms. The carbon source/sink ratio is, in part, determined by the equilibrium between methylotrophy and methanogenesis. According to in situ measurements, polar soils are a net source of CO_2 emissions but a sink for methane [41]. However, there is evidence that increased soil temperatures (thaw) result in increased methanogen diversity of both active layer and permafrost soils as well as considerable increase in methane production [41]. According to Tveit et al. [42], methane productivity changes were linked to a shift from formate- and H_2-using Methanobacteriales to Methanomicrobiales and from the acetotrophic Methanosarcinaceae to Methanosaetaceae. Very little is known concerning the distribution and abundance of methanogens in Antarctic soils. However, it is reasonable to predict that a warming climate may lead to more anaerobic soil conditions, which could ultimately result in these soils becoming net methane producers.

(ii) **Nitrogen Cycle**

Nitrogen is generally used for proteins and nucleic acid synthesis. The primary sources of bio-available nitrogen (nitrate, nitrite, ammonia and organic nitrogen) are internal remineralization; external sources like snowmelt, aerial deposition and complex organic material decomposition; as well as microbial-mediated atmospheric nitrogen fixation (by cyanobacteria or some microbial groups associated with plant roots) [43]. Additionally, certain types of sedimentary and metasedimentary bedrocks may have ecologically important amount of nitrogen, which if liberated could impact biological nitrogen cycling in soils. Most Antarctic soils are severely oligotrophic and are particularly low in organic nitrogen. Ice-free continental Antarctica is completely deprived from higher plants; thus, much of the soil nitrogen cycling is thought to be driven by microbial communities such as cyanobacteria including *Leptolyngbya*, *Phormidium*, *Oscillatoria*, *Nostoc*, *Calothrix*, *Dichothrix*, *Nodularia* and *Hydrocoryne* [44]. These are widely considered to be the central regulators of nitrogen cycling in soils and play crucial role as "ecosystem engineers" [44]. Martínez-Pérez et al. [45] demonstrated the role of small unicellular diazotrophic symbiont; UCYN-A with algal partner potentially contributes significantly to N_2 fixation from Artic to Antarctic circle. UCYN-A lacks key genes for CO_2 fixation and therefore lives in association with alga from which it receives fixed carbon. In return, UCYN-A transfers up to 95% of its recently fixed N to the algal partner [45]. Similarly, many fungi including yeasts are commonly found in Antarctic habitats such as *Rhodotorula muscorum*, *Rhodotorula mucilaginosa*, *Cryptococcus aerius* and *Cryptococcus albidus*; by producing enzymes such as urease and protease, they play an essential role in nitrogen mineralization or ammonification. In addition, nitrate and nitrite can be also used by many fungi, including yeasts, in dissimilatory pathways such as denitrification. In a study, a bacterial isolate, i.e. *Candida* sp. and a mesophilic isolate of *Trichosporon cutaneum*, is found in Antarctic soils governing the process of denitrification [46]. The genes of nitrogen cycling have been extensively studied in Antarctic soils, primarily by targeting the nitrogenase (*nif*H) gene. These soils have a high prevalence of diazotrophy, which is mostly but not entirely associated with cyanobacterial lineages. Genes implicated in nitrite oxidation and ammonia oxidation, the *nxr*A and *amo*A genes, respectively, have also been reported in Antarctic soil metagenomes [47]. Denitrification, a process that generates the "greenhouse" gas N_2O and for which the *nar*G gene is the genetic marker, is mostly linked to Actinobacteria and Proteobacteria in Antarctic soils. However, PCR-dependent and metagenomic gene surveys have suggested that the nitrogen cycle is severely truncated in these soils, with key enzymes implicated in some crucial steps (such as dissimilatory nitrate reductase and nitrous oxide reductases) either present at very low abundance or undetectable.

(iii) **Phosphorus Cycle**

Phosphorus (P) is a key limiting nutrient for organisms, and its biogeochemical cycling is believed to regulate primary productivity and ecosystem structure. Phosphorus is derived primarily from the weathering of apatite in soils following exposure of the parent substrate [48–50]. Unlike soil carbon (C) and nitrogen (N), which reside primarily in organic pools, soil P is typically dominated by inorganic pools consisting of unweathered and largely biologically unavailable material, and also its biogeochemical cycle differs from nitrogen and carbon cycle in that it does not include a gas phase. The release of bioavailable P from the lithosphere into the soil ecosystem is governed by a variety of physical, chemical and biological weathering processes. Rain and weathering cause rocks to release phosphate ions and other minerals, which initiate the phosphorous cycle [51, 52]. Inorganic phosphate is then distributed in soils and water where it is taken up by living organisms to form biomass. The organic phosphate returns to the soil when organisms decompose and die. Within the soil, organic forms of phosphate can be hydrolysed releasing phosphate to the environment in a process known as mineralization [53]. Microbial phosphatases play an important role in this process. In soils, organic phosphorus is mainly found as phytates (inositol hexa- and penta-phosphates) from which phosphate is released by the action of specific phosphatases called phytases, which remains in the surrounding in the available form [54, 55]. Several phytase-producing bacterial and yeast strain have been reported from Antarctic region including *Papiliotrema laurentii*, *Rhodotorula mucilaginosa*, *Cryptococcus laurentii* and *Pseudomonas* sp. [56].

15.5.2 Role as PGPR

Antarctic plants have developed a set of survival mechanisms including adjustment in cellular and physiological molecular responses such as modification in the membrane lipid composition and the production of antioxidants, osmoprotectants and cold-shock proteins, among others to grow and survive in this hostile environment. Thus, the ability of plants to adapt to adverse environmental conditions defines their long-term survival and geographical and environmental distribution. In addition to these biochemical and physiological changes, external factors also aid in the adaptation and dissemination of plants in Antarctica. Sometimes, such adaptations are the result of interactions between roots and soil microorganisms. For example, microorganisms produce molecules that actively cooperate in the establishment and development of plants known as plant growth-promoting microbes (PGPM) [57, 58]. PGPMs have also adapted to live and perform all their functions in extreme conditions, and they also contribute to enhance the Antarctic plant biomass [59]. Most of the microbial species found in the rhizosphere are organotrophic. Among positive effects performed by PGPM, a few microorganisms produce chemicals involved in the acquisition of nutrients (e.g. the acquisition of iron by bacterial siderophores or phosphorus by the secretion of organic acids from mycorrhizal fungi) and/or the development of roots (e.g. by the production of phytohormones) [60, 61]. Fardella et al. [62] demonstrated that fungal endophytes isolated from Antarctic plants improved the survival and water use efficiency of several tree and shrub species. Similarly, Berrios et al. [63] reported that an Antarctic strain of *Pseudomonas* sp. imparts a beneficial effect on plant *Deschampsia antarctica* growth, probably because of their contribution of hormones and nutrients (P) supplementation. Moreover, various studies have reported beneficial effect of PGPM including *Pseudomonas tolaasii* (resistance to antibiotics and heavy metals), *P. trivialis* (tolerance to temperature) and *Arthrobacter* sp. (antimicrobial properties) associated with Antarctic vascular plants *Deschampsia antarctica* [64, 65]. Thus, certain strains of rhizobacteria from Antarctica may be useful tools in promoting abiotic stress tolerance and productivity in important crop species by altering root systems.

15.5.3 Role in Bioremediation

Bioremediation is defined as a method of using microbes, plants or microbial or plant enzymes to detoxify contaminants within soil and other environments. The concept also includes biodegradation, which is defined as a partial, and occasionally complete transformation by microbes and plants [66, 67]. For at least 40 years, bioremediation has been considered as a way to promote recovery of contaminated habitats at both higher and lower temperatures. In comparison to other methods, microbial biodegradation is the most effective and economically viable method that also poses the least risk to the environment. *Rhodococcus* is one of the bacterial strains that has been isolated from the Antarctic continent and has received the most attention for its substantial role in soil ecosystems and outstanding metabolic capability. This bacterial group was able to break down aromatic molecules as well as alkanes with chain lengths ranging from C6 to C20, which are referred to as persistent fractions in Antarctic soils [68]. Another significant hydrocarbon degrader in Antarctic soil has been identified as the *Acinetobacter* genus. In a microcosm experiment, the strain *Acinetobacter* B-2-2 was combined with a *Rhodococcus* ADH strain and was able to degrade 81.1% of the oil in pristine soil that had been contaminated for the experiment, as opposed to the 75% degradation that the strain *Acinetobacter* B-2-2 achieved when used alone in a prior study [69]. One of the main types of bacteria that degrade hydrocarbons is known as *Pseudomonas*. After 6 days in the presence of 3.5% diesel oil (v/v), the Antarctic Peninsula isolate of *Pseudomonas* sp. J3 displayed impressive cellular development [70]. *Sphingomonas* is another bacterial species isolated from Antarctic environments that has been demonstrated to be able to use hydrocarbons as a special source of carbon [71]. At low temperatures ranging from 1 to 35 °C, the Scott Base-Antarctic strain Ant 17 was able to degrade the aromatic component of numerous different crude oils, although the optimal circumstance was pH 6.4 at 22 °C. Additionally, *Sphingomonas* Ant 17 demonstrated resistance to freeze-thaw cycles and UV radiation, which is highly helpful in Antarctic regions where both conditions are commonly present. Additionally, fungi and yeast from the *Aspergillus*, *Candida*, *Penicillium* and *Rhodotorula* genera have been widely reported as potential candidates in hydrocarbon degradations [72], (Table 15.1).

The Maritime Antarctic is the best area in Antarctica for successful bioremediation. In this region, temperature above 0 °C is typical in summers, as seen at King George Island. Though the Antarctica has a pH range of 6 on the island and 9 on the coast and soils with pH values exceeding 8.8 have been found to have more effective hydrocarbon biodegradation [68], bioremediation degradation is more efficient under aerobic conditions rather than anaerobic conditions, as the major aerobic hydrocarbon breakdown pathways are faster and generate more energy. Approximately 0.3 g of oxygen are required for every gram of oil oxidized [83]. Once the general biodegradation of substances takes place inside the cell, the microbial community—which promote biodegradation via aerobiosis—is also be able to internalize the substrate. The ability of microbial cells to contact and internalize a

Table 15.1 List of hydrocarbon-degrading microorganisms isolated from Antarctica

S. No	Sample isolation site	Microbial strain	Degraded substrate (s)	Reference
Bacteria				
1.	Terra Nova Bay (Ross Sea, Antarctica)	*Rhodococcus* sp. and *Alcaligenes* sp. co-culture	Diesel, n-alkane	Pini et al. [73]
2.	Scott Base –N/A	*Sphingomonas* ant 17	JP-8; m-xylene; 1-methyl naphthalene; 2-methyl naphthalene; phenanthrene; fluorene; heptane; undecane; dodecane	Aislabie et al. [74]
3.	Marambio station	*Rhodococcus* DM1-21	C12-C30, crude oil, gas oil	Ruberto et al. [75]
4.	Signy Island (South Orkney Islands, Antarctica)	*Pseudomonas* sp. ST41	Polar Blend marine gas oil	Stallwood et al. [76]
5.	Scott Base-Ross Island	Rhodococcus 43/2	C12-dodecane, C16-hexadecane, Pristane, JP5 jet fuel	Saul et al. [77]
6.	Great Wall station	*Pseudomonas* LCY16	Naphthalene, phenanthrene	Ma et al. [78]
Fungi				
7.	Macquarie Island	*Arthroderma* sp. 1	Monoaromatic hydrocarbon	Ferrari et al. [79]
8.	Ohridski Base, Livingston Island, Antarctica	*Candida antarctica* T-34	Undecane	Hua et al. [80]
9.	Macquarie Island	*Pseudeurotium bakeri*	Diesel oil	Ferrari et al. [79]
10.	Carlini Station (South Shetlands Islands, Antarctica)	*Pichia caribbica*	n-alkanes and diesel fuel	Martorell et al. [81]
11.	Rothera research station	*Mortierella* sp. HC8D	Dodecane	Hughes et al. [82]
12.	Australian Research station (Macquarie Island, Antarctica)	*Exophiala* sp. and *Pseudeurotium bakeri*	Special Antarctic Blend (SAB)	Ferrari et al. [79]

substrate is known as bioavailability and is critical for the biodegradation process. The cytoplasmic matrix can freeze, and the channels across the cell membrane may seal in conditions below the freezing point, inhibiting cell function.

Low temperatures further increase oil viscosity, and water solubility decreases evaporation thus slowing down biodegradation process. Therefore, the susceptibility of biodegradation of petroleum hydrocarbons decreases from paraffins > branched alkanes > olefins > monocyclic aromatic hydrocarbons (MAHs) > naphthenes > polycyclic aromatic hydrocarbons (PAHs) [84]. Because of the above-mentioned

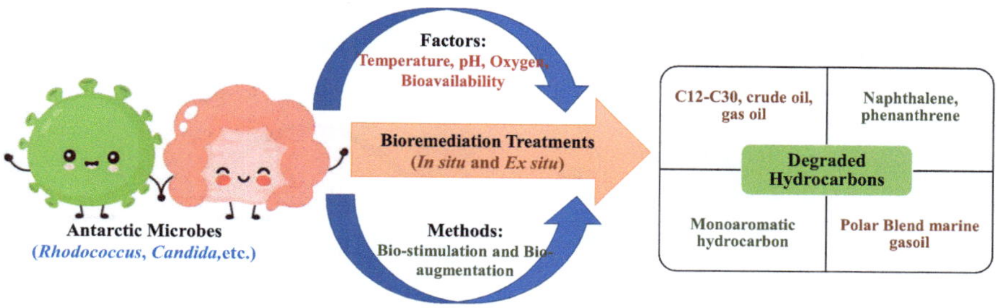

Fig. 15.1 Mechanism of bioremediation through Antarctic microbes

factors, bioremediation treatments are mostly recommended during summers when the soils are not frozen, temperatures are higher and water is readily available. Moreover, by adding nutrients, carbon sources or electron donors to the native bacteria or fungi (biostimulation, biorestoration), or by adding an enriched culture of microorganisms that have particular properties that allow them to degrade the desired contaminant more quickly (bioaugmentation), the process of bioremediation increases the rate of the natural microbial degradation of contaminants (Fig. 15.1). Low temperatures, however, can be advantageous in the case of spills since snow can contain as a containment boom and slows the spilt oil penetration by acting as an absorbent.

15.5.4 Bio-prospecting of Antarctica Microbes

Although there are numerous definitions of "bioprospecting", there isn't a single one that is universally acknowledged. Rogan [85] defines bioprospecting as "a range of activities related with the search for a novel biodiversity, whose constituent parts may be employed in a product or process and developed for commercialization". Antarctica is an important location for bioprospecting. Its microbial diversity is still poorly studied and may contain microbes with very relevant abilities for white (industrial), green (agricultural) and red (pharmaceutical) biotechnologies. Microbes have developed a variety of adaptation strategies to enable growth in the harsh Antarctic environment. These adaptations are frequently accompanied by changes to metabolic and gene regulatory systems. These adaptations include the production of antifreeze proteins and cold-active enzymes, as well as the storage of cryoprotectants such as sugars and polyols in the cell to maintain turgor pressure, high levels of unsaturated membrane phospholipids to stabilise membranes and fungal melanin for protection against freezing and UV radiation. Thus, it turns out that microbial extremophiles are a chemical reservoir of extremolytes and extremozymes that have the potential to be excellent resources for the growth of a bio-based economy [86].

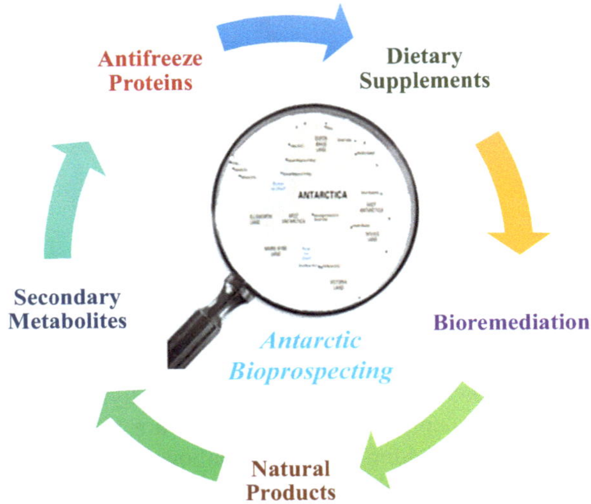

Fig. 15.2 Application of Antarctic microbes in industry

Actinobacteria represent a considerable portion of the microbial community in the majority of soils including the Antarctic region. Additionally, *Actinobacteria* found in rhizosphere soil have been linked to the development of beneficial compounds and antimicrobials [87]. The genus *Streptomyces* has demonstrated potential as a biocontrol agent for fungal diseases in commercial crops. Additionally, grape-derived *Streptomyces* spp. displayed antifungal action towards pathogenic yeast and fungus from the same habitat [88]. While the genus *Arthrobacter* is well known for producing secondary bioactive metabolites and for bioconversions, it has been regularly found in Antarctic and Arctic regions. Rojas et al. [89] searched for novel metabolites produced by Antarctic bacteria found novel molecules linked to cyclic thiazolyl peptides that were active against gram-positive pathogens and produced by *Arthrobacter agilis* from Lake Hoare and Lake Fryxell in the McMurdo Dry Valley region of Antarctica (Fig. 15.2).

15.5.5 Industrial Significance of Antarctic Dwelling Microbes

Microbes that live in Antarctica produce a number of industrially useful enzymes (Table 15.2). For example, the fungus *Cladosporium* sp. isolated from a marine sponge and the *Penicillium* species isolated from various Antarctic marine creatures (such as sea stars, mollusks and macroalgae) are both capable of producing xylanase. For instance, some yeasts from the genera *Cryptococcus*, *Leucosporidium*, *Metschnikowia*, *Candida*, *Yarrowia* and *Saccharomyces* isolated from Antarctic marine samples have been reported to produce microbial lipases, which are significant enzymes used in a variety of applications in the dairy, bakery, oil, meat and fish

Table 15.2 Industrially important cold active enzymes isolated from Antarctic region

Microbe	Enzyme/ active compound	Application	Reference
Lecanicillium muscarium CCFEE-5003; *Glaciozyma antarctica* PI12	Chitinase	Cosmetic, pharmaceutic fields, fermentation research and biomedicine	Teoh et al. [90]; Fenice et al. [91]
G. antarctica	Esterase	Degradation of natural materials and industrial pollutants	Martorell et al. [92]
G. antarctica	Proteases	Food industry	Baeza et al. [93]
Geomyces pannorum, *Bacillus subtilis* N8,	α-Amylases	Food, baking and detergent industries	He et al. [94]
Pyrococcus sp.	Cellulases	Food, ethanol and textile industry	Mao et al. [95]
Erwinia carotovora, *Candida carpophila*,	Phytases	Food and feed industry	Yu et al. [96]
Flavobacterium limicola, *Acinetobacter* sp., *Geomyces pannorum*	Proteases	Textile and leather industries, organic polymer mineralization in freshwater sediments, food and feed industry	Białkowska et al. [97]
Pseudomonas sp. LSK25	Lipase	Detergent, textile and food industries, and bioremediation	Salwoom et al. [98]
Pseudogymnoascus sp. UFMGCB 10054	Carrageenase	Biomedical, textile industry, bioethanol production and detergent additive	Furbino et al. [99]
Hymenobacter sp. UV-11	Photolyase	Formulation of dermatological cosmetics	Marizcurrena et al. [100]
Cladosporium sp.	Xylanases	Biobleaching of paper and pulps	Gil-Duran et al. [101]
Planococcus versutus L10.15	Homoserine-lactonase	Inhibition of phytopathogens	See-Too et al. [102]

processing, and beverage industries, for improving the food quality, as well as for the detergent and cosmetic industry.

These psychrophilic enzymes have been reported to have tremendous applications in industries. Some of the industrial applications of Antarctic living microorganisms have been described below:

Food Industry The use of colour to make the food appear rich and tasty is an important factor in the food industry. So far, most of the colours added to the food are synthetic and therefore unhealthy. Recently, the focus has shifted towards natural colouring agents to avoid the adverse health effects of synthetic colours. A number of microbial pigments such as Arpink red™ from *Penicillium oxalicum*, astaxanthin produced by *Xanthophyllomyces dendrorhous*, lycopene from *Fusarium sporotrichioides* and *Erwinia uredovora*, riboflavin from *Ashbya gossypii* and β-carotene produced by *Blakeslea trispora* enhance the colour of different foods when supplemented [103]. Spray-dried prodigiosin from *S. marcescens* has been reported to be an effective colouring agent in milk, yogurt and carbonated drinks [104]. In addition, cold-active enzymes are a very potent tool for the food business. Foods are increasingly being processed in industrial settings at low temperatures in an effort to preserve energy while avoiding negative impacts on flavour, texture and nutritional value. A potentially significant enzyme in the dairy business is cold-active-galactosidase, which hydrolyses lactose to glucose and galactose at refrigeration temperature. A cold-active-galactosidase from the Antarctic psychrophile *Pseudoalteromonas haloplanktis* was patented by Hoyoux et al. in [105] for its ability to hydrolyse lactose during milk storage at low temperatures. Lipase is used to degum the vegetable oil and to alter the value of oil and fats to obtain more valuable products that are rich in poly unsaturated fatty acids. Lipases are also used for producing coco butter substitutes and human milk fat substitutes through the processes of esterification. Psychrophilic lipases reported from Antarctic region include many *Moraxella* sp., *Halomonas*, *Pseudoalteromonas haloplanktis* and *Psychrobacter* sp. [106].

Textile Industries 1.3 million tonnes of synthetic dyes and chemicals are used in the textile industry, 15% of which leak as effluents after use. Unfortunately, a significant part of these colours bypass traditional wastewater treatment methods and exist as a potentially harmful environmental contaminant with negative effects on human health and the environment. As a result, there is a huge concern about using ecologically friendly colours in place of synthetic dyes in the textile business. Microorganisms can produce environmentally friendly pigments that are deemed suitable for use in the textile industry. The textile industry uses microbial pigments extensively, which could boost their market value. As a result, it is vital to investigate novel microbial sources for pigment synthesis. The cryosphere harbours immense microbial diversity that has the ability to produce pigments useful for the textile industry. The use of cold-adapted cellulases in denim finishing and fabric production could improve the smoothness and softness of tissues, decolorize textile effluents and enable the development of ecologically friendly fibre processing techniques [107].

Medical and Pharmaceutical Applications Studying psychrophilic bacteria as potential new tools for pharmacological and cosmetic applications is gaining popularity. Tomova et al. [64] observed that numerous promising psychrophilic strains were found to be a valuable source of novel active antibacterial chemicals at low temperatures. In recent years, substances produced by the halophilic actinomycete *Nocardioides* sp. Strain A-1 showed potential for use in agriculture for plant protection since it had antimicrobial activity against the bacteria *Xanthomonas oryzae*, which causes bacterial blight disease in rice. Antarticine-NF3, an antifreeze glycoprotein produced by the Antarctic bacterium *Pseudoalteromonas*, has been found to have effective results in scar treatment and is therefore being included in cosmetic regeneration creams [108]. Additionally, cold-adapted enzymes are highly effective in the field of pharmaceuticals. For example, cold-adapted dehalogenases have significant effects on the synthesis of optically pure pharmacological intermediates like halo-alkanoic acids. Cold-adapted lipases that are obtained from *Candida antarctica* can be used for several applications, such as to enhance the quality of beauty products, modifying sugars and their related compounds, synthesizing optically active drug intermediates, production of various cosmetic products, fragrance esters and pharmaceuticals [109].

Detergent and Cleaning Industry Currently, the detergent industry accounts for 30–40% of all enzyme production worldwide and needs enzymes that can function at low temperatures in order to conserve energy. Cold-adapted enzymes such as lipases, proteases and α-amylases are vigorously used in manufacturing detergent, which can improve the efficacy of detergents and also decrease the amount of chemicals used in detergents, thereby protecting the texture and colours of fabrics and reduce wear and tear during washing. Esterases and lipases are vital enzymes since they can catalyse the cleavage of ester bonds and also help in reversing the reactions in organic solvents. The inclusion of lipases in detergents or cleaning solutions can also improve the detergents' stain-removing abilities. Currently, cold-active subtilisins, isolated from Antarctic *Bacillus* species, are being used in the manufacture of cold-active detergents that can help in alkaline stability and cold activity needed for optimal washing results. Proteases, amylases and lipases are a few examples of cold-active enzymes that have shown to be quite effective in cleaning procedures at low temperatures. Wipes or other formulations containing psychrozymes can quickly clean solid objects that cannot be heated for washing. Due to their considerable market share in the enzyme industry, proteases are also shown to be important enzymes in detergents and washing powders. As detergent additions for cold washing, proteases isolated from *Acinetobacter* sp., *Bacillus* sp., *Planococcus* sp., *Pseudomonas aeruginosa* and *Serratia marcescens* can be employed [110].

Biodiesel Production The production of biodiesel at industrial scale is done mostly by using vegetable oils through a process called transesterification. However, the use of vegetable oils for this purpose is an expensive affair and also leads to competition with the food sector due to which potential alternative raw materials for the process are under investigation. A practical alternative to traditional methods of producing biodiesel is microorganisms. The most crucial enzymes employed in the manufacture of biodiesel are lipases. Long-chain triacylglycerols are hydrolysed by lipases, producing free fatty acids, glycerol and mono- and diglycerols. Lipases catalyse the two-step transesterification of vegetable, animal and algal oils in the manufacture of biodiesel. Lipases obtained from Antarctic microbes have peculiar stability characteristics. Lipase obtained from soil bacterium *Janibacter* sp. R02 presents a halophilic, alkaliphilic and thermophilic profile [111]. A cold-tolerant lipase isolated from *Pseudomonas* sp. AMS8 exhibited excellent stability when organic solvents were present. It displayed 92%, 109% and 88% activity in the presence of 25% (v/v) xylene, octane and methanol, respectively [112]. For these applications, isolated *Rhodotorula* species from Antarctica also make great candidates. They have the capacity to accumulate significant amounts (50–70%, w/w) of "single cell oils" (SCOs) with a fatty acid composition characterized by a predominance of palmitic

(16:0), oleic (18:1) and linoleic (18:2) acids, which is suitable for the biodiesel industry [113].

Nanoparticles Production Through bio-mineralization processes, some bacteria are exceptional makers of inorganic nanoparticles from organic compounds. The Antarctic microbial population has been identified as possible biomineralizing candidates, although they are still unexplored. For instance, peroxide-resistant psychrophilic strains of *Pseudomonas* spp. are able to synthesize cadmium sulphide (CdS) when grown at 15 °C in a medium with added H_2O_2 and $CdCl_2$ [114]. *Magnetotactic cocci* belonging to Alphaproteobacteria class and isolated from Antarctica marine sediments have the ability to produce elongated magnetite magnetosomes that can be used to produce iron nanoparticles. Purified magnetosomes have been used in a variety of applications, including contrast agents for magnetic resonance imaging, alternating magnetic field-induced hyperthermia agents for cancer treatments and support for enzyme immobilization. It has been discovered that bacteria from the genera *Pseudomonas*, *Psychrobacter* and *Shewanella* were able to synthesise CdS and CdTe QDs. Biosynthesized QDs post purification exhibit broad spectra of absorption and emission characteristics similar to biogenic Cd nanoparticles [114]. *Brevundimonas*, *Bacillus* and *Rhodococcus*, three recently isolated bacterial strains from a consortium linked to the psychrophilic Antarctic marine ciliate *Euplotes focardii*, were employed by John et al. [115] to synthesise silver nanoparticles (AgNPs).

15.6 Conclusion and Future Prospective

Due to the cold temperature of Antarctica region, their microbial diversity has developed various adaptations for surviving under such extreme cold environment. Nutrient mobilizing property and enzymes producing potential of these microorganisms made glory for them as they can efficiently work under low temperature and therefore have a wide application in agriculture, industry, medical, etc. Besides, they are responsible for the biogeocycling of the nutrients in the environment. So studies could be conducted to determine the role of specific microbial diversity in biogeocycling. Due to their lower richness, Antarctic lakes are excellent model systems to study how microbes affect geochemistry since it is easy to take in a significant amount of the variety and link certain taxa to specific processes. New knowledge on how microbes have adapted to the Antarctic environment under a wide range of chemical and physical circumstances is being gained through the use of molecular approaches. Managers of human activity, policy-makers, scientists and the general public can learn more about the causes, impacts and consequences of human activity, including anthropogenic climate change, by being given critical sentinel indicators.

References

1. Fretwell P, Pritchard HD, Vaughan DG, Bamber JL et al (2013) Bedmap 2: improved ice bed, surface and thickness datasets for Antarctica. Cryosphere 7:375–393
2. Cowan DA, Tow LA (2004) Endangered Antarctic environments. Annu Rev Microbiol 58: 649–690. https://doi.org/10.1146/annurev.micro.57.030502.090811
3. Chong CW, Pearce DA, Convey P (2015) Emerging spatial patterns in Antarctic prokaryotes. Front Microbiol 6:1058. https://doi.org/10.3389/fmicb.2015.01058
4. Sadaiappan B, Kannan S, Palaniappan S, Manikkam R, Ramasamy B, Anilkumar N, Subramanian M (2020) Metagenomic 16S rDNA amplicon data of microbial diversity and its predicted metabolic functions in the Southern Ocean (Antarctic). Data Br 28:104876
5. Cavicchioli R, Charlton T, Ertan H, Mohd Omar S, Siddiqui KS, Williams TJ (2011) Biotechnological uses of enzymes from psychrophiles. J Microbial Biotech 4:449
6. Bottos EM, Laughlin DC, Herbold CW, Lee CK, McDonald IR, Cary SC (2020) Abiotic factors influence patterns of bacterial diversity and community composition in the Dry Valleys of Antarctica. FEMS Microbiol Ecol 96(5):fiaa042
7. Magalhaes C, Stevens MI, Cary SC, Ball BA, Storey BC, Wall DH, Ruprecht U (2012) At limits of life: multidisciplinary insights reveal environmental constraints on biotic diversity in continental Antarctica. PLos One 7(9):e44578. https://doi.org/10.1371/journal.pone.0044578
8. Velasco-Castrillon A, Schultz MB, Colombo F, Gibson JA, Davies KA, Austin AD, Stevens MI (2014) Distribution and diversity of soil microfauna from East Antarctica: assessing the link between biotic and abiotic factors. PLoS One 9(1):e87529
9. Convey P (1997) How are the life history strategies of Antarctic terrestrial invertebrates influenced by extreme environmental conditions? J Therm Biol 22(6):429–440
10. Aronson RB, Thatje S, McClintock JB, Hughes KA (2011) Anthropogenic impacts on marine ecosystems in Antarctica. Ann N Y Acad Sci 1223(1):82–107
11. Padeiro A, Amaro E, Dos Santos MM, Araújo MF, Gomes SS, Leppe M et al (2016) Trace element contamination and availability in the Fildes Peninsula, King George Island, Antarctica. Environ Sci Process Impacts 18(6):648–657
12. Ogaki MB, Coelho LC, Vieira R, Neto AA, Zani CL, Alves TM et al (2020) Cultivable fungi present in deep-sea sediments of Antarctica: taxonomy, diversity, and bioprospecting of bioactive compounds. Extremophiles 24:227–238
13. Rosa Luiz H et al (2019) Fungi in Antarctica: diversity, ecology, effects of climate change, and bioprospection for bioactive compounds. In: Fungi of Antarctica. Springer, Cham, pp 1–17
14. Tedersoo L, Sánchez-Ramírez S, Kõljalg U, Bahram M, Döring M, Schigel D, May T, Ryberg M, Abarenkov K (2018) High-level classification of the fungi and a tool for evolutionary ecological analyses. Fungal Div 90:135–159
15. Bridge PD, Spooner BM (2012) Non-lichenized Antarctic fungi: transient visitors or members of a cryptic ecosystem? Fungal Ecol 5:381–394
16. Ruisi S, Barreca D, Selbmann L, Zucconi L, Onofri S (2007) Fungi in Antarctica. Rev Environ Sci Biotechnol 6:127–141
17. Cary SC, McDonald IR, Barrett JE, Cowan DA (2010) On the rocks: the microbiology of Antarctic Dry Valley soils. Nat Rev Microbiol 8:129–138
18. Núñez-Montero K, Barrientos L (2018) Advances in Antarctic research for antimicrobial discovery: A comprehensive narrative review of bacteria from Antarctic environments as potential sources of novel antibiotic compounds against human pathogens and microorganisms of industrial importance. Antibiotics 7:90. https://doi.org/10.3390/antibiotics7040090
19. Teixeira LCRS, Peixoto RS, Cury JC, Sul WJ, Pellizari VH, Tiedje J, Rosado AS (2010) Bacterial diversity in rhizosphere soil from Antarctic vascular plants of Admiralty Bay, maritime Antarctica. ISME J 4:989
20. O'Brien A, Sharp R, Russell NJ, Roller S (2004) Antarctic bacteria inhibit growth of foodborne microorganisms at low temperatures. FEMS Microbiol Ecol 48:157–167

21. Hultman J, Waldrop M, Mackelprang R et al (2015) Multi-omics of permafrost, active layer and thermokarst bog soil microbiomes. Nature 521:208–212. https://doi.org/10.1038/nature14238
22. Dhaulaniya AS, Balan B, Agrawal PK, Singh DK (2019) Cold survival strategies for bacteria, recent advancement and potential industrial applications. Arch Microbiol 201(1):1–16
23. Ghobakhlou AF, Johnston A, Harris L, Antoun H, Laberge S (2015) Microarray transcriptional profiling of Arctic Mesorhizobium strain N33 at low temperature provides insights into cold adaption strategies. BMC Genomics 16(1):1–14
24. Nagy G, Kerekes R (1981) Influence of cultivation temperature on the TTC-reducing capacity of psychrophilic, psychotropic, and mesophilic pseudomonas strains. ZentralblattfürBakteriologie, Parasitenkunde, Infektionskrankheiten und Hygiene. ZweiteNaturwissenschaftlicheAbteilung: Mikrobiologie der Landwirtschaft, der Technologie und des Umweltschutzes 136(3):185–188
25. Ayala-del-Río HL, Chain PS, Grzymski JJ, Ponder MA, Ivanova N, Bergholz PW et al (2010) The genome sequence of Psychrobacter arcticus 273-4, a psychoactive Siberian permafrost bacterium, reveals mechanisms for adaptation to low-temperature growth. Appl Environ Microbiol 76(7):2304–2312
26. Celik Y, Graham LA, Mok YF, Bar M, Davies PL, Braslavsky I (2010) Superheating of ice crystals in antifreeze protein solutions. Proc Natl Acad Sci 107(12):5423–5428
27. Muñoz PA, Márquez SL, González-Nilo FD, Márquez-Miranda V, Blamey JM (2017) Structure and application of antifreeze proteins from Antarctic bacteria. Microb Cell Factories 16(1):1–13
28. Woldringh CL, Jensen PR, Westerhoff HV (1995) Structure and partitioning of bacterial DNA: determined by a balance of compaction and expansion forces? FEMS Microbiol Lett 131(3):235–242
29. Shivaji S, Prakash JS (2010) How do bacteria sense and respond to low temperature? Arch Microbiol 192(2):85–95
30. Halliwell B (2006) Reactive species and antioxidants. Redox biology is a fundamental theme of aerobic life. Plant Physiol 141(2):312–322
31. Correa-Llantén DN, Amenábar MJ, Blamey JM (2012) Antioxidant capacity of novel pigments from an Antarctic bacterium. J Microbiol 50(3):374–379
32. Khan A, Singh AV, Pareek N, Arya P, Upadhayay VK, Kumar Jugran A, Kumar Mishra P, Goel R (2023a) Credibility assessment of cold adaptive Pseudomonas jesenni MP1 and P. palleroniana N26 on growth, rhizosphere dynamics, nutrient status, and yield of the kidney bean cultivated in Indian Central Himalaya. Front Plant Sci 14:1042053. https://doi.org/10.3389/fpls.2023.1042053
33. Khan A, Panthari D, Sharma RS, Punetha A, Singh AV, Upadhayay VK (2023b) Biofertilizers: a microbial-assisted strategy to improve plant growth and soil health. In: Advanced microbial techniques in agriculture, environment, and health management. Academic Press, pp 97–118
34. Bardgett RD, Van Der Putten WH (2014) Belowground biodiversity and ecosystem functioning. Nature 515(7528):505–511
35. Khan A, Singh AV (2021) Multifarious effect of ACC deaminase and EPS producing *Pseudomonas* sp. and *Serratia marcescens* to augment drought stress tolerance and nutrient status of wheat. World J Microbiol Biotechnol 37(12):1–17
36. Upadhayay VK, Singh AV, Khan A (2022a) Cross talk between zinc-solubilizing bacteria and plants: A short tale of bacterial-assisted zinc biofortification. Frontiers in Soil Science 1(788170):10–3389
37. Khan A, Upadhayay VK, Panwar M, Singh AV (2020a) Soil microbiota: A key bioagent for revitalization of soil health in hilly regions. In: Microbiological advancements for higher Altitude Agro-Ecosystems & Sustainability. Springer, Singapore, pp 183–200

38. de la Torre JR, Goebel BM, Friedmann EI, Pace NR (2003) Microbial diversity of cryptoendolithic communities from the McMurdo Dry Valleys, Antarctica. Appl Environ Microbiol 69(7):3858–3867
39. Straza TR, Cottrell MT, Ducklow HW, Kirchman DL (2009) Geographic and phylogenetic variation in bacterial biovolume as revealed by protein and nucleic acid staining. Appl Environ Microbiol 75(12):4028–4034
40. Badger MR, Bek EJ (2008) Multiple Rubisco forms in proteobacteria: their functional significance in relation to CO2 acquisition by the CBB cycle. J Exp Bot 59(7):1525–1541
41. Allan J, Ronholm J, Mykytczuk NCS, Greer CW, Onstott TC, Whyte LG (2014) Methanogen community composition and rates of methane consumption in Canadian high Arctic permafrost soils. Environ Microbiol Rep 6(2):136–144
42. Tveit AT, Urich T, Frenzel P, Svenning MM (2015) Metabolic and trophic interactions modulate methane production by Arctic peat microbiota in response to warming. Proc Natl Acad Sci 112(19):E2507–E2516
43. Khan A, Joshi M, Singh AV (2020b) Rhizospheric microbial community: ecology, methods, and functions. In: Sharma SK, Singh UB, Sahu PK, Singh HV, Sharma PK (eds) Rhizosphere microbiology. Springer, Singapore, pp 127–148
44. Ortiz M, Bosch J, Coclet C, Johnson J, Lebre P, Salawu-Rotimi A et al (2020) Microbial nitrogen cycling in Antarctic soils. Microorganisms 8(9):1442
45. Martínez-Pérez C, Mohr W, Löscher CR, Dekaezemacker J, Littmann S, Yilmaz P et al (2016) The small unicellular diazotrophic symbiont, UCYN-A, is a key player in the marine nitrogen cycle. Nat Microbiol 1(11):1–7
46. Vishniac HS (2006) Yeast biodiversity in the Antarctic. In: Biodiversity and ecophysiology of yeasts. Springer, Berlin, Heidelberg, pp 419–440
47. Chan Y, Van Nostrand JD, Zhou J, Pointing SB, Farrell RL (2013) Functional ecology of an Antarctic dry valley. Proc Natl Acad Sci 110(22):8990–8995
48. Singh J, Singh AV, Upadhayay VK, Khan A, Chandra R (2022) Prolific contribution of Pseudomonas protegens in Zn biofortification of wheat by modulating multifaceted physiological response under saline and non-saline conditions. World J Microbiol Biotechnol 38(12):1–20
49. Khan A, Singh AV, Upadhayay VK, Ballabh A, Prasad B (2022a) Influence of PGPR on growth and yield of oat (Avena sativa L.) under field conditions. Indian. J Ecol 49(4):1351–1356
50. Upadhayay VK, Singh AV, Khan A, Singh J, Pareek N, Raghav A (2022b) FE-SEM/EDX based zinc mobilization analysis of Burkholderia cepacia and Pantoea rodasii and their functional annotation in crop productivity, soil quality, and zinc biofortification of Paddy. Front Microbiol 13. https://doi.org/10.3389/fmicb.2022.852192
51. Khan A, Singh AV, Kumar R, Kukreti B, Bundela V, Upadhayay VK (2022b) Relative impact of PGPR inoculation on biofortification and yield of wheat under field conditions and their performance assessment through statistical tools. J Pharm Innov 11(8):490–495
52. Kaviya N, Upadhayay VK, Singh J, Khan A, Panwar M, Singh AV (2019) Role of microorganisms in soil genesis and functions. In: Mycorrhizosphere and Pedogenesis. Springer, Singapore, pp 25–52
53. Khan A, Singh J, Upadhayay VK, Singh AV, Shah S (2019) Microbial biofortification: A green technology through plant growth promoting microorganisms. In: Sustainable green technologies for environmental management. Springer, Singapore, pp 255–269
54. Punetha A, Punetha S, Khan A (2022) Soil community composition and ecosystem processes. In: Rukhsana, Alam A (eds) Agriculture, environment and sustainable development. Springer, Cham, pp 217–236
55. Upadhayay VK, Singh J, Khan A, Lohani S, Singh AV (2019) Mycorrhizal mediated micronutrients transportation in food based plants: A biofortification strategy. In: Mycorrhizosphere and Pedogenesis. Springer, Singapore, pp 1–24

56. Duarte AWF, Dos Santos JA, Vianna MV, Vieira JMF, Mallagutti VH, Inforsato FJ et al (2018) Cold-adapted enzymes produced by fungi from terrestrial and marine Antarctic environments. Crit Rev Biotechnol 38(4):600–619
57. Upadhayay VK, Khan A, Singh J, Singh AV (2021a) Bacterial assisted improved Zn consignment in root and shoot of rice plant by zinc solubilizing *Serratia marcescens* bearing plant probiotic traits. Adv Biores 13(1):1–8
58. Upadhayay VK, Singh AV, Khan A, Pareek N (2021b) Influence of zinc solubilizing bacterial co-inoculation with zinc oxide supplement on rice plant growth and Zn uptake. J Pharm Innov 10:113–116
59. Jorquera MA, Shaharoona B, Nadeem SM, de la Luz Mora M, Crowley DE (2012) Plant growth-promoting rhizobacteria associated with ancient clones of creosote bush (Larrea tridentata). Microb Ecol 64(4):1008–1017
60. Roshani KA, Singh AV, Upadhayay VK, Prasad B (2020) Development of potential microbial consortia and their assessment on wheat (Triticum aestivum) seed germination. Environ Ecol 38(1):6–16
61. Parveen H, Singh AV, Khan A, Prasad B, Pareek N (2018) Influence of plant growth promoting rhizobacteria on seed germination and seedling vigor of green gram. Int J Chem Stud 6(4):611–618
62. Fardella C, Oses R, Torres-Díaz C, Molina-Montenegro MA (2014) Antarctic fungal endophytes as tool for the reintroduction of native plant species in arid zones. Bosque 35:235–239. https://doi.org/10.4067/S0717-92002014000200011
63. Berríos G, Cabrera G, Gidekel M, Gutiérrez-Moraga A (2013) Characterization of a novel antarctic plant growth-promoting bacterial strain and its interaction with antarctic hair grass (Deschampsia Antarctica Desv). Polar Biol 36(3):349–362
64. Tomova I, Stoilova-Disheva M, Lazarkevich I, Vasileva-Tonkova E (2015) Antimicrobial activity and resistance to heavy metals and antibiotics of heterotrophic bacteria isolated from sediment and soil samples collected from two Antarctic islands. Front Life Sci 8(4):348–357
65. Presta L, Inzucchi I, Bosi E, Fondi M, Perrin E, Miceli E et al (2016) Draft genome sequence of Flavobacterium sp. strain TAB 87, able to inhibit the growth of cystic fibrosis bacterial pathogens belonging to the Burkholderia cepacia complex. Genome Announc 4(3):e00410–e00416
66. Khan A, Kukreti B, Makarana G, Suyal DC, Singh AV, Kumar S (2023c) Role of microorganisms in alleviation of arsenic toxicity in plants. In: Unravelling plant-microbe synergy. Academic Press, pp 263–281
67. Khan A, Sharma RS, Panthari D, Kukreti B, Singh AV, Upadhayay VK (2023d) Bioremediation of heavy metals by soil dwelling microbes: an environment survival approach. In: Pandey SC, Pande V, Sati D, Samant M (eds) Advanced microbial techniques in agriculture, environment and health management. Elsevier, London, pp 167–190
68. Ismail W, Gescher J (2012) Epoxy coenzyme A thioester pathways for degradation of aromatic compounds. Appl Environ Microbiol 78(15):5043–5051
69. Dias RL, Ruberto L, Hernández E, Vázquez SC, Balbo AL, Del Panno MT, Mac Cormack WP (2012) Bioremediation of an aged diesel oil-contaminated Antarctic soil: evaluation of the "on site" biostimulation strategy using different nutrient sources. Int Biodeterior Biodegradation 75:96–103
70. Pérez-de-Mora A, Engel M, Schloter M (2011) Abundance and diversity of n-alkane-degrading bacteria in a forest soil co-contaminated with hydrocarbons and metals: a molecular study on alkB homologous genes. Microb Ecol 62:959–972
71. Powell SM, Ferguson SH, Bowman JP, Snape I (2006) Using real-time PCR to assess changes in the hydrocarbon-degrading microbial community in Antarctic soil during bioremediation. Microb Ecol 52:523–532
72. Reyes-César A, Absalon AE, Fernández FJ, González JM, Cortés-Espinosa DV (2014) Biodegradation of a mixture of PAHs by non-ligninolytic fungal strains isolated from crude oil-contaminated soil. World J Microbiol Biotechnol 30:999–1009

73. Pini F, Grossi C, Nereo S, Michaud L, Giudice AL, Bruni V et al (2007) Molecular and physiological characterisation of psychotropic hydrocarbon-degrading bacteria isolated from Terra Nova Bay (Antarctica). Eur J Soil Biol 43(5–6):368–379
74. Aislabie JM, Balks MR, Foght JM, Waterhouse EJ (2004) Hydrocarbon spills on Antarctic soils: effects and management. Environ Sci Technol 38(5):1265–1274
75. Ruberto LA, Vazquez S, Lobalbo A, Mac Cormack WP (2005) Psychrotolerant hydrocarbon-degrading Rhodococcus strains isolated from polluted Antarctic soils. Antarct Sci 17(1):47–56
76. Stallwood B, Shears J, Williams PA, Hughes KA (2005) Low temperature bioremediation of oil-contaminated soil using biostimulation and bioaugmentation with a Pseudomonas sp. from maritime Antarctica. J Appl Microbiol 99(4):794–802
77. Saul DJ, Aislabie JM, Brown CE, Harris L, Foght JM (2005) Hydrocarbon contamination changes the bacterial diversity of soil from around Scott Base, Antarctica. FEMS Microbiol Ecol 53(1):141–155
78. Ma Y, Wang L, Shao Z (2006) Pseudomonas, the dominant polycyclic aromatic hydrocarbon-degrading bacteria isolated from Antarctic soils and the role of large plasmids in horizontal gene transfer. Environ Microbiol 8(3):455–465
79. Ferrari BC, Zhang C, Van Dorst J (2011) Recovering greater fungal diversity from pristine and diesel fuel contaminated sub-Antarctic soil through cultivation using both a high and a low nutrient media approach. Front Microbiol 2:217
80. Hua Z, Chen J, Lun S, Wang X (2003) Influence of biosurfactants produced by Candida Antarctica on surface properties of microorganism and biodegradation of n-alkanes. Water Res 37(17):4143–4150
81. Martorell MM, Ruberto LAM, Fernández PM, Castellanos de Figueroa LI, Mac Cormack WP (2017) Bioprospection of cold-adapted yeasts with biotechnological potential from Antarctica. J Basic Microbiol 57(6):504–516
82. Hughes KA, Bridge P, Clark MS (2007) Tolerance of Antarctic soil fungi to hydrocarbons. Sci Total Environ 372(2–3):539–548
83. Confalonieri F, Sommer S (2011) Bacterial and archaeal resistance to ionizing radiation. J Phys Conf Ser 261(1):012005. IOP Publishing
84. Das N, Chandran P (2011) Microbial degradation of petroleum hydrocarbon contaminants: an overview. Biotechnol Res Int 2011:941810. https://doi.org/10.4061/2011/941810
85. Rogan-Finnemore M (2005) The legal implications of bioprospecting in the Antarctic region. UC Research Repository
86. Zucconi L, Canini F, Temporiti ME, Tosi S (2020) Extracellular enzymes and bioactive compounds from Antarctic terrestrial fungi for bioprospecting. Int J Environ Res Public Health 17(18):6459
87. Zainal N, Ser HL, Yin WF, Tee KK, Lee LH, Chan KG (2016) Streptomyces humi sp. nov., an actinobacterium isolated from soil of a mangrove forest. Antonie Van Leeuwenhoek 109:467–474
88. Vercesi A, Nasini G, Locci R (1992) Biological and chemical characterization of the antibiotic activity of Streptomyces species isolated from grapevine carposphere. Actinomycetes 3(1)
89. Rojas JL, Martín J, Tormo JR, Vicente F, Brunati M, Ciciliato I et al (2009) Bacterial diversity from benthic mats of Antarctic lakes as a source of new bioactive metabolites. Mar Genomics 2(1):33–41
90. Teoh CP, Koh SP, Ling CMWV (2020) Characterisation of an Antarctic yeast, Glaciozyma Antarctica PI12. Borneo Int J Biotechnol 1:89
91. Fenice M (2016) The Psychrotolerant Antarctic fungus Lecanicillium muscarium CCFEE 5003: A powerful producer of cold-tolerant Chitinolytic enzymes. Molecules 21:447
92. Martorell MM, Ruberto LAM, de Figueroa LIC, Mac Cormack WP (2019) Antarctic Yeasts as a Source of Enzymes for Biotechnological Applications. In: Rosa LH (ed) Fungi of Antarctica: diversity, ecology and biotechnological applications. Springer International Publishing, Cham, pp 285–304

93. Baeza M, Alcaíno J, Cifuentes V, Turchetti B, Buzzini P (2017) Cold-active enzymes from cold-adapted yeasts. In: Sibirny AA (ed) Biotechnology of yeasts and filamentous fungi. Springer International Publishing, Cham, pp 297–324
94. He L, Mao Y, Zhang L, Wang H, Alias SA, Gao B (2017) Functional expression of a novel α-amylase from Antarctic psychrotolerant fungus for the baking industry and its magnetic immobilization. BMC Biotechnol 17:22
95. Mao Y, Yin Y, Zhang L, Alias SS, Gao B, Wei D (2015) Development of a novel Aspergillus uracil deficient expression system and its application in expressing a cold-adapted α-amylase gene from Antarctic fungi Geomyces pannorum. Process Biochem 50:1581–1590
96. Yu P, Wang XT, Liu JW (2015) Purification and characterization of a novel cold-adapted phytase from strain JMUY14 isolated from the Antarctic: characterization of a novel cold-adapted phytase. J Basic Microbiol 55:1029–1039
97. Białkowska AM, Szulczewska KM, Krysiak J, Florczak T, Gromek E, Kassassir H, Kur J, Turkiewicz M (2017) Genetic and biochemical characterization of yeasts isolated from Antarctic soil samples. Polar Biol 40:1787–1803
98. Salwoom L, Raja Abd Rahman RNZ, Salleh AB, Mohd Shariff F, Convey P, Pearce D, Mohamad Ali MS (2019) Isolation, characterisation, and lipase production of a cold-adapted bacterial strain Pseudomonas sp. LSK25 isolated from Signy Island, Antarctica. Molecules 24(4):715
99. Furbino LE, Pellizzari FM, Neto PC, Rosa CA, Rosa LH (2018) Isolation of fungi associated with macroalgae from maritime Antarctica and their production of agarolytic and carrageenolytic activities. Polar Biol 41:527–535
100. Marizcurrena JJ, Martı'nez-Lo'pez W, Ma H, Lamparter T, Castro-Sowinski S (2018) A highly efficient and cost-effective recombinant production of a bacterial photolyase from the Antarctic isolate Hymenobacter sp. UV11. Extremophiles 23:49
101. Gil-Dura'n C, Ravanal MC, Ubilla P, Vaca I, Cha'vez R (2018) Heterologous expression, purification and characterization of a highly thermolabile endoxylanase from the Antarctic fungus Cladosporium sp. Fungal Biol 122:87582
102. See-Too WS, Convey P, Pearce DA, Chan KG (2018) Characterization of a novel N-*acyl homoserine lactonase*, AidP, from *Antarctic Planococcus* sp. Microb Cell Factories 17:179
103. Dharmaraj S, Ashokkumar B, Dhevendaran K (2009) Fermentative production of carotenoids from marine actinomycetes. Iran J Microbiol 1(4):36–41
104. Namazkar S, Ahmad WA (2013) Spray-dried prodigiosin from Serratia marcescens as a colorant. Biosci Biotechnol Res Asia 10(1):69
105. Hoyoux A, Jennes I, Dubois P, Genicot S, Dubail F, François JM et al (2001) Cold-adapted β-galactosidase from the Antarctic psychrophile Pseudoalteromonas haloplanktis. Appl Environ Microbiol 67(4):1529–1535
106. Kavitha M (2016) Cold active lipases—an update. Front Life Sci 9:226–238
107. Araujo R, Casal M, Cavaco-Paulo A (2008) Application of enzymes for textile fibres processing. Biocatal Biotransformation 26:332–349. https://doi.org/10.1080/10242420802390457
108. Bisht S (2011) Cold active proteins in food and pharmaceutical industry [article on the Internet]. Biotech Articles. [cited 2017 Dec. 15]. Available from http://www.biotecharticles.com/Biotechnology-products-Article/ColdActive-Proteins-in-Food-and-Pharmaceutical-Industry-719
109. Hamid B, Bashir Z, Yatoo AM, Mohiddin F, Majeed N, Bansal M, Poczai P, Almalki WH, Sayyed RZ, Shati AA et al (2022) Cold-active enzymes and their potential industrial applications—a review. Molecules 27:5885. https://doi.org/10.3390/molecules27185885
110. Furhan J, Awasthi P, Sharma S (2019) Biochemical characterization and homology modeling of cold-active alkophilic protease from Northwestern Himalayas and its application in detergent industry. Biocatal Agric Biotechnol 17:726–735

111. Castilla A, Giordano SR, Irazoqui G (2022) Extremophilic lipases and esterases: characteristics and industrial applications. In: Kuddus M (ed) Microbial Extremozymes. Academic Press, pp 207–222
112. Ganasen M, Yaacob N, Rahman RN, Leow AT, Basri M, Salleh AB et al (2016) Cold-adapted organic solvent tolerant alkalophilic family I.3 lipase from an Antarctic pseudomonas. Int J Biol Macromol 92:1266–1276
113. Viñarta SC, Maza DD, Fernández PM, Aybar MJ, de Figueroa LIC (2022) Integrated production of biodiesel and industrial wastewater treatment by culturing oleaginous microorganisms. In: Kumar V, Kumar M (eds) Integrated environmental technologies for wastewater treatment and sustainable development. Elsevier, Amsterdam, pp 81–101
114. Plaza DO, Gallardo C, Straub YD, Bravo D, Pérez-Donoso JM (2016) Biological synthesis of fluorescent nanoparticles by cadmium and tellurite resistant Antarctic bacteria: exploring novel natural nanofactories. Microb Cell Factories 15(1):1–11
115. John MS, Nagoth JA, Ramasamy KP, Ballarini P, Mozzicafreddo M, Mancini A et al (2020) Horizontal gene transfer and silver nanoparticles production in a new Marinomonas strain isolated from the Antarctic psychrophilic ciliate Euplotes focardii. Sci Rep 10(1):1–14

Chapter 16
Microbial Pigments: Overview and Industrial Perspective

Anita Mishra, Pragati Srivastava, Manali Singh, Divya Joshi, Ravindra Soni, and Deep Chandra Suyal

Abstract Natural colors derived from plants, animals, and microorganism are being used since ancient time. Keeping in mind the harmful effects of synthetic dyes, these eco-friendly pigments are of high demand. Among these, microbial pigments are widely used in various industries including food, textile, pharmaceuticals, etc. They have attracted scientific attention due to their easy harvesting, cost-effectivity, higher stability, and high production value. However, the availability and diversity of microbial pigments are not expanding in the same ratio their demand is increasing. It may increase the undesirable use of synthetic colorants to meet commercial demands. Therefore, besides identifying the new microbial sources, scientific efforts are urgently needed to develop novel strategies for easy extraction and monitoring processes. This review provides an overview of microbial pigments and their characteristics features. Moreover, it summarizes their recent applications in various field and industries.

Keywords Pigments · Canthaxanthin · Astaxanthin · Riboflavin · Beta-carotene · Sustainability

A. Mishra · D. C. Suyal (✉)
Vidyadayini Institute of Science, Management and Technology, Bhopal, Madhya Pradesh, India

P. Srivastava · D. Joshi
Department of Microbiology, College of Basic Sciences and Humanities, G. B. Pant University of Agriculture and Technology, Pantnagar, India

M. Singh
Department of Life Sciences, Parul Institute of Applied Sciences, Parul University, Vododara, India

R. Soni
Department of Agricultural Microbiology, Indira Gandhi Krishi Vishwavidyalaya, Raipur, Chhattisgarh, India

R. Soni et al. (eds.), *Microbial Bioactive Compounds*,
https://doi.org/10.1007/978-3-031-40082-7_16

16.1 Introduction

Pigments are molecules that take in a selected wavelength of light and replicate the rest of the pulchritude visible spectrum (380–750 nm). They are widely used in the industry for various applications in cosmetics, pharmaceuticals, textiles food, and other industries [1]. They are known to possess antioxidant, cytotoxic, antimalarial, antimicrobial, antifouling, antitumor, and anticancer activities.

Pigment manufacturing is one of the charismatic tendencies of microbes. Generally, microbial pigments aren't simply colors; however, they possess a combination of numerous chemical components with multifaceted biological activities [2]. In recent years, studies on pigmented microorganisms from terrestrial and marine ecosystems have incredibly improved and thus utilized in advanced research. Microbial-pigmented molecules include bacteriochlorophylls, flavins, indigoids, β-carotene, carotenoids, phenazines, monascus, phenazostatin, melanins, pheomelanin, prodigiosin, glaukothalin, quinone precursors, canthaxanthin, violacein, phycocyanin, xanthomonadin, phenazine, astaxanthin, and others [3, 4]. They can be produced directly or generated as a byproduct of their metabolism.

16.2 Commercially Important Microbial Pigments

The major microbial pigments that can be utilized as food colorants are canthaxanthin, prodigiosin, phycocyanin, astaxanthin, lycopene, riboflavin, violacein, melanin, and beta-carotene (Fig. 16.1). They may be inorganic or organic in nature. However, organic pigments are considered more useful in the food industry. Some of them are described below:

1. Canthaxanthin—It is a reddish-orange color pigment of the trans-carotenoid category. It is a potent antioxidant and generally lipid soluble. It can be isolated from several species of *Bradyrhizobium*. It has been approved as a food colorant and is widely utilized for poultry and salmon feed.
2. Prodigiosin—It is a reddish pigment having antimalarial, antineoplastic, antibiotic, and antibacterial properties. It can be isolated from *Serratia marcescens*. It is reported to use as a food colorant in carbonated drinks, yogurt, and flavored milk.
3. Astaxanthin—It is an pinkish-red colored microbial pigment. It can be obtained from the bacteria, yeasts, salmon, red shrimp, crayfish, and microalgae, crustaceans, and bird's feathers. It is agency-approved colorant for feeding purposes to animals and fishes.
4. Riboflavin—It is yellow-colored compound and is generally produced by several microorganisms. It is utilized in the dairy items, baby foods, energy drinks, sauces, cereals, and packaged juices.

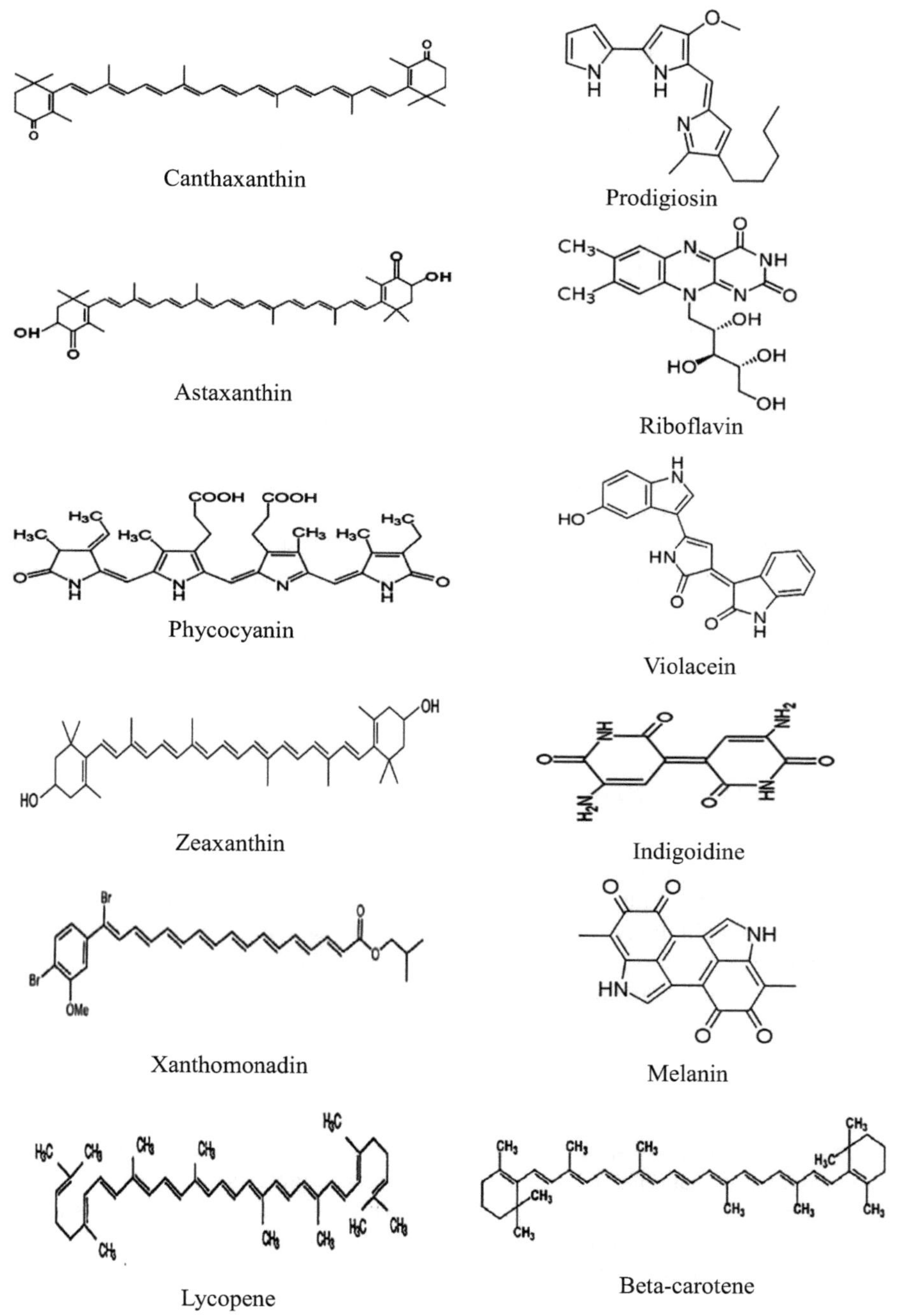

Fig. 16.1 Microbial pigments and their chemical structures

5. Phycocyanin—These are bluish-colored pigments that are produced by cyanobacteria having chlorophyll A. Commercial phycocyanin can be extracted from *Aphanizomenon flos-aquae* and *Spirulina*. These are used in packaged foods, beverages, ice creams, and sweets.
6. Violacein—It is a purple-colored pigment that exhibits antifungal, antibiotic, antitumor, and antibacterial properties. It is frequently obtained from *Chromobacterium violaceum* and has a useful role in cosmetic, food, and textile and beverage industries.
7. Lycopene—This pigment is naturally present in tomatoes. It is red carotenoid pigment and can be extracted from several microbes, viz., *Sporotrichioides*, *Blakeslea trispora*, and *Fusarium*. It has been reported to prevent the disease including coronary heart disease and cancers. Several countries, viz., New Zealand, the USA, the UK, Italy, and Australia use it for the meat coloring.
8. Melanin—These are natural occurring pigments found in several plants, animals, and microbes. They are frequently used in cosmetic items, food and feed purposes, eye glasses, sunscreen creams, and several pharma compounds.
9. Beta-carotene—It is a red- or orange-colored pigment and is organic in nature. It is generally obtained from *Dunaliella salina* and a few other algae. It is an agency-approved colorant that can be utilized in packaged food and beverages.
10. Anthocyanin—These are primarily the plant's generated pigments. However, in recent years, genetically engineered bacteria are used to produce such plant-derived pigments. They are water soluble and flavonoid compounds.

16.3 Ecology of Pigment-Producing Microorganisms

Pigment-producing microbes are found in diverse geographical conditions, ranging from polar environments to tropical regions to deep-sea ecosystems. Pigment production is thought to be an adaptive feature of such microbes. Few terrestrial microbes, viz., *Stenotrophomonas* and *Rhodotorula*, are proven to adapt to marine environments by producing pigments. Therefore, pigmented microbes can also be classified as true marine-pigmented microbes and adaptive pigmented microbes. However, true marine-pigmented microbes are gaining more attention of the research community due to their diverse portfolio of bioactive compounds. Besides these, pigmented microbes are also reported from ice cores, sea surface, salt lakes, air-water interfaces, hot springs, lava caves, human skin, plant surfaces, rhizosphere soils, algal mats, lagoons, microbial mats, corals, and glaciers.

Several microorganisms are reported to have polyextremophilic features as per their habitats, viz., xerophilic (*Penicillium purpurogenum*), psychrophilic (*Kocuria polaris*), alkaliphilic (*Microbacterium arborescens*), barophilic (*Halomonas salaria*), radioresistance (*Deinococcus grandis*), color mimic (*Cellulophaga lytica*), pleomorphic (*Arthrobacter*), dimorphic (*Metschnikowia laotica*), acidophilic (*Acidobacterium*), polyextremophile (*Halorubrum*), and halophilic (*Salinibacter*).

16.4 Significance of Microbial Pigments

It is well reported that a maximum of the microbial pigments discovered are known to act as protecting systems against harmful radiations [5–7]. It, therefore, enhances their survivability to the encircling environmental conditions as compared to nonpigmented microbes. Further, it is also known that it protects their respective host from their pathogens. Furthermore, extremophile microbes, viz., *Thermus filiformis*, *Halobacterium salinarum*, and *Halococcus morrhuae* are reported to produce pigments that stabilize their cell membrane and thus help them in their survival. These pigments also act as antioxidants and neutralize the harmful effects of reactive species of oxygen, nitrogen, and other free radicles. Some pigments are known to protect the lipid membrane of microbes, viz., violacein. In another report, phenazines are observed to regulate bacterial gene regulation [8].

16.5 Industrial Applications

16.5.1 Food Industry

Generally microbial pigments are eco-friendly, noncarcinogenic, nonallergic, and nontoxic. Therefore, they are considered safe for human consumption and preferred to be used in food industry [9]. It has been estimated that their use in food and associated industries will increase with the rate of 7% annually. The important microbial pigments that are preferred in food industry include phycocyanin, riboflavin, lycopene, canthaxanthin, prodigiosin, violacein, melanin, astaxanthin, and beta-carotene (Table 16.1). Further, the organic pigments are found more useful than the inorganic ones. Phycocyanin is widely used in beverages, packaged foods, ice creams, and sweets. Riboflavin is utilized in baby foods, fruit drinks, dairy products,

Table 16.1 Important microbial pigments and their appearance

Pigments	Appearance	Associated microorganisms
Canthaxanthin	Reddish orange	*Bradyrhizobium, Haloferax*
Prodigiosin	Reddish	*Serratia*
Astaxanthin	Pinkish red	*Agrobacterium, Paracoccus*
Riboflavin	Yellow	*Bacillus, Ashbya*
Phycocyanin	Blue	*Aphanizomenon, Spirulina*
Zeaxanthin	Yellow	*Flavobacterium*
Xanthomonadin	Yellow	*Xanthomonas*
Violacein	Purple	*Janthinobacterium, Chromobacterium*
Indigoidine	Blue	*Corynebacterium*
Lycopene	Red	*Sporotrichioides, Blakeslea, Fusarium*
Melanin	Dark brown	*Bacillus, Aeromonas, Rhizobium*
Beta-carotene	Orange	*Dunaliella*

sauces, energy drinks, and breakfast cereals. Canthaxanthin and astaxanthin are approved by regulatory agencies and used in animal and fish foods. Prodigiosin is a red-colored pigment that is used in carbonated drinks, ice creams, sweets, and flavored milk [10]. Violacein is an important ingredient for several packaged foods. Beta-carotene can be utilized in various sweets, flavored items, and ice creams due to its reddish yellow color. Furthermore, melanin and lycopene are commercially important pigments that are widely used in foods and beverages.

16.5.2 In Pharmaceutical Industry

Microbial pigments are also used in pharmaceutical industries due to their bioactive nature. Moreover, in recent years, several researches are going on to identify the major roles of these pigments in the treatment of leukemia, diabetes, cancer, and autoimmune diseases [11]. These pigments are found to possess anticancer, immunosuppressant, antipyretic, antibiotic, and antiproliferative properties [12, 13]. The most important pigments in this category include anthocyanin, prodigiosin, violacein, and melanin.

Anthocyanins are used as antioxidants, anticarcinogen, and immunosuppressant. Prodigiosin is a tripyrrole pigment and known to possess cytotoxic property [14]. Moreover, it has also shown immunosuppressant and antiproliferative activities. It has also been used for the treatment of diabetes mellitus. Similarly, violacein possesses antiparasitic, anticarcinogenic, antimicrobial, and antiprotozoal abilities [15].

16.5.3 Textile Industry

Microbial pigments are frequently used in textile industries for coloring the cloths [16]. Due to their eco-friendly and noncarcinogenic nature, they are being now preferred over the synthetic dyes. Prodigiosin and violacein are important microbial pigments that are used to dye polyesters, acrylic, nylon, silk, cotton vinylon, rayon, and microfibers. Further, iron mordants are used to fix microbial pigments over the cloths and to provide the variable patterns.

16.6 Conclusion

In recent years, commercial requirement of microbial pigments has increased enormously and continue to be in demand. However, the available range of these pigments is limited and proven insufficient, especially food and beverages industries. Therefore, the discovery of novel and unique pigments is becoming important nowadays. Moreover, new methodologies and technologies are required to extract the pigments from microorganisms through nondestructive and high-throughput manner. Additionally, it must be cost-effective, easy to use, and easy to monitor. Therefore, detailed and in-depth investigations are urgently required in this field.

References

1. Paillié-Jiménez ME, Stincone P, Brandelli A (2020) Natural pigments of microbial origin. Front Sustain Food Syst 4:590439
2. Chatragadda R, Dufossé L (2021) Ecological and biotechnological aspects of pigmented microbes: a way forward in development of food and pharmaceutical grade pigments. Microorganisms 9:637
3. Venil CK, Dufossé L, Renuka Devi P (2020) Bacterial pigments: sustainable compounds with market potential for pharma and food industry. Front Sustain Food Syst 4:100. https://doi.org/10.3389/fsufs.2020.00100
4. Egea MB, Dantas LA, Sousa TL, Lima AG, Lemes AC (2023) The potential, strategies, and challenges of Monascus pigment for food application. Front Sustain Food Syst 7:1141644
5. Minghui Z, Chen Y, Fang X, Li W, Zhang Y (2023) Isolation and identification of pigment-producing filamentous fungus DBFL05 and its pigment characteristics and chemical structure, CyTA. J Food 21(1):374–385
6. Agarwal H, Bajpai S, Mishra A, Kohli I, Varma A, Fouillaud M, Dufossé L, Joshi NC (2023) Bacterial pigments and their multifaceted roles in contemporary biotechnology and pharmacological applications. Microorganisms 11(3):614
7. Rastogi RP, Sinha RP (2013) Incharoensakdi, A characterization, UV-induction and photoprotective function of sunscreen pigment, scytonemin from Rivularia sp. HKAR-4. Chemosphere 93:1874–1878
8. Sorrels CM, Proteau PJ, Gerwick WH (2009) Organization, evolution, and expression analysis of the biosynthetic gene cluster for scytonemin, a cyanobacterial UV-absorbing pigment. Appl Environ Microbiol 75:4861–4869
9. Mohammadi M, Ahangari H, Mousazadeh S, Hosseini S (2021) Microbial pigments as an alternative to synthetic dyes and food additives: a brief review of recent studies. Bioprocess Biosyst Eng 45:1–12
10. Sajjad W, Ahmad S, Aziz I, Azam SS, Hasan F, Shah AA (2018) Antiproliferative, antioxidant and binding mechanism analysis of prodigiosin from newly isolated radio-resistant Streptomyces sp. Strain WMA-LM31. Mol Biol Rep 45:1787–1798
11. Celedón R, Díaz L (2021) Natural pigments of bacterial origin and their possible biomedical applications. Microorganisms 9:739
12. Montaner B, Prez-Toms R (2003) The prodigiosins: a new family of anticancer drugs. Curr Cancer Drug Targets 3:57–65
13. Siegel RL, Miller KD, Jemal A (2019) Cancer statistics. CA Cancer J Clin 69:7–34
14. Cao W, Chen W, Sun S, Guo P, Song J, Tian C (2007) Investigating the antioxidant mechanism of violacein by density functional theory method. J Mol Struct Theochem 817:1–4
15. Gutierrez-dalmau A, Campistol JM (2007) Immunosuppressive therapy and malignancy in organ a systematic review. Drugs 67:1167–1198
16. Heer K, Sharma S (2017) Microbial pigments as a natural colour—a review. Int J Pharm Sci Res 8:1913–1922

MIX
Papier aus verantwortungsvollen Quellen
Paper from responsible sources
FSC® C105338

If you have any concerns about our products,
you can contact us on
ProductSafety@springernature.com

In case Publisher is established outside the EU,
the EU authorized representative is:
Springer Nature Customer Service Center GmbH
Europaplatz 3, 69115 Heidelberg, Germany

Printed by Libri Plureos GmbH
in Hamburg, Germany